Physics of Rare Earth Solids

Physics of
Rare Earth Solids

K. N. R. TAYLOR
Senior Lecturer, Department of Physics,
University of Durham

M. I. DARBY
Lecturer, Department of Pure and Applied Physics,
University of Salford

CHAPMAN AND HALL LTD
11 NEW FETTER LANE, LONDON EC4

First published 1972
by Chapman and Hall Ltd,
11 New Fetter Lane, London EC4P 4EE
Filmset by Keyspools Ltd, Golborne, Lancs.
Printed in Great Britain by
Redwood Press, Trowbridge, Wiltshire

SBN 412 10160 2

Contents

Preface

During the past fifteen years the rare earths have become one of the most intensively studied groups of elements in the periodic table. The elements, alternatively known as the lanthanides, the fraternal fifteen and the step children of the periodic table, lie between lanthanum ($Z = 57$) and lutetium ($Z = 71$). Strictly speaking, lanthanum is not a rare earth element and is often omitted in discussions of their properties. In this work we have taken an intermediate view and have considered its properties where these can contribute to an understanding of the behaviour of the other elements.

The series of elements is characterized by a gradual filling of the $4f$ shell. In most of the elements this shell is deeply buried within the atom and the magnetic moment which is associated with it is highly localized. These features of the electronic structure give rise to a large variety of interesting physical properties when the ions are situated in a crystalline lattice. As a result the rare earths have become the subject of multi-discipline research in which physicist, chemist and metallurgist alike are contributing to the increasing insight into their behaviour.

Prior to World War II, the availability of the rare earths in a pure metallic form made their name justifiable, although from a geological viewpoint it is a misnomer, as their abundances are in general comparable to many of the better known elements, for example, silver, tin, tungsten and mercury. The development of ion exchange, solvent extraction and metal reduction techniques following the work of F. H. Spedding and his group at Ames, Iowa, has drastically changed the situation however, and made available the quantities of pure metal which are necessary for the research in this branch of science to develop to its present state. Although increasing purity will allow certain fundamental and sophisticated experiments to be performed we believe that a point has been reached at which the physical properties of the rare earth elements can be considered in a single volume and where the work of the next decade is going to contribute significantly less to the basic knowledge and understanding of their physical properties than the work of the past ten years.

The volume of published work is sufficiently large however, to make a complete text unpublishably large and we have prepared this manuscript more in the light of an extended review. Each one of the chapters could be expanded into a separate treatise and in some cases this has already been done. At no point would we claim to have made a complete coverage of all the features of interest or of all the workers who have contributed independently to the subject· We hope however, that the referencing is sufficiently complete to allow the bibliographers and tabulators access to the many hundreds of papers applicable to each topic and to give the new worker in this field sufficient information for his purpose. Omissions are perhaps most noticeable in the first and last chapters and here we have tried to discuss typical behaviour rather than the particular. In Chapter 6 this has meant that some classes of compound receive no mention, e.g. the garnets which, although technologically important, contribute relatively little to an understanding of the rare earths.

In discussing the rare earth metals, the elements are conveniently divided into two groups, the light rare earths (La, Ce–Eu) and the heavy rare earths (Gd–Yb), which have distinctively different properties. The metal cerium exhibits behaviour unlike the other elements and is frequently treated separately. Promethium is a natural radioactive element of relatively short half life and little work has been done on it, consequently it will receive little mention in the following. On the other hand the elements scandium ($Z = 21$) and yttrium ($Z = 39$) whose atoms have similar electronic configurations to the heavy rare earths, with a single d-electron inside a closed s-shell, are often included in the discussion. Their importance to the rare earths has usually been in the role of an ideal diluent, by which the effective atomic spacing could be varied without causing too drastic a change in the band structure of the material.

Concerning units, the Gaussian c.g.s. system is employed here, together with special units where relevant. These units have been universally used in the literature in the past and at present there is little indication of their replacement by, for example, SI units. Conversions between the various units are given in an appendix. Regarding notation for crystal point groups and space groups, the Schönflies symbols are used throughout, but in Chapter 2 on crystal structures the international symbols are also given.

It is a pleasure to thank Dr W. Jones, Dr M. A. H. McCausland, Mr D. E. O'Connor and Dr D. J. Saunders for critically reading various sections of the book. We are also greatly indebted to Mrs D. Anson, Miss M. Brawn, Mrs B. Collins, Miss H. Farrand, Mrs J. Moore and Mrs S. Naylor for their help in preparing the manuscript. Finally our thanks to Miss J. M. Brown for her extensive assistance in surveying the literature.

We are grateful to the publishers and editors of the following journals and books for permission to reproduce a number of figures: Comptes Rendus de l'Academie des Sciences, Journal of Applied Physics, Journal of Physics and Chemistry of Solids, Leiden Communications, Physical Review, Physical Review Letters, Physics Letters, Proceedings of the International Conference on Magnetism (Nottingham 1964), Proceedings of the Physical Society (Journal of Physics), Proceedings of the Royal Society, Progress in the Science and Technology of the Rare Earths, Progress in Solid State Chemistry, Solid State Communications, Solid State Physics (ed. F. Seitz and D. Turnbull), Technical Report AFM2-TF-68-159(1968), Transactions of the American Institute of Metallurgical Engineers, Transactions of the American Society of Metals; 'Spectra and Energy Levels of Rare Earth Ions in Crystals' by G. H. Dieke (Interscience Publishers, 1968), 'Physical Chemistry of Metals', by L. S. Darken and R. W. Gurry (McGraw Hill Publishing Co., 1953).

K. N. R. Taylor
M. I. Darby

1 October 1970

TO

Our long suffering wives

1

Ionic Properties

The purpose of this first chapter is to outline the theory of the energy level structure of ions as it applies to the rare earths. In particular, attention will be focused upon ions in solids, as our interest is primarily concerned with the physics of the solid state, and we shall be concerned with the lowest electronic states, which are accessible to electrons at thermal energy. Excitations of the order of 10,000 cm^{-1}, i.e. optical transitions, are not considered here as they have been reviewed in detail elsewhere, e.g. Wybourne [1], Dieke [2], and McClure [3]. However, it is sometimes necessary to consider in the theory the effect of these much higher states upon the lowest ones, and, of course, optical spectroscopy can yield very accurate experimental information about the lowest levels. Some of these results will be quoted where relevant.

The energy levels of ions may be investigated experimentally by spectroscopy. Having disregarded optical measurements, the most important measurements on paramagnetic ions in crystals are those of electron spin resonance (esr) at low temperatures ($< 20°K$). The interpretation of such experiments on rare earth ions, which historically stimulated much interest in their properties, will be discussed in Section 1.4. In such experiments the interaction of the ions with the lattice is important and, since these effects are reasonably well understood for the rare earths, they are considered at some length in Section 1.5. Recently esr has been used to investigate the interaction between pairs of ions in crystals at very low temperatures (4·2°K). These experiments promise to be important in helping to understand the exchange interactions between ions in non-conducting crystals, and their interpretation is discussed in Section 1.6. In the final sections some bulk properties of salts are considered.

The nuclear properties of the ions will only be considered in so far as they determine the hyperfine structure of the electronic states, and then only briefly in connection with esr. Frequently nuclear magnetic resonance (nmr) is impossible in paramagnetic salts.

1.1 Theory of free ions

The theory of rare earth ions in solids shows two simplifying features compared with the theory of most other ions. In solids the rare earth ions are usually triply ionized. Consequently their electrons are in the closed shell configurations of the xenon structure, except for the $4f$ electrons. The first point is that the $4f$ electrons, which determine the electronic states of interest, are shielded by the closed shells $5s^2$ $5p^6$ from perturbations due to neighbouring ions, so that the effects of the latter are small. The second feature is that the spin-orbit interaction for the $4f$ electrons, which couples the angular momenta L and S to give a total angular momentum J, is larger than other interactions. As a result J is considered to be a good quantum number in almost all circumstances, and the lowest levels are usually taken to be determined by a single value of J. The effect of other interactions, such as those with the nucleus or crystal fields, is to lift the $(2J+1)$ degeneracy of the level, and the small splittings produced may be treated by perturbation theory. In contrast, consider the ions of the iron transition elements, where the $3d$ electrons are not screened, so that they may take part in bonding, etc., and where the spin-orbit interaction is generally not larger than the crystalline field, so that one must employ L and S quantization.

As would be expected, the simplest theories of rare earth ions do not give exact agreement with experiment, and various small corrections have been made. Some of these are due to screening of the $4f$ electrons, either from neighbouring ions by the outer shells, or from the nucleus by inner shells. There are relativistic effects other than the spin-orbit interaction which are not negligible for such heavy ions. Further, some small part of higher energy levels (J') may be mixed into the ground state J and, in solids, there may be some small mixing of the outer orbitals of the rare earth ion with those of neighbouring anions, i.e. covalency. Each of these effects will be mentioned in the relevant sections. It must be emphasized however, that these are small corrections. The balance between them may be delicate, as will be seen, for example, in determining the crystal field splittings of Gd^{3+} [34], and an exact treatment of them appears impossible. For this reason the simple theory is often employed and the theoretical parameters are replaced by empirical ones. This is the approach which will be adopted in most of this and subsequent chapters.

1.1.1 *The central field approximation*

In the simplest theory of the atom [4] each electron is assumed to move independently in a spherically symmetric potential arising from the nuclear charge Ze and from the charge of the other electrons. It is well known that the appropriate single particle wave functions have the form

$$\psi_{nl}(\mathbf{r}) = R_{nl}(r)\, Y_l^m(\theta, \phi) \tag{1.1}$$

where the Y_l^m are spherical harmonics and n, l, m are quantum numbers.

In the approximate *Hartree method* electron spin is ignored and the total wave function of the atom Ψ is taken to be a simple product of the ψ_i. Unfortunately the Pauli exclusion principle has to be imposed separately. This difficulty is removed in the *Hartree–Fock* (HF) approximation where Ψ is constructed as a determinant of orbitals which include spin: $\phi_i = \psi_i(\mathbf{r}_i)\chi_i)m_s)$, where χ_i is a Pauli spin function and m_s is the spin quantum number. The best ϕ_i are chosen to minimize the total energy of the atom and they are found to satisfy a set of coupled equations:

$$\left\{-\frac{\hbar^2}{2m}\nabla_1^2 - \frac{Ze^2}{r_1} + e^2\sum_j \int \frac{|\phi_j(\mathbf{r}_2)|^2}{|\mathbf{r}_1 - \mathbf{r}_2|}\mathrm{d}\mathbf{r}_2\right\}\phi_i(\mathbf{r}_1)$$

$$-e^2\sum_{\substack{j \\ (m_{s_j}\|m_{s_i})}}\left\{\int \frac{\phi_j^*(\mathbf{r}_2)\phi_i(\mathbf{r}_2)}{|\mathbf{r}_1 - \mathbf{r}_2|}\mathrm{d}\mathbf{r}_2\right\}\phi_j(\mathbf{r}_1) = \varepsilon_i\phi_i(\mathbf{r}_1) \tag{1.2}$$

The third term in the Hamiltonian is an average potential due to the other electrons and is not present in the Hartree scheme. The final term is an exchange potential, V_{ex}, which is non-local in the sense that the wave functions operated upon enter under an integral. It is not present in the Hartree method.

Often V_{ex} is approximated by taking suitable averages of the ϕ_j, and the resulting local potential spherically averaged, so that the solutions for ϕ again have the form (1.1). The effect of such an exchange potential is to keep electrons of like spin apart. Around a given electron the density of electrons with spins parallel to it is reduced, and there is said to exist an exchange or *Fermi hole*. In the case of a uniform free electron gas of (constant) density $\rho(\mathbf{r})$ the appropriate average potential is

$$V_{ex}(\mathbf{r}) = -6\left\{\frac{3}{8\pi}|\rho(\mathbf{r})|\right\}^{1/3} \quad \text{a.u.} \tag{1.3}$$

Slater first applied this form of potential in the single particle equation for atoms, see [5], employing for $\rho(\mathbf{r})$ the spherically averaged charge density

$$\rho(\mathbf{r}) = (4\pi r^2)^{-1}\textstyle\sum_{nl} w_{nl} r^2 |R_{nl}(r)|^2, \tag{1.4}$$

where w_{nl} is a weighting factor for shell nl. The resulting equation, known as the Hartree–Fock–Slater equation, has been solved for all atoms up to $Z = 103$ by Herman and Skillman [5].

The HF method would be expected to give superior results to the Hartree scheme, but comparison with experiment usually indicates that this is not so, and in some cases it predicts unphysical results, e.g. for the density of states of a free electron gas. In all schemes independent motion is assumed. In fact the motions will be correlated, so that electrons avoid one another. In the HF scheme for electrons with parallel spins the effect of exchange is to ensure just this through the Pauli condition, i.e. the Fermi hole, but there is no similar effect for electrons with antiparallel spins, which are allowed to be at the same place. Thus in one sense the inclusion of exchange leads to an unbalanced correction of the Hartree method, and correlation effects must be included to redress the balance. It is therefore often better to employ the Hartree scheme, where neither effect contributes.

1.1.2 *The radial wavefunction*

The configuration of the rare earth ions have as basis the xenon structure ($1s^2 2s^2 2p^6 3s^2 3p^6 3d^{10} 4s^2 4p^6 4d^{10} 5s^2 5p^6$), with electrons at least partly filling the $4f$ shell, and with either two or three outer electrons ($6s^2$ or $5d6s^2$). The appearance of the $4f$ electrons at the beginning of the lanthanide series clearly means that the nuclear charge becomes sufficiently large for them to be bound at that point. However, it might be supposed that the $4f$ electrons would occupy the outermost orbitals, whereas all the evidence indicates that they lie inside the $5s$ shells.

The binding of the $4f$ electrons must in some way be understood in terms of the screening of the nucleus by the other electrons. Unfortunately a Hartree-type calculation of the (self-consistent) potential of the other electrons may become unreliable when the dependence of the orbitals upon Z is not smooth. However, this is not the case for the statistical Thomas–Fermi method, in which the central potentials are always smooth functions of Z, and the method has been applied to the $4f$ shells by Latter [7] and Coulson and Sharma [8]. Some results are shown in Figs. 1.1 and 1.2.

One expects a $4f$ electron to be bound when its energy lies below its ionization potential, and this occurs at $Z = 60$ which is in reasonable

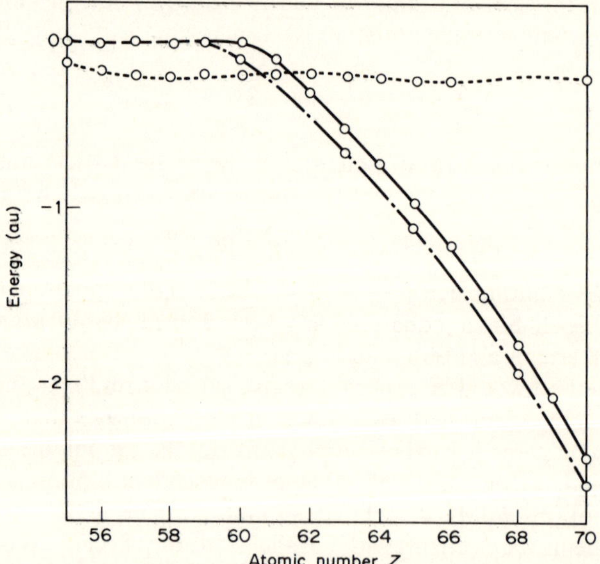

Fig. 1.1 Energies of $4f$ electrons as a function of atomic number, Z, calculated using the Thomas Fermi method. The solid curve is a calculation by Coulson and Sharma [8]. The broken curve is a calculation by Latter [7] and includes relativistic effects. The dotted line is the experimental ionization potential.

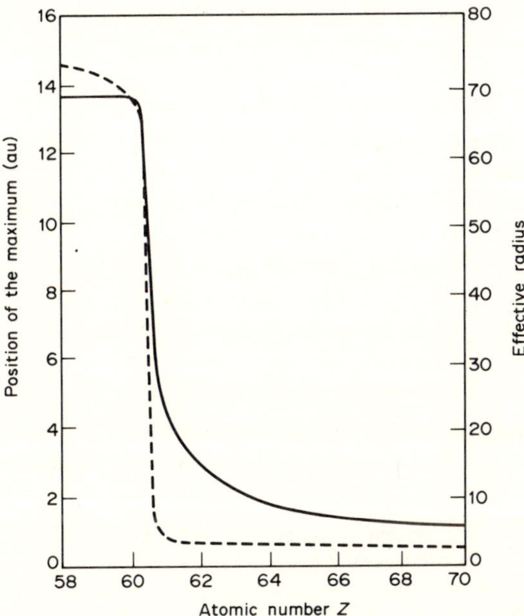

Fig. 1.2 Thomas Fermi calculations of the radius of the $4f$ shell as a function of atomic number [8]. The broken curve gives the radius at which $rR_{4f}(r)$ has its maximum value. The solid curve gives the radius beyond which $rR_{4f}(r)$ becomes effectively zero.

agreement with the beginning of the lanthanide series. The calculated radial charge distributions, Fig. 1.2, show a maximum which moves dramatically from outside to inside the $5s^2 5p^6 5d^1 6s^2$ electrons between $Z = 60$ and 61, again in reasonable agreement with the observed results. It is interesting that the Hartree–Fock–Slater calculations of Herman and Skillman [5] differ very little from these Thomas–Fermi ones.

Because the $4f$ electrons are so well embedded within the core orbitals the valence electrons are those in the $5d^1 6s^2$ states. In salts these electrons are readily transferred to other ions, while in metals they form the conduction band. Hence in most practical circumstances the rare earth ions are tripositive, in which case all electrons are in closed shells, and therefore inactive, except for the $4f$ electrons.

In order to determine the energy levels of the ions theoretically, detailed radial wavefunctions R_{4f} are needed in order to evaluate expectation integrals of the various interactions. These integrals are usually treated as empirical parameters, but the calculation of the R_{4f} functions for the rare earths is of considerable interest. There are three basic methods: (i) hydrogen-type functions with empirical parameters, (ii) Hartree, (iii) Hartree–Fock.

Hydrogen-like $4f$ radial wavefunctions of the form

$$R_{4f} = (Z-\sigma)^{3/2} r^3 e^{-r/2} (96\sqrt{35})^{-1}$$

have been used [9] in which σ is a screening parameter which takes values $32 \leq \sigma \leq 36$ to obtain agreement with the energy integrals. More often a representation first due to Slater [10] is employed; see also Burns [11].

The Hartree equations have been solved by Ridley [12] for Pr^{3+} and Tm^{3+}, and Judd and Lindgren [13] have fitted an empirical function to her results for large r.

Solutions of the HF equations have been computed by Freeman and Watson [14] for Ce^{3+}, Pr^{3+}, Nd^{3+}, Sm^{3+}, Eu^{3+}, Gd^{3+}, Er^{3+}, and Yb^{3+}. They present normalized wavefunctions in the form

$$R_{4f}(r) = \sum_{i=1}^{4} C_i r^3 e^{-Z_i r}, \tag{1.5}$$

where the C_i and Z_i are tabulated. Sovers [15] has interpolated their results for the remaining rare earth ions, and values of C_i and Z_i are given for all ions in Table 1.1. Fig. 1.3 shows a plot of the radial charge density (rR_{4f})

Table 1.1 Coefficients in an analytical expansion for Hartree–Fock radial wavefunctions of the $4f^n$ electrons of the tripositive rare earth ions, equation (1.5), taken from [14] and [15].

	n	C_1	C_2	C_3	C_4	Z_1	Z_2	Z_3	Z_4
Ce^{3+}	1	752·34	103·114	16·9079	0·443299	9·815	5·585	3·723	2·034
Pr^{3+}	2	902·18	129·677	20·0061	0·527930	10·271	5·828	3·885	2·125
Nd^{3+}	3	1068·89	159·822	23·5767	0·645797	10·727	6·071	4·047	2·216
Pm^{3+}	4	1251·81	194·289	27·6555	0·798327	11·183	6·314	4·209	2·307
Sm^{3+}	5	1453·86	233·105	32·1353	1·01200	11·639	6·557	4·371	2·398
Eu^{3+}	6	1676·72	278·194	37·3965	1·23048	12·095	6·800	4·533	2·489
Eu^{2+}	7	1562·47	238·626	30·6217	1·31648	11·764	6·603	4·402	2·416
Gd^{3+}	7	1923·82	329·667	43·2748	1·50474	12·554	7·046	4·697	2·578
Tb^{3+}	8	2188·66	385·383	49·2579	1·89821	13·007	7·286	4·857	2·671
Dy^{3+}	9	2480·40	448·837	55·9670	2·35247	13·463	7·529	5·019	2·762
Ho^{3+}	10	2797·67	520·930	63·5658	2·84272	13·919	7·772	5·181	2·853
Er^{3+}	11	3141·61	601·335	71·8456	3·41911	14·375	8·015	5·343	2·944
Tm^{3+}	12	3513·46	690·919	80·9179	4·07972	14·831	8·258	5·505	3·035
Yb^{3+}	13	3914·44	790·999	90·9984	4·80641	15·287	8·501	5·667	3·126

against r for the $4f$, $5s$, $5p$ and $6s$ electrons of Gd^{3+} which was obtained by Freeman and Watson. It clearly demonstrates that the $4f$ electrons are well embedded within the ion.

The $4f$ radial wavefunctions for Pr^{3+} computed by various methods are compared in Fig. 1.4. It can be seen that both Hartree and HF functions predict greater charge density at large r than the hydrogenic functions. The Hartree–Fock–Slater function for the neutral atom is intermediate between Hartree and HF results. In general the effect of exchange is to contract the HF functions relative to the Hartree ones, and this is again apparent from the

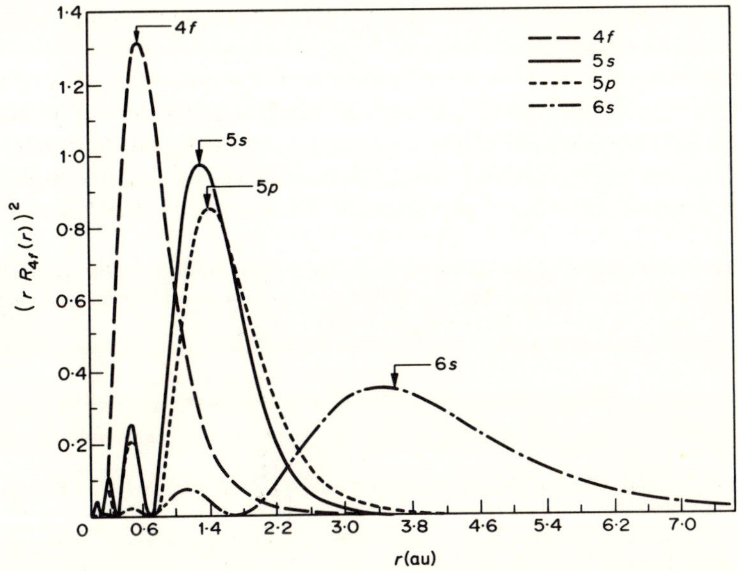

Fig. 1.3 The Hartree–Fock radial densities for the 4*f*, 5*s*, 5*p* and 6*s* electrons of Gd^{3+} [14].

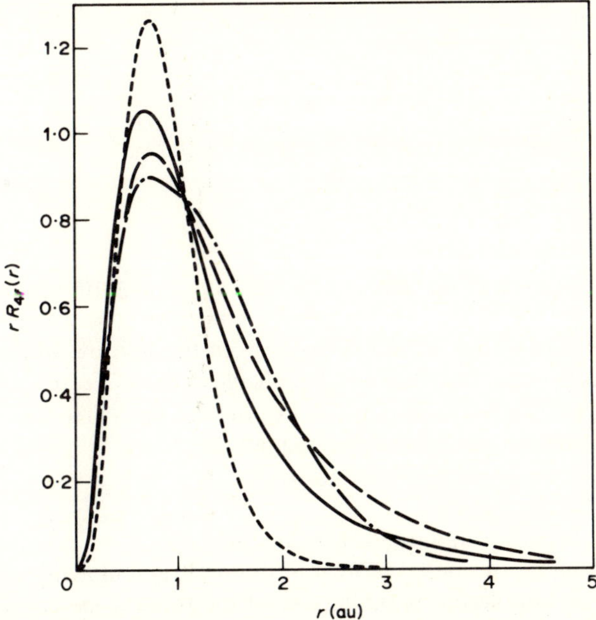

Fig. 1.4 A comparison of the radial 4*f* ;avefunctions of Pr^{3+} calculated by various methods: ---- hydrogenic, —— Hartree–Fock [14], — — Hartree [12]. The curve ·—·— is for the neutral atom [5].

figure. In principle, but see Section 1.1.1, the HF wavefunction should be the best of those obtained.

As the nuclear charge Z increases throughout the rare earth series the potential seen by the $4f$ electrons deepens and their orbitals show a systematic contraction in radius; this effect is known as the *lanthanide* contraction [6–8]. This is shown in Table 1.2 where the positions of the maxima in R_{4f}, computed using the functions of Freeman and Watson, are given for the series.

Table 1.2 The configurations, ground states, and sizes of the rare earth ions. The quantities $(\langle r^2\rangle_{4f})$, which are a measure of the radius of the $4f$ shell, have been computed from the theoretical Hartree–Fock values of $\langle r^2\rangle$ [14]. The ionic radii are after [2].

	Z	R	R^{3+}	$(\langle r^2\rangle_{4f})^{\frac{1}{2}}$ Å	Ground state R^{3+}	Ionic radius Å
La	57	$5d^16s^2$	$5s^25p^6$	0·610	$^1S_0$	1·061
Ce	58	$4f^26s^2$	$4f^15s^25p^6$	0·578	$^2F_{5/2}$	1·034
Pr	59	$4f^36s^2$	$4f^25s^25p^6$	0·550	$^3H_4$	1·013
Nd	60	$4f^46s^2$	$4f^35s^25p^6$	0·528	$^4I_{9/2}$	0·995
Pm	61	$4f^56s^2$	$4f^45s^25p^6$	0·511	$^5I_4$	0·979
Sm	62	$4f^66s^2$	$4f^55s^25p^6$	0·496	$^6H_{5/2}$	0·964
Eu	63	$4f^76s^2$	$4f^65s^25p^6$	0·480	$^7F_0$	0·950
Gd	64	$4f^75d^16s^2$	$4f^75s^25p^6$	0·468	$^8S_{7/2}$	0·938
Tb	65	$4f^96s^2$	$4f^85s^25p^6$	0·458	$^7F_6$	0·923
Dy	66	$4f^{10}6s^2$	$4f^95s^25p^6$	0·450	$^6H_{15/2}$	0·908
Ho	67	$4f^{11}6s^2$	$4f^{10}5s^25p^6$	0·440	$^5I_8$	0·894
Er	68	$4f^{12}6s^2$	$4f^{11}5s^25p^6$	0·431	$^4I_{15/2}$	0·881
Tm	69	$4f^{13}6s^2$	$4f^{12}5s^25p^6$	0·421	$^3H_6$	0·869
Yb	70	$4f^{14}6s^2$	$4f^{13}5s^25p^6$	0·413	$^2F_{7/2}$	0·858
Lu	71	$4f^{14}5d^16s^2$	$4f^{14}5s^25p^6$	0·405	$^1S_0$	0·848

1.1.3 *The Coulomb interaction*

The basic energy levels of free ions are those of the configurations, which are $2(2l_1+1)\ 2(2l_2+1)\ \dots\ 2(2l_n+1)$ degenerate, and have separation $\sim 5\times 10^4$ cm^{-1}. The detailed energy structure arises from the splitting of these levels by interactions between the electrons, here between the $4f$ electrons involved, and also between electrons and nucleus. For the free rare earth ions the three dominant interactions are, in order of magnitude

(i) Coulomb repulsion between electrons
(ii) spin-orbit interaction
(iii) nuclear hyperfine and quadruple interactions

The effect of each is considered in turn in this and the following sections.

The Coulomb interaction is the largest of the three to be considered. It splits each configuration energy level into *term* levels which are specified by values of the total angular momentum quantum number L, and total spin quantum number S.

There are two important rules, known as *Hund's rules*, for predicting which of the terms is the ground state. These are: (i) the lowest state has the maximum number of electrons with parallel spins, i.e. the largest possible S; (ii) subject to (i) the lowest level also has the largest possible L. The first rule is reasonable in that the Pauli exclusion principle favours parallel spins. In some cases the ground state configuration may often be modified to obtain the maximum number of parallel spins, e.g. $Eu^{3+}(4f^6)$ to $4f^7$. In the case of a half-filled shell, $Gd^{3+}(4f^7)$, the first rule is sufficient to determine L, because all $(2l+1)$ orbital states are occupied by one electron, so that for every $+m_l$ there is a $-m_l$, i.e. $L = 0$. The second rule can also be made plausible. For L to be a maximum all the l_i must be parallel, so that all electrons go around the nucleus in the same direction, which is the most favourable condition for electrons to be able to keep apart. Both rules appear to be valid for all ions. The lowest terms, and their energy separations for the rare earths are given in Fig. 1.5. The theoretical calculation of the energy splitting between terms involves evaluation of the Slater radial integrals $F^k(4f, 4f)$ [16, 3].

A wavefunction $\psi_0(L, S, M_L, M_S)$ which has the correct eigenvalues L and S for a given term can be constructed by taking suitable linear combinations of the wavefunctions of the configuration from which that term comes. In doing this it is found, and this is a general rule, that configurations $4f^{14-n}$ and $4f^n$ give rise to identical terms.

Other wavefunctions $\psi_n(L, S, M_L, M_S)$ may be constructed, with the same L and S, from other configurations, and in general the best term eigenfunction is a linear combination of all such functions, i.e.

$$\Psi(L, S, M_L, M_S) = \alpha_0\psi_0 + \sum_n \alpha_n\psi_n. \tag{1.6}$$

The mixing of the ψ_n's with ψ_0 in this way is called *configurational interaction.* For rare earth ions the energy separations between configurations is much larger than the Coulomb interaction, and consequently the mixing is small. In this case second order perturbation theory is adequate for calculating the mixing and energy shifts due to the interaction [17]. However, for many purposes configurational interaction is neglected.

In constructing wavefunctions for different configurations the concept of *fractional parentage* is useful [3]. This involves expanding the wavefunction of a configuration f^n in terms of products of states of the wavefunctions for configurations f^{n-1} with those of the nth particle, i.e.

$$\psi(f^n; L, S, M_L, M_S) = \sum_{S'L'} \beta \sum_{M'_S M'_L m_s m_l} \gamma\psi(f^{n-1}; L', S', M'_L, M'_S)\phi(f; m_l, m_s) \tag{1.7}$$

Here the β are known as the coefficients of fractional parentage, and the γ are complicated constants which correctly couple together the spins.

1.1.4 *The spin-orbit interaction*

The coupling of the individual orbital angular momenta $\mathbf{l}_i$ and $\mathbf{s}_i$ between

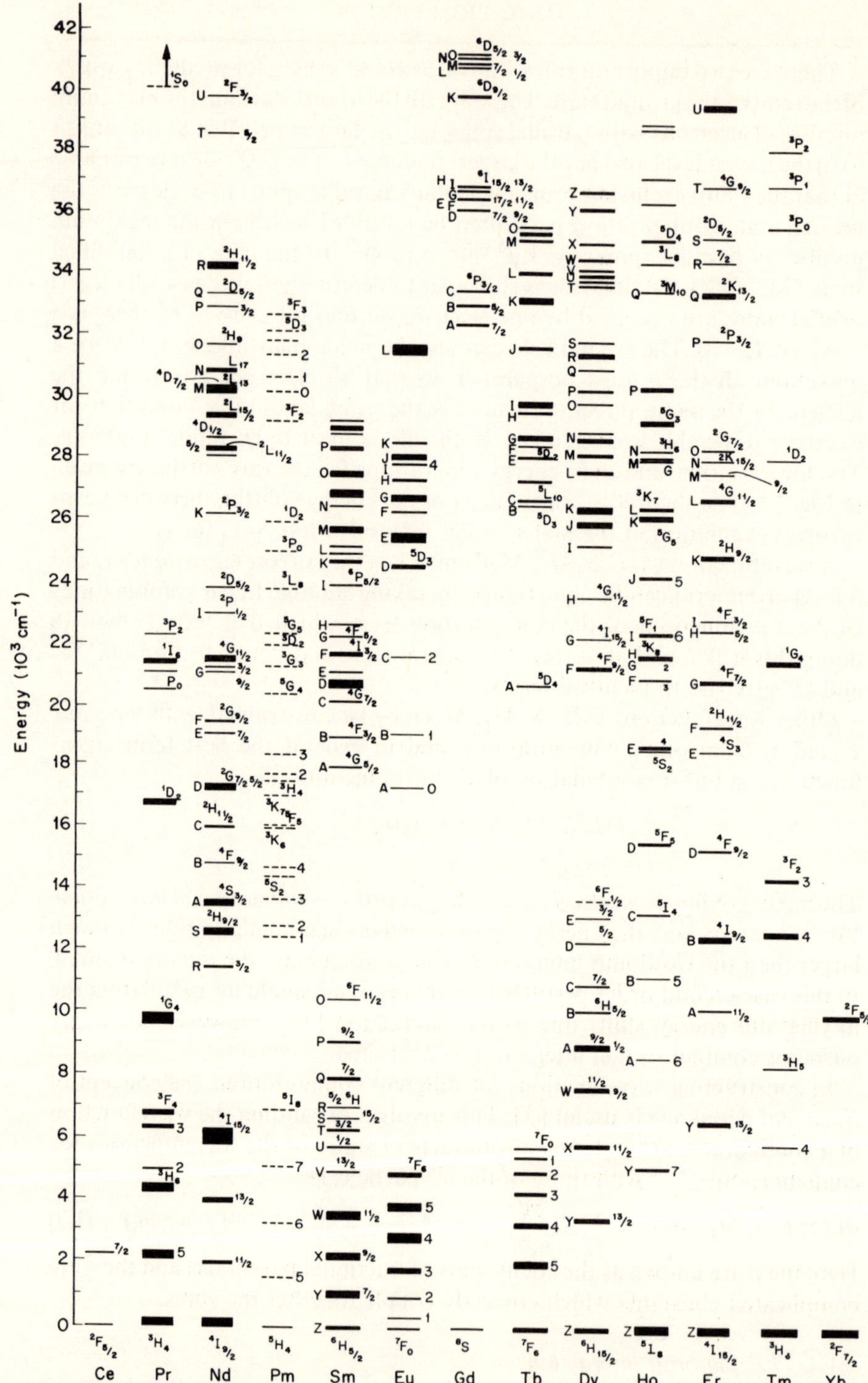

Fig. 1.5 The observed multiplet energy levels of the rare earth ions in $LaCl_3$ (after ref. [2]).

electrons is a relativistic effect; see Section 1.1.7. Within the central field approximation the interaction takes the form

$$\mathscr{H}_{sl} = \sum_{i=1}^{n} \xi(r_i)\mathbf{l}_i.\mathbf{s}_i, \tag{1.8}$$

where

$$\xi(r_i) = \frac{\hbar^2}{2m^2c^2}\frac{1}{r_i}\frac{\partial V(r_i)}{\partial r}. \tag{1.9}$$

For the rare earths the Coulomb interaction is much stronger than (1.8) so that the individual $\mathbf{l}_i$ and $\mathbf{s}_i$ always couple to give $\mathbf{L}$ and $\mathbf{S}$. The effect of the spin-orbit interaction is then to couple $\mathbf{L}$ and $\mathbf{S}$ according to the Russell–Saunders scheme, $\mathbf{L}+\mathbf{S} = \mathbf{J}$, and partially lift the degeneracy of the term; J is the total angular momentum. The resulting multiplet of levels, which are each $(2J+1)$ degenerate, is the starting point of most theoretical considerations of the rare earths. L and S are still good quantum numbers and, since the overall multiplet splitting due to $\mathscr{H}_{sl}$ is usually much less than the energy separation of the terms, wavefunctions $\Psi(L, S, J, M_J)$ may be constructed from the $\Psi(L, S, M_L, M_S)$ of a single term, i.e. mixing from other terms can usually be neglected:

$$\Psi(L, S, J, M_J) = \sum_{M_L', M_S'} C\Psi(L, S, M_L, M_S) \tag{1.10}$$

where the constants C are known as the *Clebsch–Gordon* or *Wigner coefficients.*

The multiplet splittings are given by evaluating matrix elements of $\mathscr{H}_{sl}$ between single particle wavefunctions $|n_i, l_i, m_{l_i}, m_{s_i}\rangle$. For ions with a single unfilled shell, here the $4f$ shell, the matrix elements reduce to

$$\zeta_{4f} \sum_{i=1}^{n} m_{l_i} m_{s_i},$$

where

$$\zeta_{4f} = \hbar^2 \int_0^\infty r^2 R_{4f}^2(r)\xi(r)\mathrm{d}r \tag{1.11}$$

is *the spin-orbit radial integral.* It can be shown [18] that assuming Russell–Saunders coupling, identical matrix elements are obtained between term wavefunctions $|L, S, M_L, M_S\rangle$ if the operator

$$\mathscr{H}_{SL} = \zeta(LS)\mathbf{L}.\mathbf{S} \tag{1.12}$$

is employed for the spin-orbit coupling, where $\zeta(LS)$ is to be determined. This is useful since the matrix elements of (1.12) are much easier to calculate than those of (1.8). In particular, using first order perturbation theory it is readily shown that $\mathscr{H}_{SL}$ splits a single term level E_T into levels

$$E_J = E_T + \langle L, S, M_L, M_S | \mathscr{H}_{SL} | L, S, M_L, M_S \rangle$$
$$= E_T + \tfrac{1}{2}\lambda\zeta_{4f}(LS)[(J(J+1) - L(L+1) - S(S+1)], \tag{1.13}$$

where $\lambda = \pm\frac{1}{2}S$ for states of maximum multiplicity; see Elliott *et al* for other states. For a given L and S the *Landé interval rule* for the separation of multiplet levels follows immediately:

$$E_J - E_{J-1} = \lambda\zeta_{4f}(LS)J. \tag{1.14}$$

For ions with their $4f$ shells less than half full the lowest energy level has the smallest J value, i.e. $|L-S|$, so that $\lambda > 0$, while if the shell is more than half full it has the largest possible J, i.e. $|L+J|$, so that $\lambda < 0$. The Landé rule is a reasonable approximation for the rare earths.

Departures from the Landé rule have been discussed by Judd [19]. There is a correction due to interactions between the magnetic moments of the $4f$ electrons, but this is very small. Deviations also arise from a partial breakdown of Russell–Saunders coupling which occurs if the spin-orbit coupling is sufficiently strong to mix other terms into the ground state term. This situation is described as *intermediate coupling*. In practice these corrections are neglected when considering the lowest multiplet levels of the ions.

The spin-orbit coupling constant ζ_{4f} is most often treated as an empirical parameter, and values are shown for the lanthanides in Table 1.3. From (1.11)

Table 1.3 The spin–orbit coupling constants ζ_{4f} and radial integrals $\langle r^{-3}\rangle$ for tripositive ions. The table is based upon results given by Judd [20]. The Hartree–Fock values are from [14] and the Lindgren (Hartree) values are from [13]. $\langle r^{-3}\rangle_{exp}$ data marked * is from hyperfine measurements and independent determination of μ_I; the other values have been interpolated. $\langle r^{-3}\rangle_{eff}$ is discussed in the text. H = Hartree.

Ion	ζ_{4f} (cm^{-1})			$\langle r^{-3}\rangle$ (a_0^{-3})			
	H	HF	expt	HF	expt	eff	Lindgren
Ce^{3+}		830	640	4·72		4·17	3·66
Pr^{3+}	785	980	750	5·37	5·06	4·76	4·26
Nd^{3+}		1130	900	6·03	5·64*	5·35	4·86
Pm^{3+}			1070		6·18	5·95	5·46
Sm^{3+}		1480	1200	7·36	6·72	6·56	6·07
Eu^{3+}			1320		7·28	7·22	6·70
Gd^{3+}			1620	8·84	7·89	7·90	7·35
Tb^{3+}			1700		8·53	8·60	8·03
Dy^{3+}		2310	1900	10·34	9·20	9·32	8·74
Ho^{3+}			2160		9·91	10·18	9·50
Er^{3+}		2830	2440	12·01	10·60*	10·89	10·32
Tm^{3+}	2740		2640		11·72	11·73	11·20
Yb^{3+}		3400	2880	13·83	12·50*	12·63	12·18

and (1.9) it can be seen that ζ_{4f} depends upon the radial wavefunctions and the central potential $V(r)$. The theoretical values obtained using Hartree–Fock [14] and Hartree [12] wavefunctions, but in both cases a Hartree potential, are shown in Table 1.3. The experimental values are mostly those obtained for the trivalent ions in salts or solutions, but it is thought that their magnitudes are little affected by the ions' environment. It will be observed that the Hartree ζ_{4f} are in better agreement with experiment than the HF. Because of exchange the HF radial functions are more contracted, and this leads to larger ζ_{4f} values. However, as discussed in Section 1.1.7, relativistic effects will partly cancel those of exchange. Judd [20] has suggested that configuration interaction is important and Blume *et al* [72] have found that two body corrections to $\mathscr{H}_{SL}$ modify the HF values in the right direction. Therefore, again it appears that the Hartree method is better because of a cancellation of errors.

It is also possible to obtain an empirical $\langle r^{-3}\rangle_{\text{so}}$ *parameter* from the multiplet splittings. If the central field potential is assumed to be Coulombic i.e. $V(r) = -e^2 Z^*_{4f}/r$, where Z^*_{4f} is an effective nuclear charge seen by a $4f$ electron, then by (1.9) and (1.11)

$$\zeta_{4f} = \frac{\hbar^2 e^2}{2m^2c^2} Z^*_{4f}\langle r^{-3}\rangle_{\text{so}}. \tag{1.15}$$

Since HF calculations predict ζ_{4f} values which are too large, then the $\langle r^{-3}\rangle_{\text{HF}}$ values will also be too large. Judd [20] has represented the effects of configurational interaction, which essentially lead to a further screening of the nucleus, by a small correction Δ_{so} to $\langle r^{-3}\rangle_{\text{HF}}$ defined such that the effective theoretical average ($\langle r^{-3}\rangle_{\text{eff}}$) is given by $(1+\Delta_{\text{so}})\langle r^{-3}\rangle_{\text{HF}}$. Empirical formulae like (1.15) have been employed [21, 22] to estimate nuclear magnetic moments from hyperfine splittings.

1.1.5 *Hyperfine structure*

If the nucleus has spin $\mathbf{I}$ then associated with it is a magnetic moment $\boldsymbol{\mu}_I = g_I \mu_{\text{N}} \mathbf{I}$, where g_I is a splitting factor and μ_{N} the nuclear magneton. Values for the rare earths are shown in Table 1.4. Due to their orbital and spin angular momenta, the $4f$ electrons produce an effective magnetic (hyperfine) field $\mathbf{H}_{\text{eff}}$ at the nucleus which interacts with $\boldsymbol{\mu}_I$. The complete Hamiltonian is [23]

$$\begin{aligned}\mathscr{H}_{\text{N}} &= -\boldsymbol{\mu}_I.\mathbf{H}_{\text{eff}} \\ &= 2g_I\mu_{\text{B}}\mu_{\text{N}}\left\{\sum_{i=1}^{n}\frac{8\pi}{3}\,\delta(\mathbf{r}_i)\mathbf{s}_i + \frac{\mathbf{N}}{r_i^3}\right\}.\mathbf{I},\end{aligned} \tag{1.16}$$

where μ_{B} is the Bohr magneton $= e\hbar/2mc$, and

$$\mathbf{N} = \sum_{i=1}^{n}\left\{\mathbf{l}_i - \mathbf{s}_i + \frac{3(\mathbf{r}_i.\mathbf{s}_i)\mathbf{r}_i}{r_i^2}\right\}. \tag{1.17}$$

Table 1.4 The nuclear properties of some rare earth isotopes, indicating the method of determination. The values are based upon those quoted by Low [98].

Isotope		Nuclear spin	Nuclear magnetic moments (μ_N)		Quadrupole moment (barns)	
Ce	141	7/2	0·968	esr		
Pr	141	5/2	4·28	AB	−0·0589	AB
	142	2	0·250	AB	0·0297	AB
Nd	143	7/2	−1·0644	AB	−0·482	AB
			−1·085	ENDOR	0·0207	ENDOR
	145	7/2	−0·6534	AB	−0·255	
			−0·675	ENDOR	0·0105	ENDOR
	147	5/2	0·579	esr		
Pm	147	7/2	2·77	AB		
Sm	147	7/2	−0·796	AB	−0·208	AB
			0·812	esr		
	149	7/2	−0·643	AB	0·060	AB
			0·656	esr		
Eu	151	5/2	3·419	AB		
	152	3	1·931	esr		
	153	5/2	1·507	AB		
	154	3	2·001	esr		
Gd	155	3/2	−0·2567	ENDOR		
	157	3/2	−0·3363	ENDOR		
Tb	156	3	1·41		1·40	
	159	3/2	1·60	esr	1·32	
	160	3	1·56		1·87	
Dy	161	5/2	−0·455	esr	1·35	esr
	163	5/2	0·635	esr	1·62	esr
Ho	165	7/2	4·01	AB	2·82	AB
Er	167	7/2	−0·564	AB	2·82	AB
			0·563	esr		
	171	5/2	2·96	AB	6·4	AB
Tm	166	2	0·0465	AB	4·36	AB
	169	1/2	−0·229	AB		
	170	1	0·245	AB	0·574	AB
Yb	171	1/2	0·4930	nmr		
	173	5/2	−0·673	esr	3·0	

The δ-function term is a *contact term* which is only non-zero for *s*-electrons, in which case the other term is zero. The effect of (1.16) is to give rise to hyperfine splittings of the multiplet levels. Because μ_N is only $(1/1836)\mu_B$ the interaction is much smaller than those between the electrons. In the rare earths the splittings are $\sim 10^{-2}-10^{-3}$ cm^{-1}.

Assuming that $\mathbf{H}_{\text{eff}} \propto \mathbf{J}$, equation (1.16) may be written as

$$\mathscr{H}_N = a_J \mathbf{I}.\mathbf{J}, \tag{1.18}$$

where a_J is the *magnetic hyperfine structure constant.* In the absence of external fields, $\mathscr{H}_N$ couples $\mathbf{I}$ and $\mathbf{J}$ to give states characterized by $\mathbf{F} = \mathbf{I}+\mathbf{J}$, and F, I, L and J are good quantum numbers. Each multiplet energy is split into levels which are $(2F+1)$ degenerate. The hyperfine splittings depend upon a_J in the same way that the multiplet structure depends upon ζ_{4f} in (1.13), i.e.

$$\begin{aligned} E_F - E_J &= \tfrac{1}{2}a_J[F(F+1) - I(I+1) - J(J+1)] \\ &= \tfrac{1}{2}a_J K \qquad \text{(say).} \end{aligned} \tag{1.19}$$

Explicit expressions for a_J are obtained by equating matrix elements of (1.16) and (1.18). For a single electron it may be shown that

$$\left.\begin{aligned} a_J &= \frac{2g_I\mu_N\mu_B}{J(J+1)} L(L+1)\langle r^{-3}\rangle_h, \text{ for a non } s\text{-electron,} \\ &= \frac{16\pi}{3} g_I\mu_N\mu_B|\psi(0)|^2 \qquad , \text{ for an } s\text{-electron,} \end{aligned}\right\} \tag{1.20}$$

where $\langle r^{-3}\rangle_h$ is an average of r^{-3} between radial wavefunctions, and $|\psi(0)|^2$ is the electron density at the nucleus. For more than one (non s-) electron, and with Russell–Saunders coupling [24],

$$\begin{aligned} a_J &= \frac{2g_I\mu_N\mu_B}{J(J+1)}[\mathbf{L}.\mathbf{S} + \xi'\{L(L+1) - 3(\mathbf{L}.\mathbf{J})(\mathbf{L}.\mathbf{S})\}]\langle r^{-3}\rangle_h \\ &= 2g_I\mu_N\mu_B\langle r^{-3}\rangle_h\langle J\|N\|J\rangle, \end{aligned} \tag{1.21}$$

where the numbers $\langle J\|N\|J\rangle$ are given by Elliott and Stevens [21] for the rare earths, and are reproduced in Table 1.8. ξ' is a constant.

A direct comparison of theoretical and experimental hyperfine splittings for the rare earths is not particularly rewarding because of a lack of knowledge about the values of the nuclear magnetic moments. In principle the structure constant a_J is amenable to experimental determination in several ways: (1) nuclear magnetic resonance, (2) atomic beam magnetic resonance (AB), (3) optical spectroscopy, (4) electron spin resonance, (5) electron nuclear double resonance (ENDOR), and (6) nuclear alignment measurements. Most of these methods involve the application of a large magnetic field, with the result that it is the Zeeman levels of the ion which are split into $(2I+1)$ hyperfine states. A discussion of the effect of a magnetic field is deferred until Section 1.2. In the past the hyperfine splittings have been used together with theoretical estimates of $\mathbf{H}_{\text{eff}}$ to obtain the nuclear moments. The value of $\mathbf{H}_{\text{eff}}$ will depend upon the ions environment and this will be considered later. For the deeply embedded $4f$ electrons the effects of environment are minimized, but even so it is found that it is not possible to correlate experimental hyperfine field estimates made from different measurements. Consequently a common procedure is to use theoretical hyperfine fields calculated from the constant a_J, in which the critical quantity is $\langle r^{-3}\rangle_h$. On theoretical grounds it is expected that the $\langle r^{-3}\rangle_h$ involved here is not exactly the same as $\langle r^{-3}\rangle_{\text{so}}$, if only for the obvious reason that the averages are taken within the matrix elements of the different Hamiltonians $\mathscr{H}_N$ and $\mathscr{H}_{SL}$. These differences are observed experimentally. For the calculated radial integrals $\langle r^{-3}\rangle_h$ it should be noted that the Hartree values [13] are not any better than the HF ones. A correction Δ_h to the HF value $\langle r^{-3}\rangle_{HF}$ arising due to configurational interaction has been calculated by Judd [20]; explicitly $\langle r^{-3}\rangle_h = (1+\Delta_h)\langle r^{-3}\rangle_{HF}$. However, Watson and Freeman [25] have suggested that $\Delta_h \sim \Delta_{so}$, and that configuration interaction is not so important.

For an S-state ion such as Gd^{3+} or Eu^{2+} it can be seen from (1.20) that there is no orbital contribution to the hyperfine field. In the case of Gd^{3+} there are no unpaired s-electrons, so that there should also be no contact term. In fact hyperfine fields of $\sim -3\times 10^5$ gauss are observed for Gd [81] and Eu [82], and these are explained by assuming that the densities ρ_+ and ρ_- of up and down s-electrons do not cancel at the nucleus because of *exchange* (or *core*) *polarization.* The exchange interaction between the s-electrons and the net $4f$ electron spin, **S**, of a single ion, favours more s-electron spins opposed to **S**. An approximate calculation of its effect has been made for Gd [25] using an *unrestricted Hartree–Fock* method. The restriction of the HF method that each spatial orbital $\psi(\mathbf{r})$ has two electrons of opposite spin, is removed. Results of the calculation for $\rho_+ - \rho_-$ at the nucleus give a hyperfine field in good agreement with experiment.

1.1.6 *The nuclear quadrupole interaction*

If the nucleus has a quadrupole moment Q there is also an electrostatic interaction between it and the $4f$ electrons of the form

$$\mathscr{H}_Q = e^2 q_J Q\left[\frac{3(\mathbf{I}.\mathbf{J})^2+\frac{3}{2}(\mathbf{I}.\mathbf{J})-I(I+1)J(J+1)}{2I(2I-1)J(2J-1)}\right], \tag{1.22}$$

where Q, the scalar quadrupole moment, is given in terms of the nuclear charge density ρ_n and coordinates $\mathbf{r}_n$ by

$$Q = \int \rho_n(\mathbf{r}_n)_{M_I=I}(3z_n^2 - r_n^2)\mathrm{d}\mathbf{r}_n, \tag{1.23}$$

and

$$\begin{aligned} eq_J &= \int \rho_e(\mathbf{r}_e)_{M_J=J}\left(\frac{3\cos^2\theta-1}{r_e^3}\right)\mathrm{d}\mathbf{r}_e \\ &= -eJ(2J-1)\alpha_J\langle r^{-3}\rangle_q \end{aligned} \tag{1.24}$$

is the field gradient at the nucleus due to the $4f$ electrons. The energy level splittings of states $|I, J, F, M_F\rangle$ are

$$E_Q = e^2 q_J Q\left[\frac{\frac{3}{4}K(K+1)-I(I+1)J(J+1)}{2I(2I-1)J(2J-1)}\right], \tag{1.25}$$

and hence again depend upon $\langle r^{-3}\rangle_q$. In general $\langle r^{-3}\rangle_q$ differs from values obtained from either ζ_{4f} or a_J.

The quadrupole moment gives rise to a potential having symmetry described by the spherical harmonic Y_2^0. The effect of this non-central potential is to cause configurational interaction so that orbitals ψ_l', ψ_{l+2} and ψ_{l-2} are mixed into the central field orbital ψ_l. Physically the angular distribution of the electronic charge density may be distorted without radial change, which leads to screening of Q, or else there may be radial distortions, which lead to antishielding. In either case it is common to account for the effects by introducing a correction to $\langle r^{-3}\rangle_q$ and hence to eq_J, which is known as the *Sternheimer shielding factor* [21] R_q, i.e.

$$eq_{J,\mathrm{eff}} = eq_J(1-R_q). \tag{1.26}$$

Calculations [39] of R_q for Pr^{3+} and Tm^{3+} using Hartree wavefunctions [12] give values of 0·2 and 0·15 respectively, and the value for Tm is in good agreement with experimental results on the ethylsulphate. In both cases it is found that the angular (shielding) distortion dominates. For the radial distortions, the inner orbitals produce an antishielding, but this is largely cancelled by shieldings due to the outer orbitals, so that the net effect of radial perturbations is weak antishielding. A Thomas–Fermi calculation [22] for the Eu atom indicates that $R_q \sim 0{\cdot}4$, but because of the method, this is probably too high by a factor of 1·5. For ions in crystals there is the added complication of field gradients arising from neighbouring ions, and this will be discussed in Section 1.4.2.

1.1.7 *Relativistic effects*

In general relativistic effects become appreciable when the potential energy of an electron is not negligible compared with its rest mass. For hydrogen-like atoms, for which $V = Ze^2/r$, it is clear that such effects are most important for heavy atoms $Z \gtrsim 55$, and especially so for the electrons of such atoms with the smallest quantum numbers. Since the rare earths are in the second half of the periodic table, and since the spin-orbit coupling is a relativistic effect, and it is large for the rare earths, it might be expected that other relativistic effects are also important.

Dirac's relativistic wave equation for a single particle has the form

$$\{c\boldsymbol{\alpha}\,.\,\mathbf{p}+\beta mc^2+V(\mathbf{r})\}\psi = E\psi, \tag{1.27}$$

where $\mathbf{p}$ is the momentum, and β and the three components of $\boldsymbol{\alpha}$ are all 4×4 matrices. ψ has four components, forming a column vector, and for electrons the two upper components are much larger than the lower ones, while the converse is true for positrons. In the case of one electron in a central potential

$V(r)$ it may be shown that for low energies (1.27) approximates to the following equation for the two large components [4]:

$$\left[\frac{\mathbf{p}^2}{2m}+V(r)+\frac{1}{2m^2c^2r}\frac{\mathrm{d}V}{\mathrm{d}r}\mathbf{s}.\mathbf{l}-\frac{\mathbf{p}^4}{8m^3c^2}\right.$$
$$\left.-\frac{\hbar^2}{4m^2c^2}\frac{\mathrm{d}V}{\mathrm{d}r}\frac{\partial}{\partial r}\right]\psi(\mathbf{r})\begin{pmatrix}\chi_{+\frac{1}{2}}\\ \chi_{-\frac{1}{2}}\end{pmatrix}=E\psi(\mathbf{r})\begin{pmatrix}\chi_{+\frac{1}{2}}\\ \chi_{-\frac{1}{2}}\end{pmatrix}, \tag{1.28}$$

where the electronic spin $\mathbf{s}$ is defined in terms of the Pauli matrices $\boldsymbol{\sigma}$, $\mathbf{s}=\frac{1}{2}\boldsymbol{\sigma}$, and the $\chi_{\pm\frac{1}{2}}$ are the Pauli spin functions which satisfy $s_z\chi_{\pm\frac{1}{2}}=\pm\frac{1}{2}\chi_{\pm\frac{1}{2}}$. Here $\mathbf{l}=\mathbf{r}\times\mathbf{p}$ is the orbital angular momentum of the electron. The first two terms give the Schrödinger equation, while the third term is just the spin-orbit interaction, and the fourth term represents a relativistic mass correction. The final term, which is known as the Darwin term, depends upon the divergence of the electric field produced by the nucleus, and is only non-zero for *s*-state electrons. The mass and Darwin corrections shift the non-relativistic energies, but do not lift any degeneracies. Herman and Skillman [5] have used first order perturbation theory to calculate the effect of the final three terms of (1.28) upon their Hartree–Fock–Slater eigenvalues. Typically for the $4f$ electrons of the rare earths they find that the mass correction is $\sim 5\%$, while the Darwin correction is only $\sim 0{\cdot}1\%$.

The Dirac equation for a central field can be separated in spherical coordinates without approximation; see for example Schiff [4] or, for a more detailed exposition, Rose [95]. The four component equation becomes

$$\left\{-ic\alpha_r\left(\frac{\partial}{\partial r}+\frac{1}{r}\right)+\frac{ic}{r}\alpha_r\beta^2K+\beta mc^2+V\right\}\psi=W\psi, \tag{1.29}$$

where $\alpha_r=r^{-1}(\boldsymbol{\alpha}.\mathbf{r})$ and $K=(\boldsymbol{\sigma}.\mathbf{l}+1)$. The separated wavefunction may be written as

$$\psi_\kappa^\mu=\frac{1}{(2l+1)^{\frac{1}{2}}}\left\{\begin{matrix}g_\kappa(r)\begin{bmatrix}(1+\mu+\frac{1}{2})^{\frac{1}{2}}\chi_{+\frac{1}{2}}Y_l^{\mu-\frac{1}{2}}\\(1-\mu+\frac{1}{2})^{\frac{1}{2}}\chi_{-\frac{1}{2}}Y_l^{\mu+\frac{1}{2}}\end{bmatrix}\\ if_\kappa(r)\begin{bmatrix}-(1-\mu+\frac{1}{2})^{\frac{1}{2}}\chi_{+\frac{1}{2}}Y_{l'}^{\mu-\frac{1}{2}}\\(1+\mu+\frac{1}{2})^{\frac{1}{2}}\chi_{-\frac{1}{2}}Y_{l'}^{\mu+\frac{1}{2}}\end{bmatrix}\end{matrix}\right\}, \tag{1.30}$$

This particular form is chosen because, besides being an eigenfunction of the Hamiltonian, it is simultaneously an eigenfunction of the three operators $\mathbf{j}^2=(\mathbf{l}+\mathbf{s})^2$, j_z, and K, with eigenvalues of $j(j+1)$, μ and $-\kappa$ say, respectively. Explicitly

$$\kappa=l(l+1)-(j+\tfrac{1}{2})^2=\pm1,\pm2,\cdots, \tag{1.31}$$

so that its value specifies both j and l, and therefore is a very convenient quantum number. Substituting ψ_κ^μ into (1.29) it may be shown that the radial functions g and f satisfy

$$\left.\begin{aligned}\frac{\mathrm{d}f}{\mathrm{d}r} &= \frac{\kappa-1}{r}f-(W-1-V)g\\ \frac{\mathrm{d}g}{\mathrm{d}r} &= (W-V+1)f-\frac{\kappa+1}{r}g\end{aligned}\right\}, \tag{1.32}$$

where the units are (see Appendix 1) $m = c = \hbar = 1$, $e^2 = 1/137$, and W contains the electron's rest mass $mc^2 = 1$. Normalization requires

$$\int_0^\infty r^2(g^2+f^2)\mathrm{d}\mathbf{r} = 1,$$

and the electronic charge density at $\mathbf{r}$ is the sum $\rho(\mathbf{r}) = \Sigma_\kappa(g_\kappa^2(r)+f_\kappa^2(r))$ over all occupied orbitals.

Liberman *et al* [96] have obtained self-consistent solutions of equations (1.32) for atoms and ions of a large number of elements, employing the Slater $\rho^{1/3}$ approximation for the exchange potential. A general relativistic effect is that electrons with small quantum numbers are more concentrated near the nucleus than non-relativistic calculations predict. These electrons therefore screen the nucleus more effectively, with the result that electrons with higher quantum numbers, $l = 2, 3$, are less tightly bound. On the other hand, the effect of exchange is to contract the orbitals of the outer electrons, thereby tending to cancel the relativistic effect for such electrons, and the simple Hartree method would therefore be expected to be the best of the non-relativistic treatments. Unfortunately the results of Liberman *et al* for the rare earths are not readily accessible. For Cu ($Z = 29$) they find that the relativistic results differ little from the simple Hartree ones, but for Hg ($Z = 80$) they find appreciable deviations from previous non-relativistic calculations and obtain much better agreement with experimental X-ray data.

Without doubt the relativistic solutions should be employed for calculations involving the rare earths. However, these are only known numerically and hence their use is not expected to lead to a better understanding of the important physical factors which determine the properties of the ions. In the rest of this chapter relativistic effects, except spin-orbit coupling, will therefore be neglected. However, they will be considered again in Section 3.4 in a discussion of the band structures of rare earth metals, where the small relativistic energy shifts can be very important.

1.2 Free ions in a magnetic field

Ignoring hyperfine structure, the basic energy levels of a free rare earth ion are those of the multiplets (E_J). The application of an external magnetic field H_z lifts the $(2J+1)$ degeneracy of these levels. Due to the unfilled $4f$ shell an ion has a permanent magnetic moment $\boldsymbol{\mu} = -\mu_B(\mathbf{L}+2\mathbf{S})$ which interacts with H_z:

$$\mathscr{H}_H = -\boldsymbol{\mu}.\mathbf{H} = \mu_B H_z(L_z+2S_z). \tag{1.33}$$

The Zeeman splittings are much smaller than the multiplet energy separations, except for Sm^{3+} and Eu^{2+}, so that first order perturbation theory is adequate for their calculation. The matrix elements required are readily achieved using

$$\langle J, M_J|\mathbf{L}+2\mathbf{S}|J, M'_J\rangle = g_J\langle J, M_J|\mathbf{J}|J, M'_J\rangle, \tag{1.34}$$

where q_J is the Landé splitting factor. This is a simple example of a more general result known as the Wigner–Eckart theorem. Hence the $(2J+1)$ Zeeman levels are given by

$$\begin{aligned} E_{J,M_J} - E_J &= g_J\mu_B H_z\langle J, M_J|J_z|J, M_J\rangle \\ &= g_J\mu_B H_z M_J = E^{(1)}H_z \quad \text{say.} \end{aligned} \tag{1.35}$$

1.2.1 *The paramagnetic susceptibility*

A knowledge of the energy levels E_i in a magnetic field enables the paramagnetic susceptibility χ of an ensemble of N non-interacting ions to be calculated from

$$\chi = -\frac{N}{H}\left\{\sum_{i=1}\frac{\partial E_i}{\partial H}e^{-E_i/kT}\right\}\bigg/\sum_{i=1}e^{-E_i/kT}. \tag{1.36}$$

Excepting Sm^{3+} and Eu^{2+}, the first excited multiplet levels of the rare earths have energies much greater than kT at room temperature and are not populated. Employing energies E_{J,M_J} in (1.36), and summing over M_J for the ground state multiplet, *Hunds' formula* is obtained,

$$\chi = Ng_J^2\mu_B^2 J(J+1)/3kT. \tag{1.37}$$

This is in good agreement with the experimental susceptibilities of most rare earth ions in salts, as may be seen from Table 1.5.

The separation of multiplet levels for Sm^{3+} and Eu^{2+} is comparable with kT so that not only do levels other than the ground state contribute to χ, but the second order correction to E_{J,M_J}, of the form $E^{(2)}H^2$ say, is important. Expanding the Boltzmann factor for $\mu H \ll kT$ and since there is no ordered moment, so that $\sum_{J,M_J} E^{(1)}e^{-E_{J,M_J}/kT} = 0$, (1.36) becomes

$$\chi = -N\sum_{J,M_J}\left\{2E^{(2)} - \frac{[E^{(1)}]^2}{kT}\right\}e^{-E_J/kT}\bigg/\sum_{J,M_J}e^{-E_J/kT}. \tag{1.38}$$

For the summation over M_J, the temperature-dependent term gives Hund's formula, while the temperature-independent term, called the *Van Vleck*

Table 1.5 Theoretical and experimental data for the susceptibilities of rare earth ions. The susceptibility per atom is given by $\mu_B p^2/3kT$. The values in parentheses for Sm and Eu are calculated using the Van Vleck formula, equation (1.41).

Ion	Ground state	S	L	J	g_J	$p^2 = g^2{}_J(J+1)$	p^2 expt.
La^{3+}	$^1S_0$	0	0	0		0	0
Ce^{3+}	$^2F_{5/2}$	1/2	3	5/2	6/7	6·43	6
Pr^{3+}	$^3H_4$	1	5	4	4/5	12·8	12
Nd^{3+}	$^4I_{9/2}$	3/2	6	9/2	8/11	13·1	12
Pm^{3+}	$^5I_4$	2	6	4	3/5	7·2	—
Sm^{3+}	$^6H_{5/2}$	5/2	5	5/2	2/7	0·71 (2·5)	2·4
Eu^{3+}	$^7F_0$	3	3	0		0 (12)	12·6
Gd^{3+}	$^8S_{7/2}$	7/2	0	7/2	2	63	63
Tb^{3+}	$^7F_6$	3	3	6	3/2	94·5	92
Dy^{3+}	$^6H_{15/2}$	5/2	5	15/2	4/3	113	110
Ho^{3+}	$^5I_8$	2	6	8	5/4	112	110
Er^{3+}	$^4I_{15/2}$	3/2	6	15/2	6/5	92	90
Tm^{3+}	$^3H_6$	1	5	6	7/6	57	52
Yb^{3+}	$^2F_{7/2}$	1/2	3	7/2	8/7	20·6	19

term, may be written

$$\frac{2N}{(2J+1)}\sum_{M_J} E^{(2)} = \frac{2N}{(2J+1)}\sum_{M_J}\left\{\frac{|\langle J|\mathscr{H}_H|J+1\rangle|^2}{E_J-E_{J+1}}+\frac{|\langle J|\mathscr{H}_H|J-1\rangle|^2}{E_J-E_{J-1}}\right\}$$

$$= \frac{N\mu_B^2}{6(2J+1)}\left\{\frac{F(J+1)}{E_J-E_{J+1}}-\frac{F(J)}{E_J-E_{J-1}}\right\}$$

$$= N\alpha \quad \text{say,} \tag{1.39}$$

with

$$F(J) = J^{-1}[(S+L+1)^2-J^2][J^2-(S-L)^2]. \tag{1.40}$$

Finally, summing the Hund and Van Vleck terms over allowed J, noting that each J state is $(2J+1)$ degenerate,

$$\chi = \frac{N\Sigma_J\{g_J^2\mu_B^2 J(J+1)/kT-\alpha\}(2J+1)e^{-E_J/kT}}{\Sigma_J(2J+1)e^{-E_J/kT}}, \tag{1.41}$$

a result which is in good agreement with the experimental susceptibilities of Sm^{3+} and Eu^{2+} salts at room temperature (Table 1.5).

1.2.2 *Effects upon hyperfine structure*

A magnetic field, be it external $\mathbf{H}_0$, or one which arises internally from other ions, $\mathbf{H}_{int}$, will also lift the degeneracy of the hyperfine levels.

The Hamiltonian is approximately

$$\mathscr{H} = a_J\mathbf{I}.\mathbf{J}+g_J\mu_B\mathbf{J}.\mathbf{H}_0, \tag{1.42}$$

where the term $\boldsymbol{\mu}_I \cdot \mathbf{H}_0$ is assumed negligible, and the two limiting cases of (a) weak $\mathbf{H}_0$ and (b) strong $\mathbf{H}_0$ are of particular interest.

(a) For weak $\mathbf{H}_0$ ($\lesssim 100$ gauss), $\mu_B H_0 \ll a_J$ and the field perturbs the hyperfine levels E_F. F is still a good quantum number so that $\mathbf{H}_0$ splits each hyperfine level into the $2F+1$ levels:

$$E_{F,M_F} = E_F + g_J \mu_B \langle I, J, F, M_F | \mathbf{J} \cdot \mathbf{H}_0 | I, J, F, M_F \rangle$$
$$= \tfrac{1}{2} a_J K + g_J \mu_B \mathbf{H}_0 \frac{K}{2F(F+1)}. \tag{1.43}$$

(b) For strong $\mathbf{H}_0$ ($\gtrsim 10^4$ gauss), $\mu_B \mathbf{H}_0 \gg a_J$, the angular momenta I and J are decoupled so that F is no longer a good quantum number, and the basis states $|I, M_I, J, M_J\rangle$ must be employed. The hyperfine field now perturbs the Zeeman levels, and if $\mu_B \mathbf{H}_0 \ll \zeta_{4f}$, so that mixing of multiplet levels can be neglected, then:

$$E_{M_J,M_I} = g_J \mu_B \mathbf{H}_0 M_J + a_J M_I M_J. \tag{1.44}$$

Each Zeeman level is split into $(2I+1)$ states and this provides a means of determining I.

Expressions (1.43) and (1.44) are of fundamental importance in the experimental determination of the hyperfine structure constant, using the methods listed in Section 1.1.5, and hence in obtaining nuclear magnetic moments.

1.3 Ions in crystals

Experiments on rare earth ions are usually made on crystalline salts. Because the $4f$ electrons are so well shielded it is expected that the effect of the ion's environment is not very great and the agreement of Hund's formula with experimental susceptibilities supports this view. However, over the last two decades experimental resonance techniques have been developed which are able to determine the fine structure of ionic energy levels in crystals [26]. Such experiments are valuable in investigating (i) the electronic structure of ions and (ii) the environment in which the ions are situated. The most important technique is that of electron spin resonance and this is discussed in Section 1.4.

1.3.1 *The crystal potential*

A rare earth ion in a crystalline salt is situated in a potential $V(\mathbf{r})$, the crystalline field, which arises from the charges eZ_i on neighbouring ions at positions $\mathbf{R}_i$. Approximating the other ions by point charges we have $V(\mathbf{r}) = \sum_i eZ_i / |\mathbf{r} - \mathbf{R}_i|$. This potential partially lifts the $(2J+1)$ degeneracy of the multiplet levels of the free rare earth ion. From susceptibility measurements it is clear that these splittings due to $V(r)$ are much smaller than the multiplet separations, so that mixing of different multiplets can usually be neglected. In this respect the rare earths differ from the iron transition elements where the crystal field splittings are greater than those of the multiplets. The smallness of the splittings for the rare earths is due (i) to the $4f$ electrons being well shielded from $V(\mathbf{r})$ by the outer electrons, and (ii) to the ions being well separated.

The number of new levels arising depends upon the symmetry of the crystal field, there being fewer levels the greater the symmetry. For most rare earth salts the symmetry is relatively low, but by introducing rare earth ions into other host lattices, e.g. CaF_2, the effects of higher symmetries can be investigated. There are two general theorems of group theory which are invaluable in understanding the lowest energy level structure.

(*i*) *Kramers' theorem.* In the absence of an applied magnetic field the levels of an ion with an *odd* number of $4f$ electrons can at most be split into levels which are doubly degenerate. This applies to rare earths with half integral J ground states, e.g. Ce^{3+}.

(*ii*) The *Jahn–Teller effect.* The environment of an ion with a degenerate ground state, Kramers' doublets excepted, spontaneously distorts to a lower symmetry so as to remove the degeneracy. It implies that ions with an *even* number of $4f$ electrons always have singlet ground states, e.g. Pr^{3+}. This rule does not apply to excited states.

If it is assumed that the $4f$ electrons do not overlap the neighbouring ions, then $V(\mathbf{r})$ satisfies Laplace's equation and can be expanded in spherical harmonics:

$$V(\mathbf{r}) = \sum_{l} \sum_{m=-l}^{+l} A_l^m r^l Y_l^m(\mathbf{r}) = \sum_{l,|m| \leq l} V_l^m \quad \text{say,}$$

where for the point charge model

$$A_l^m = \sum_j \frac{4\pi}{(2l+1)} \frac{Z_j}{R_j^{(l+1)}} (-1)^m Y_l^{-m}(\hat{\mathbf{R}}_j). \tag{1.45}$$

For n $4f$ electrons the potential energy is written $\sum_i V(\mathbf{r}_i) = V_c$. Usually the A_l^m are treated as empirical parameters. The number of terms in V_c is restricted by symmetry as follows:

(i) If the z-axis is an m-fold rotation axis then V_c contains V_l^m.

(ii) If there is a centre of inversion there are no odd l terms.

Further simplication comes when the matrix elements $\langle 4f | V_c | 4f \rangle$ are computed. The required integrals are $\int \psi_{4f}^* r^l V_l^m \psi_{4f} \mathrm{d}\mathbf{r}$, and the radial integrals $\langle r^l \rangle$ factor out. Since $\psi_{4f} \propto V_3^\mu$ the angular integrals involved are $\int (V_3^\mu)^* V_l^m V_3^{\mu'} \sin\theta \mathrm{d}\theta$, and these vanish unless $l \leq 6$. The non-zero terms for crystal fields of commonly occurring symmetries are given in Table 1.6.

1.3.2 *Operator equivalents*

In evaluating matrix elements of V_c between multiplet states $|J, M_J\rangle$ the angular integrations are most readily achieved by using the method of

Table 1.6 Some crystal fields of commonly occurring symmetries.

Salt	Symmetry of rare earth ion site	Non-zero terms in V_c
Ethylsulphate; some trichlorides	C_{3h}	$V_2^0, V_4^0, V_6^0, V_6^6, (V_6^{-6})$
Double nitrate	C_{3v}	$V_2^0, V_4^0, V_4^3, V_6^0, V_6^3, V_6^6$
CaF_2, ThO_2, etc.	Cubic	$V_4^0, V_4^4, V_6^0, V_6^4$

operator equivalents [27]. For a given multiplet the matrix elements of V_c are proportional to those of 'equivalent' operators which contain J_x, J_y, and J_z in place of x, y, and z respectively. For example in V_2^0,

$$\sum_i \left(\frac{3z_i^2 - r_i^2}{r_i^2}\right) \rightarrow \alpha_J\{3J_z^2 - J(J+1)\} \equiv \alpha_J O_2^0 \quad \text{say}$$

where α_J is a proportionality constant which need only be calculated once.

Table 1.7 Equivalent operators for the rare earths [27, 28]. The constants α_J, β_J, and γ_J are given in Table 1.8.

Term in electrostatic potential	Operator equivalent	Standard notation
$\sum(3z^2-r^2)$	$\alpha_J\langle r^2\rangle[3J_z^2-J(J+1)]$	$\alpha_J\langle r^2\rangle O_2^0$
$\sum(x^2-y^2)$	$\alpha_J\langle r^2\rangle\frac{1}{2}[J_+^2+J_-^2]$	$\alpha_J\langle r^2\rangle O_2^2$
$\sum(35z^4-30r^2z^2+3r^4)$	$\beta_J\langle r^4\rangle[35J_z^4-30J(J+1)J_z^2+25J_z^2-$ $-6J(J+1)+3J^2(J+1)^2]$	$\beta_J\langle r^4\rangle O_4^0$
$\sum(7z^2-r^2)(x^2-y^2)$	$\beta_J\langle r^4\rangle\frac{1}{4}[\{7J_z^2-J(J+1)-5\}\{J_+^2+J_-^2\}+$ $+\{J_+^2+J_-^2\}\{7J_z^2-J(J+1)-5\}]$	$\beta_J\langle r^4\rangle O_4^2$
$\sum z(x^3-3xy^2)$	$\beta_J\langle r^4\rangle\frac{1}{4}[J_z(J_+^3+J_-^3)+(J_+^3+J_-^3)J_z]$	$\beta_J\langle r^4\rangle O_4^3$
$\sum(x^4-6x^2y^2+y^4)$	$\beta_J\langle r^4\rangle\frac{1}{2}[J_+^4+J_-^4]$	$\beta_J\langle r^4\rangle O_4^4$
$\sum(231z^6-315z^4r^2+105z^2r^4-5r^6)$	$\gamma_J\langle r^6\rangle[231J_z^6-315J(J+1)J_z^4+735J_z^4+$ $+105J^2(J+1)^2J_z^2-525J(J+1)J_z^2+$ $+294J_z^2-5J^3(J+1)^3+40J^2(J+1)^2-$ $-60J(J+1)]$	$\gamma_J\langle r^6\rangle O_6^0$
$\sum\{16z^4-16(x^2+y^2)z^2+$ $+(x^2+y^2)^2\}\{x^2-y^2\}$	$\gamma_J\langle r^6\rangle\frac{1}{4}[\{33J_z^4-(18J(J+1)+123)J_z^2+$ $+J^2(J+1)^2+10J(J+1)+102\}\times$ $\times(J_+^2+J_-^2)+(J_+^2+J_-^2)\{33J_z^4-\text{etc.}\}]$	$\gamma_J\langle r^6\rangle O_6^2$
$\sum(11z^3-3zr^2)(x^3-3xy^2)$	$\gamma_J\langle r^6\rangle\frac{1}{4}[\{11J_z^3-3J(J+1)J_z-59J_z\}\times$ $\times(J_+^3+J_-^3)+(J_+^3+J_-^3)\{11J_z^3-\text{etc.}\}]$	$\gamma_J\langle r^6\rangle O_6^3$
$\sum(11z^2-r^2)(x^4-6x^2y^2+y^4)$	$\gamma_J\langle r^6\rangle\frac{1}{4}[\{11J_z^2-J(J+1)-38\}\times$ $\times(J_+^4+J_-^4)+(J_+^4+J_-^4)\{11J_z^2-\text{etc.}\}]$	$\gamma_J\langle r^6\rangle O_6^4$
$\sum(x^6-15x^4y^2+15x^2y^4-y^6)$	$\gamma_J\langle r^6\rangle\frac{1}{2}[J_+^6+J_-^6]$	$\gamma_J\langle r^6\rangle O_6^6$

The matrix elements of the equivalent operator O_2^0 between states $|J, M_J\rangle$ are easily obtained (see Appendix 2). Some commonly occurring operators and their equivalents are given in Table 1.7, and values of the proportionality constants α_J, β_J, and γ_J are presented in Table 1.8. In the following, Hutchings' [28] notation will be used for which the operator equivalent crystal field Hamiltonian is written:

$$\mathscr{H}_c = \sum_{l,m} B_l^m O_l^m, \tag{1.46}$$

where the B_l^m's are related to the crystal field parameters through $B_l^m = A_l^m \langle r^l \rangle \theta_l$, and $\theta_2 = \alpha_J$, $\theta_4 = \beta_J$, and $\theta_6 = \gamma_J$. Hutchings has tabulated the matrix elements for all these operators between states $|J = \text{const}, M_J\rangle$ for J values appropriate to the rare earths. Use of these tables enables the energy

Table 1.8 Multiplicative factors associated with equivalent operators for the ground states of the ions [21], and calculated Hartree–Fock radial integrals $\langle r^n \rangle$ in atomic units of length a_0^{-n} [14].

Ion	Ground state	α_J	β_J	γ_J	$\langle J\|N\|J\rangle$	$\langle r^2\rangle$	$\langle r^4\rangle$	$\langle r^6\rangle$
Ce^{3+}	$^2F_{5/2}$	$\frac{-2}{5\cdot7}$	$\frac{2}{3^2\cdot5\cdot7}$	0	$\frac{2^4\cdot3}{5\cdot7}$	1·200	3·455	21·226
Pr^{3+}	$^3H_4$	$\frac{-2\cdot13}{3^2\cdot5^2\cdot11}$	$\frac{-2^2}{3^2\cdot5\cdot11^2}$	$\frac{2^4\cdot17}{3^4\cdot5\cdot7\cdot11^2\cdot13}$	$\frac{2^3\cdot37}{3^2\cdot5^2}$	1·086	2·822	15·726
Nd^{3+}	$^4I_{9/2}$	$\frac{-7}{3^2\cdot11^2}$	$\frac{-2^3\cdot17}{3^3\cdot11^3\cdot13}$	$\frac{-5\cdot17\cdot19}{3^3\cdot7\cdot11^3\cdot13^2}$	$\frac{2^2\cdot7\cdot17}{3\cdot11^2}$	1·001	2·401	12·396
Pm^{3+}	$^5I_4$	$\frac{2\cdot7}{3\cdot5\cdot11^2}$	$\frac{2^3\cdot7\cdot17}{3^3\cdot5\cdot11^3\cdot13}$	$\frac{2^3\cdot17\cdot19}{3^3\cdot7\cdot11^2\cdot13^2}$	$\frac{2^5\cdot7}{3\cdot5\cdot11}$	0·935	2·13	10·4
Sm^{3+}	$^6H_{5/2}$	$\frac{13}{3^2\cdot5\cdot7}$	$\frac{2\cdot13}{3^3\cdot5\cdot7\cdot11}$	0	$\frac{2^3\cdot61}{3^2\cdot5\cdot7}$	0·883	1·897	8·775
Tb^{3+}	$^7F_6$	$\frac{-1}{3^2\cdot11}$	$\frac{2}{3^3\cdot5\cdot11^2}$	$\frac{-1}{3^4\cdot7\cdot11^2\cdot13}$	$\frac{7^2}{2\cdot3^2\cdot5}$	0·758	1·44	5·8
Dy^{3+}	$^6H_{15/2}$	$\frac{-2}{3^2\cdot5\cdot7}$	$\frac{-2^3}{3^3\cdot5\cdot7\cdot11\cdot13}$	$\frac{2^2}{3^3\cdot7\cdot11^2\cdot13^2}$	$\frac{2^5}{3^2\cdot5}$	0·726	1·322	5·102
Ho^{3+}	$^5I_8$	$\frac{-1}{2\cdot3^2\cdot5^2}$	$\frac{-1}{2\cdot3\cdot5\cdot7\cdot11\cdot13}$	$\frac{-5}{3^3\cdot7\cdot11^2\cdot13^2}$	$\frac{23}{2\cdot3\cdot5}$	0·695	1·22	4·5
Er^{3+}	$^4I_{15/2}$	$\frac{2^2}{3^2\cdot5^2\cdot7}$	$\frac{2}{3^2\cdot5\cdot7\cdot11\cdot13}$	$\frac{2^3}{3^3\cdot7\cdot11^2\cdot13^2}$	$\frac{2^4\cdot11}{3^2\cdot5^2}$	0·666	1·126	3·978
Tm^{3+}	$^3H_6$	$\frac{1}{3^2\cdot11}$	$\frac{2^3}{3^4\cdot5\cdot11^2}$	$\frac{-5}{3^4\cdot7\cdot11^2\cdot13}$	$\frac{7}{3^2}$	0·640	1·03	3·45
Yb^{3+}	$^2F_{7/2}$	$\frac{2}{3^2\cdot7}$	$\frac{-2}{3\cdot5\cdot7\cdot11}$	$\frac{2^2}{3^3\cdot7\cdot11\cdot13}$	$\frac{2^4}{3\cdot7}$	0·613	0·960	3·104

matrix to be written down immediately, and the energy levels are obtained by diagonalizing it. The corresponding wavefunctions are mixtures of the free ion levels, being appropriate linear combinations of the free ion states, i.e. $\sum_{M_J} C_{M_J} |J, M_J\rangle$, where the C_{M_J} are constants.

As an explicit example of the procedure for determining the crystal field splittings, the effect of the cubic field

$$V_c = B_4^0[O_4^0 + 5O_4^4]$$

upon a sixfold degenerate ground state for which $J = \frac{5}{2}$, is considered in Appendix 2.

The energy level splittings of all multiplets $2 \leq J \leq 8$ by a cubic crystal field have been tabulated in general form by Lea *et al* [29]. With a four-fold axis as reference, and in their notation,

$$V_c = W\left\{\frac{x}{F(4)}[O_4^0 + 5\,O_4^4] + \frac{(1-|x|)}{F(6)}[O_6^0 - 21\,O_6^6]\right\}, \tag{1.47}$$

where W and x are defined by

$$\frac{Wx}{F(4)} = B_4^0 \quad , \quad \frac{W(1-|x|)}{F(6)} = B_6^0, \tag{1.48}$$

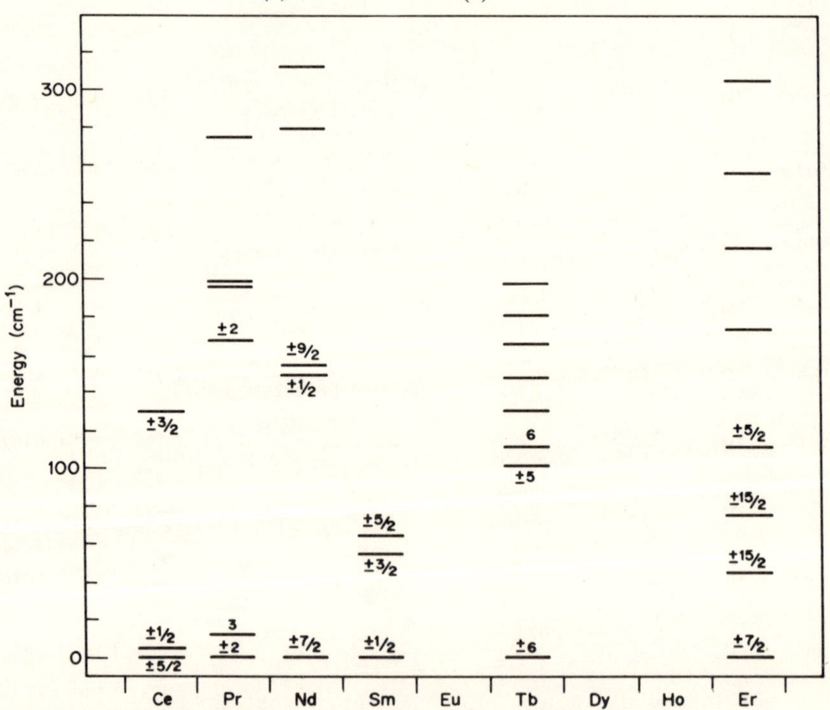

Fig. 1.6 The observed crystal field splittings of the lowest multiplet levels of some rare earth ions in the concentrated ethylsulphates (after ref. [53]).

and $F(4)$ and $F(6)$ are factors common to the matrix elements of O_4^m and O_6^n respectively. Lea *et al* have tabulated energies and associated wavefunctions as functions of x and W.

The crystal field splittings of some rare earth ions in the concentrated ethylsulphate salts are shown in Fig. 1.6. In Figs. 1.7 and 1.8 are shown results for rare earth ions in $LaCl_3$ and CaF_2 respectively.

1.3.3 *Perturbation of a Kramers' doublet*

Having discussed how, in principle, the crystal field levels are obtained, it is necessary to consider the effect of a magnetic field **H** on these levels, since most experiments require the application of such a field. The magnetic moment of the 4*f* electrons interacts with **H** to lift any remaining degeneracy in the level structure. The particularly interesting case for the rare earths is the splitting of a ground state Kramers' doublet. The two initially degenerate levels may be written symbolically as

$$
\begin{aligned}
+\rangle &= \sum_{M_J} C_{M_J} |J, M_J\rangle \\
-\rangle &= \sum_{M_J} (-1)^{J+M_J} C_{M_J} |J, -M_J\rangle,
\end{aligned}
\qquad (1.49)
$$

where the coefficients C_{M_J} are assumed known from a crystal field calculation. If the z-axis is reference axis for the crystal field, then the splitting

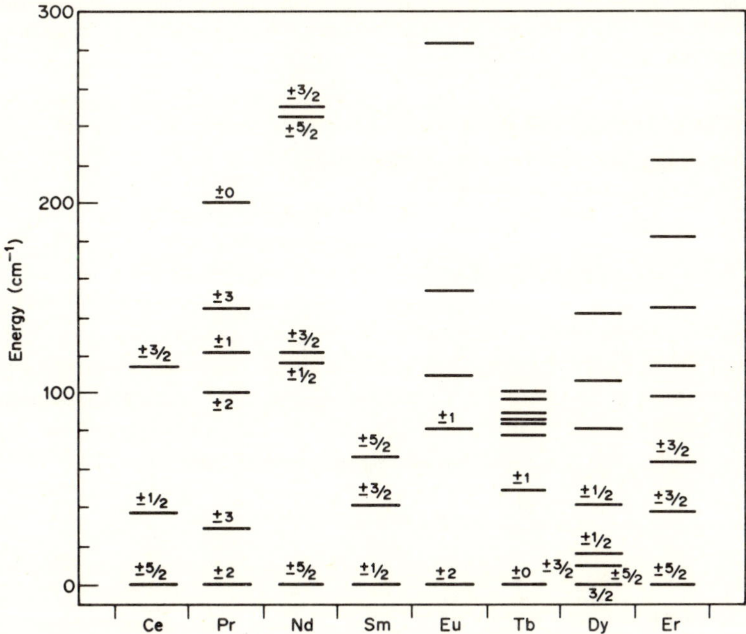

Fig. 1.7 The observed crystal field splittings of the lowest multiplet levels of some rare earth ions in $LaCl_3$ (after ref. [2]).

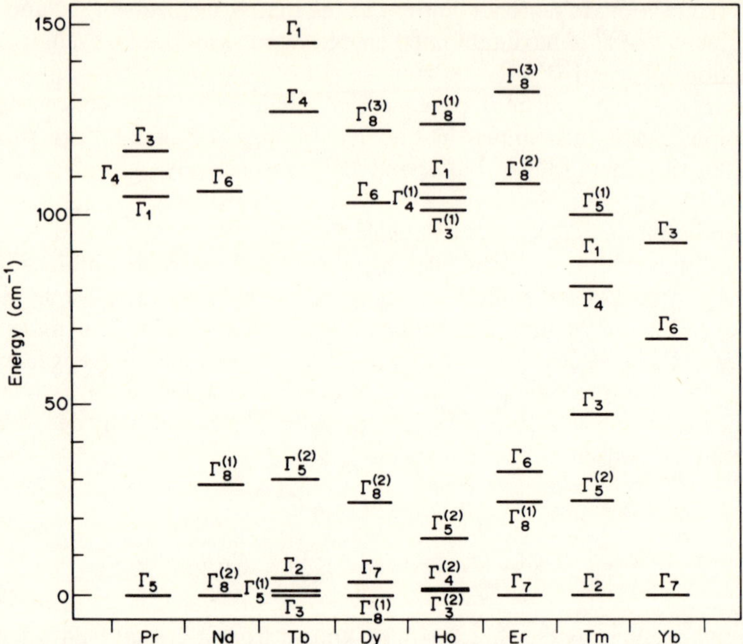

Fig. 1.8 The calculated crystal field splittings of the lowest multiplet levels of some rare earth ions occupying sites of cubic symmetry in CaF_2. The results are from Ranon [41]. The values of the parameters x which are used in the calculations differ from values given in Table 1.11 by a few percent.

$\Delta E_{\|}$ produced by a field **H** parallel to the z-axis depends only upon the matrix elements of J_z, as for the free ion. In first order

$$\Delta E_{\|} = 2g_J\mu_B|\langle+|J_z|+\rangle|H = 2g_J\mu_B H\sum_{M_J}(C_{M_J})^2 M_J$$
$$= g_{\|}\mu_B H, \quad \text{say.} \tag{1.50}$$

On the other hand, if **H** is perpendicular to the z-axis the situation differs from that of the free ion in that the ion, and hence its moment, is fixed and cannot rotate so as to bring the moment along the field. The required matrix elements are those of J_x, and it is readily shown that the diagonal elements are zero, so that the 2×2 secular equation is trivial, and

$$\Delta E_{\perp} = 2g_J\mu_B|\langle+|J_x|-\rangle|H$$
$$= g_{\perp}\mu_B H, \quad \text{say.} \tag{1.51}$$

For intermediate directions at angle θ to the z-axis, $g^2(\theta) = g_{\|}^2\cos^2\theta + g_{\perp}^2\sin^2\theta$. For the simple example given in Appendix 2 it is readily shown that the g-values of the Γ_7 Kramers' doublet are

$$g_{\parallel} = 2g_J \times \{(0{\cdot}4083)^2(\tfrac{5}{2}) + (0{\cdot}9129)^2(-\tfrac{3}{2})\}$$

and

$$g_{\perp} = 2g_J \times \sqrt{5}(0{\cdot}4083)(0{\cdot}9129).$$

The calculation of the magnetic hyperfine splittings is similar. In first order, for non *s*-electrons, one finds that a doublet is split into $(2I+1)$ equally spaced levels at energies

$$E_{n\parallel} = 2g_I\mu_N\mu_B\langle r^{-3}\rangle\langle +|N_z|+\rangle I_z\, J_z \tag{1.52}$$

and

$$E_{n\perp} = 2g_I\mu_N\mu_B\langle r^{-3}\rangle\langle +|N_x|-\rangle I_x, J_x \tag{1.53}$$

where N is defined in equation (1.17). It may also be shown that the interaction of the nuclear electric quadrupole with an axial field gradient parallel to z produces splittings which have energies given by [24]:

$$E_Q = -\frac{3e^2Q}{4I(2I-1)}\langle r^{-3}\rangle\alpha_J\langle \pm|3J_z^2 - J(J+1)|\pm\rangle\{I_z^2 - \tfrac{1}{3}I(I+1)\}. \tag{1.54}$$

The expressions (1.50) to (1.54) are invaluable in analysing esr data.

1.4 Electron spin resonance

Electron spin resonance experiments provide direct information about the lowest energy levels of ions in crystals. Basically a large static magnetic field is applied to the specimen, and dipolar transitions between Zeeman levels, with separations $\Delta E \sim 1\ \text{cm}^{-1}$, are induced by a small radio frequency field. In the absence of nuclear spin the selection rules are $\Delta M_J = \pm 1$, and if there is hyperfine splitting, the added condition $\Delta M_I = 0$ also applies. It should be noted that the presence of a nuclear quadrupole interaction can give rise to transitions with $\Delta M_I = 1, \pm 2$, which are normally forbidden. Radio frequency energy is absorbed within a band of frequencies $\Delta\nu$, centred on a frequency ν_0 given by $h\nu_0 = \Delta E$, provided a population imbalance is maintained between the levels by downward transitions involving the emission of phonons. A set of absorption lines corresponding to the allowed transitions can then be detected. In practice the frequency is fixed and the static magnetic field is slowly varied to find resonance.

The inverse probability of downward transitions between levels with phonon emission is called the *spin-lattice relaxation time* τ_{sl}. It is important in determining the width $\Delta\nu$ of the absorption lines, i.e. $\Delta\nu \propto \tau_{sl}^{-1}$. In particular, if τ_{sl} is very short $\Delta\nu$ may be so broad that no resonance can be observed. Since τ_{sl} increases as the temperature is decreased, measurements are often only possible at low temperatures, generally below 20°K for rare earths in salts, and in some cases only below 4°K. At such low temperature no resonances might be expected for ions with singlet ground states, but in fact the first excited state is sometimes sufficiently close to the ground state for them to be observed, e.g. for Pr^{3+}, Tb^{3+}, and Ho^{3+}. If, on the other hand, τ_{sl} is very long, so that the probability of downward transitions with phonon emission is very small, the populations of the two levels may become equal at quite low applied rf power. When this happens no further power absorption is possible and the resonance signal vanishes. Under these conditions saturation is said to have occurred.

The mechanisms which determine the spin-lattice relaxation rates of rare earth ions in salts are considered in detail in the next section. Of course there are also factors other than τ_{sl} which contribute to the line width, such as inhomogeneities in both the crystal structure and in the internal magnetic field, but these are not discussed here.

If $\Delta\nu$ is too large to resolve the hyperfine structure the technique of electron nuclear double resonance may be useful (e.g. ref. [79]). The method is as follows. Consider the simplest case, $I = \frac{1}{2}$, so that each Zeeman level is split into two, as illustrated in Fig. 1.9, and suppose that levels 1 and 4 are

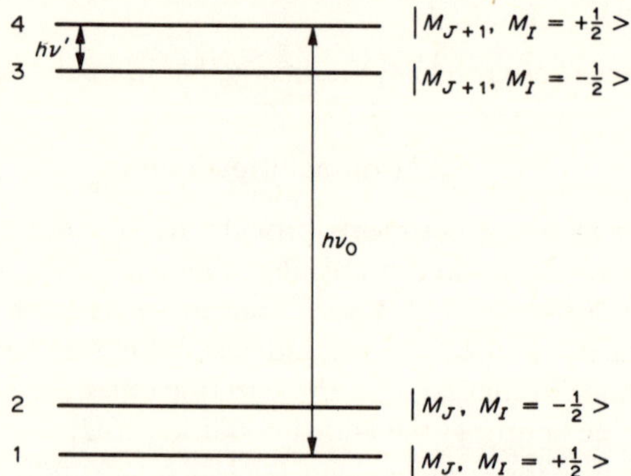

Fig. 1.9 Hyperfine splittings of Zeeman levels by a nuclear spin $I = \frac{1}{2}$.

saturated in an esr experiment. If a variable rf field is then also applied resonant transitions occur between the hyperfine levels 4 and 3 when the frequency ν is equal to ν'. This upsets the saturation conditions and the esr reappears. The line width $\Delta\nu'$ of a hyperfine transition is much narrower than a conventional esr line and allows the hyperfine structure to be determined with higher precision. Another advantage of the method is that nuclear magnetic moment measurements may be determined directly, as in ordinary nmr experiments; see Table 1.4.

1.4.1 *The spin Hamiltonian*

A convenient way of describing the separations of the energy levels deduced from esr experiments is by the use of a *spin Hamiltonian*, $\mathscr{H}_{S'}$, in which a fictitious spin S' is defined such that $(2S'+1)$ equals the number of observed levels. For a Kramers' doublet $S' = \frac{1}{2}$, and the spin Hamiltonian has the general form:

$$\mathscr{H}_{S'} = g_{\parallel}\mu_B H_z S'_z + g_{\perp}\mu_B(H_x S'_x + H_y S'_y) + A_{\parallel} I_z S'_z + A_{\perp}(I_x S'_x + I_y S'_y) + P\{I_z^2 - \tfrac{1}{3}I(I+1)\}, \tag{1.55}$$

where $g_{\parallel}$ and $g_{\perp}$ are constants of the Zeeman splitting, $A_{\parallel}$ and $A_{\perp}$ are constants of the magnetic hyperfine splittings, and P is a constant for the nuclear quadrupole interaction. Theoretical values of $g_{\parallel}$, $g_{\perp}$, $A_{\parallel}$, $A_{\perp}$ and P, are given by equations (1.50) to (1.54) in terms of the doublet states $|+\rangle$ and $|-\rangle$, which in turn depend upon the crystal field parameters B_l^m. The latter parameters must therefore be chosen to fit the experimental values of $g_{\parallel}$, $g_{\perp}$ and the ratio $A_{\parallel}/A_{\perp}$. Comparison is not made with $A_{\parallel}$ and $A_{\perp}$ individually since both contain the nuclear magnetic moment which is usually not accurately known. In principle μ_I can be obtained from $A_{\parallel}$ and $A_{\perp}$ (equations (1.52) and (1.53)), using estimates of $\langle r^{-3}\rangle$, and the calculated states $|+\rangle$ and $|-\rangle$, as discussed in Section 1.1.5. On the other hand $\langle r^{-3}\rangle$ values may be obtained if μ_I is known. The ratio $A_{\perp}g_{\parallel}/A_{\parallel}g_{\perp}$ is also of interest. Assuming that the crystal field splittings are very much less than the multiplet separations, then as a zero order approximation the Zeeman splittings may be considered to be the same as for the free ion. In this case $|\pm\rangle_{M_J} = |J, \pm M_J\rangle$, and it can easily be shown that $A_{\perp}g_{\parallel}/A_{\parallel}g_{\perp} = 1$. Deviations from unity therefore indicate mixing of other levels, e.g. of type $|J, M'_J\rangle$ and $|J', M_{J'}\rangle$ into $|J, \pm M_J\rangle$.

For ions with an even number of electrons the symmetry of V_c is distorted to produce a singlet ground state. For such ions which have two low lying singlet states, e.g. Pr^{3+}, the splitting in a magnetic field may be represented by the spin Hamiltonian:

$$\mathscr{H}_S = g_{\parallel}\mu_B H_z S_z + A_z I_z S_z + A_x I_x S_x + A_y I_y S_y + P\{I_z^2 - \tfrac{1}{3}I(I+1)\}. \tag{1.56}$$

Extensive experimental work has been done on rare earth ions in salts of the ethylsulphates and the chlorides in which the site symmetry is C_{3h}. Some results of these experiments are summarized in Tables 1.9 and 1.10. Note that the crystal field parameters $A_l^m\langle r^l\rangle$ given there are from optical measurements [2]. Numerous experiments have also been carried out on rare earth ions in other host lattices, particularly in CaF_2 which can provide sites of cubic symmetry [30]. For cubic fields it is possible for the ground state to be four-fold degenerate, a Γ_8 quartet, and the Zeeman splittings are not necessarily equally spaced. Bleaney [31] has proposed a suitable spin Hamiltonian to describe this behaviour. Table 1.11 summarizes some results for rare earth ions in CaF_2.

The ions Gd^{3+} and Eu^{2+} are in S-states, i.e. $L = 0$, so that there can be no $\mathbf{L}\cdot\mathbf{S}$ coupling and, since V_c acts only upon the orbital part of the wavefunction, and this is non-degenerate $|L = 0, M_L = 0\rangle$, first order perturbation theory gives no crystal field splitting. Theory would predict $g = 2$, but

Table 1.9 Experimental results for the rare earth ethylsulphates $RE(C_2H_5SO_4)_3.9H_2O$. The esr data is after Low [26]. The crystal field parameters have been deduced from optical measurements and are taken from Dieke [2]. The constants θ are to be chosen to fit experiment.

	$g_\parallel$ (Dilute Salts)	$g_\perp$	$\dfrac{A_\perp g_\parallel}{A_\parallel g_\perp}$	$A_2^0\langle r^2\rangle$	$A_4^0\langle r^4\rangle$ (cm^{-1})	$A_6^0\langle r^6\rangle$	$A_6^6\langle r^6\rangle$	Ground states (theory)
Ce^{3+}	0·955	2·185		9	−42	−45	680	complex [80] $\left\{\begin{array}{l}\lvert\pm\frac{1}{2}\rangle\\ \lvert\pm\frac{5}{2}\rangle\end{array}\right.$
	3·725	0·20						
Pr^{3+}	1·525			15	−88	−49	548	$\cos\theta\lvert\pm4\rangle+\sin\theta\lvert\mp2\rangle$
Nd^{3+}	3·535	2·072	0·892	58	−68	−43	595	$\cos\theta\lvert\pm\frac{7}{2}\rangle+\sin\theta\lvert\mp\frac{5}{2}\rangle$
Sm^{3+}	0·596	0·604	4·2	77	−48	−39	550	$\cos\theta\lvert\frac{5}{2},\pm\frac{1}{2}\rangle\pm\sin\theta\lvert\frac{7}{2},\pm\frac{1}{2}\rangle$
Eu^{3+}	1·991	1·991		80	−63	−39	510	
Gd^{3+}	1·991	1·991		96				
Tb^{3+}	17·72	0·3		110	−75	−34	465	$\cos\theta\lvert\pm6\rangle+\sin\theta\lvert0\rangle$
Dy^{3+}	no resonance observed			119	−26	−31	473	$a\lvert\pm\frac{15}{2}\rangle+b\lvert\pm\frac{3}{2}\rangle+c\lvert\mp\frac{9}{2}\rangle$
Ho^{3+}	15·36 (complex)			125	−79	−30	391	$a\lvert\pm7\rangle+b\lvert\pm1\rangle+c\lvert\mp5\rangle$
Er^{3+}	1·47	8·85	1·00	126	−81	−31	387	$\cos\theta\lvert\pm\frac{7}{2}\rangle+\sin\theta\lvert\mp\frac{5}{2}\rangle$
Tm^{3+}	no resonance detected			130	−71	−29	433	$a\lvert+6\rangle+b\lvert0\rangle+a\lvert-6\rangle$

Table 1.10 Experimental results for rare earth ions in $LaCl_3$, from references quoted in Table 1.9. More complete esr data is given by Hutchinson and Wong [73]. The groundstates are the same as for the ethylsulphates.

	$g_\parallel$	$g_\perp$	$A_\parallel$ ($\times10^{-4}$ cm^{-1})	$A_\perp$	$A_2^0\langle r^2\rangle$	$A_4^0\langle r^4\rangle$ (cm^{-1})	$A_6^0\langle r^6\rangle$	$A_6^6\langle r^6\rangle$
Ce^{3+}	4·0366	0·17			64	−41	−62	383
$^{141}Pr^{3+}$	1·035	0·1	502		47	−41	−40	427
$^{143}Nd^{3+}$	3·996	1·713	425	167	98	−39	−44	443
$^{147}Sm^{3+}$	0·5841	0·6127	60·7	245	81	−23	−44	426
Eu^{3+}					89	−38	−51	495
$^{155}Gd^{3+}$	1·991	1·991	3·8		97	−42	−30	290
$^{159}Tb^{3+}$	17·78	20·1	2120		92	−40	−30	290
Dy^{3+}		no resonance detected			91	−39	−23	258
$^{165}Ho^{3+}$	16·01	~0	3500		114	−34	−39	277
Er^{3+}	1·989	8·757	66·4	304	94	−37	−27	265
Tm^{3+}		no resonance detected						

Table 1.11 Date for rare earth ions occupying cubic sites in CaF_2 [30]. The parameters x, which are defined in equation (1.48), are best fit values and the groundstates are deduced from them using the results of Lea, Leask and Wolf [29]. The states Γ_1, Γ_2, and Γ_3 are singlets and hence show no esr. Results for ions in tetragonal and trigonal sites may be found in [30].

	g	$A_4\langle r^4\rangle$ (cm^{-1})	$A_6\langle r^6\rangle$ (cm^{-1})	x	Ground state
Ce^{3+}	$\Gamma_8: 2{\cdot}0; 3{\cdot}1$ $\Gamma_7: 1{\cdot}297$			1·0	Γ_8
Pr^{3+}				0·9	$\Gamma_5\,(\Gamma_1)$
Nd^{3+}		−415	33	−0·7	Γ_8
Sm^{3+}				1·0	Γ_8
Eu^{3+}					Γ_1
Tb^{3+}				0·9	$\Gamma_3\,(\Gamma_2)$
Dy^{3+}	$\Gamma_8: 2{\cdot}63; 5{\cdot}48; 13{\cdot}7$ $\Gamma_7: 7{\cdot}47$	−242	41	0·6	Γ_8
Ho^{3+}				−0·5	$\Gamma_3\,(\Gamma_5)$
Er^{3+}	6·785	−310	44	−0·4	Γ_7
Tm^{3+}				0·6	$\Gamma_2\,(\Gamma_3)$
Yb^{3+}	3·443	−280		0·7	Γ_7

Table 1.12 Some esr results for Gd^{3+} in cubic crystalline fields, taken from Abraham *et al* [75]. The constants are related to those employed in equation (1.57) by $b_4 = 60\ B_4$ and $b_6 = 1260\ B_6$.

Crystal	Lattice constant (Å)	g	$\lvert b_4\rvert$ ($\times 10^4$ cm^{-1})	$\lvert b_6\rvert$ ($\times 10^4$ cm^{-1})
CaF_2	5·46	1·991	46	1·0
BaF_2	6·20	1·9921	36	0·2
SrF_2	5·79	1·9923	41	0·5
CdF_2	5·39	1·992	48	0·0
$SrCl_2$	6·99	1·9906	10	0·0
SrO	5·14	1·991	5·8	
CaO	4·81	1·9922	12	1·0
CeO_2	5·41	1·9921	162	1·1
ThO_2	5·60	1·9917	166	1·2

in fact complex resonance spectra are found which may be described by spin Hamiltonians of the form [33]

$$\mathscr{H}_S = g\mu_B \mathbf{H}\cdot\mathbf{S} + A\mathbf{S}\cdot\mathbf{I} + \sum_{l,m} B_l^m O_l^m, \tag{1.57}$$

with effective spin $S = \frac{7}{2}$. Experimental results are shown in Table 1.12.

Wybourne [34] has considered in detail many of the second order effects which might be expected to split the ground state of Gd^{3+} in lanthanum ethylsulphate, but with little success. Several mechanisms mix in part of the excited $^6P_{7/2}$ state, but in doing so predict that the $\pm\frac{7}{2}$ Kramers' doublet is lowest, whereas experiment [35] shows that it is highest. Other corrections, e.g. relativistic, spin–spin, and configurational mixing, are important but are too small to explain the experimental observations. It would appear that covalency effects, referred to below, are dominant.

1.4.2 *Crystal field parameters*

Theoretical calculations of the crystal field parameters A_l^m have mainly been attempted for Pr^{3+} in $PrCl_3$, as Hartree wavefunctions are available for all orbitals of Pr^{3+} [12], and the structure is relatively simple with only one kind of anion. The point charge values, equation (1.45), are in considerable disagreement with the experimental ones. Because the $4f$ electrons are spatially compact, their overlap with neighbouring ions would be expected to be small, so that an ionic description would be expected to be valid. Ray and co-workers [36–40] and others [42], adopting this view, have considered two electrostatic contributions to the crystal field seen by $4f$ electrons on the rare earth ion. The first is the interaction between the multipole moment of the $4f$ electrons of the selected ion with those of neighbouring rare earth ions. The moments should then be determined self-consistently since they are induced by the crystal field. Unfortunately the modified results are worse than those of the point charge model, see Table 1.13. The second is the

Table 1.13 Theoretical estimates of contributions to the crystal field parameters for $PrCl_3$. Rows (i) and (ii) are from reference [40] and rows (iii)–(v) from reference [47]. The polarization effect in (iv) differs from that in (ii) in that explicit wavefunctions have been used for the outer Cl^- electrons, but only nearest-neighbour ions have been considered.

	$A_2^0\langle r^2\rangle$	$A_4^0\langle r^4\rangle$	$A_6^0\langle r^6\rangle$	$A_6^6\langle r^6\rangle$
		(cm^{-1})		
(*i*) *Point charge model*	560	−35·4	−3·07	36·9
(*ii*) *Electrostatic corrections* ($\sigma_l \times$ point charge values)	−392	1·02		
(*iii*) *Charge penetration*	−38·9	17·4	6·8	−81·9
(*iv*) *Polarization contribution*	171	−44	−30	359
(*v*) *Covalency*	6·8	−13·5	−13·1	159·6
Total theoretical value (*iii*)+(*iv*)+(*v*)	138·9	−40·1	−36·3	436·7
Experiment	47·3	−40·6	−39·6	405·4

contribution of the charge distribution of the selected ion itself. The crystal field distorts the $4f$ and closed shell orbitals, which in turn affects the crystal field and, for the outer electrons in particular, results in the shielding of the $4f$ electrons from the external crystalline field. Introducing an electronic shielding factor σ_l, the effective $A_{l\,\mathrm{eff}}^m$ are

$$A_{l\,\mathrm{eff}}^m = A_l^m(1-\sigma_l). \tag{1.58}$$

Calculations [38] for Pr^{3+} and Tm^{3+} indicate that shielding is only important for A_2^0, for which $\sigma_2 = 0{\cdot}59$ and $0{\cdot}71$ respectively. It is clear that a purely ionic description is inadequate, and in fact there is appreciable overlap between the $3p$ electrons of Cl^- and the rare earth $4f$ electrons [36]; Fig. 1.10. As a result there is an electrostatic interaction between the $4f$

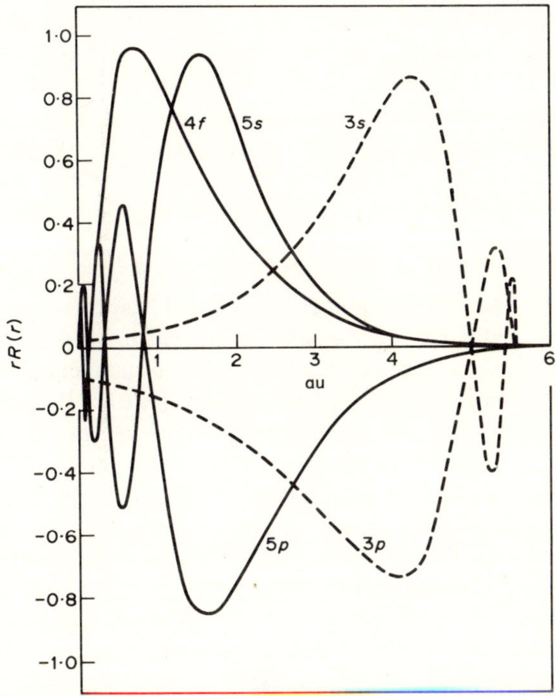

Fig. 1.10 Hartree radial wavefunctions for Pr^{3+} (full curves), and Hartree–Fock wavefunctions for Cl^- (broken curves) assuming that the Cl^- ion is at the nearest-neighbour distance from the Pr^{3+} ion [36].

electrons and the charge distribution of the Cl^- electrons giving an overlap, or *charge penetration* correction to V_c. Theoretical estimates of the correction [40, 42] indicate that for A_2^0 they are negligible while for A_4^0, A_6^0 and A_6^6 they are comparable or larger than the electrostatic contributions, but are in the wrong direction, and too small, to account for the experimental values; see Table 1.13. The situation is therefore more complex, and it is necessary to consider the mixing, or hybridization, of the $4f$ electrons with those of their neighbouring anions, i.e. true *covalency effects.* The earliest calculation by Jorgensen [43], suggested that the effect was one of antibonding, in which the

$4f$ orbitals mixed with the (ligand) orbitals of the anions along the internuclei direction (sigma orbitals). This was confirmed by a semi-empirical molecular orbital calculation by Axe and Burns [44] for Tm^{2+} in CaF_2, who found that covalency, including overlap, can increase the values of the A_l^m by as much as 50%. However, a subsequent calculation by Watson and Freeman [45] indicated that overlap is dominant, yielding a total contribution of $\sim 10\%$. The covalency effects for Pr^{3+} in $PrCl_3$ have been evaluated by Ellis and Newman [42, 47] and, with overlap included, they obtain values of A_4^0, A_6^0 and A_6^6 in such good agreement with experiment that they must be partly fortuitous; Table 1.13. The same authors [46] have also made calculations for Sm^{3+} and Er^{3+} in $LaCl_3$, without overlap, and investigated the effects of small changes in lattice constants upon the crystal field, which they find are appreciable.

The crystal field also produces an electric field gradient q_L which, together with q_J of the $4f$ electrons, has an interaction with the quadrupole moment Q of the nucleus of the form

$$e^2Q\{q_J(1-R_q)+q_L(1-\gamma_\infty)\}, \tag{1.59}$$

where γ_∞ corrects for the screening of the crystal field by the electrons and is related to σ_2. Usually $q_L \ll q_J$, but for the rare earths γ_∞ is found to be large (~ -100) so that the lattice contributions can be significant. For some ions where q_J is very small, e.g. Eu^{3+} in the ethylsulphate [58], it is the dominant term [59, 60]. Theoretical calculations [39] of γ_∞ for Pr^{3+} and Tm^{3+} indicate that angular distortions are only of the order of 1% of the radial ones; this may be compared to calculations of R_q for which the angular contribution is the larger. Further, since all the screening (orbital) charge is assumed to be inside the perturbing neighbouring ionic charges, the radial distortions produce antishielding. The major contributions are from the outer shells, particularly the distortion $5p \rightarrow np$. As would be expected from the earlier discussion, the results of these point charge model calculations do not agree well with experimental values of γ_∞. For example, the value for La^{3+} in $LaCl_3$ is 4·5 times smaller [59] than the predicted estimate for Pr^{3+} or Tm^{3+}. Overlap and covalency effects must therefore be included. The effect of charge penetration inside the $4f$ shell will certainly be in the correct direction, since it will reduce γ_∞ by producing radial shielding.

In concluding this brief discussion of the parameters $A_l^m\langle r^l\rangle$, it must be emphasized that we are dealing with small effects, and that the observed values arise from a delicate balance between them.

1.5 Spin-lattice relaxation

The loss of energy associated with the reorientation of the spins in a changing magnetic field takes place over a finite time with the emission of phonons, and may be characterized by the spin-lattice relaxation rate τ_{sl}^{-1}. The interaction between an ionic spin and the phonons takes place via the spin-orbit coupling and the crystal field $V_c(t)$, which is assumed to be oscillating and temperature dependent. The perturbing effect of the crystal field is expected to be larger at higher temperatures, so that τ_{sl} becomes shorter, as stated in the previous section. For many salts a quantitative analysis of τ_{sl} as a function of temperature is difficult, but for rare earth salts the problem is simplified by the fact that the spin orbit coupling is stronger than the crystal field so that the difficult term $V_c(t)$ may be treated as the final perturbation, and comparison of theory with experimental results has been profitable.

A common method of determining spin-lattice relaxation rates experimentally is to place the specimen in an esr spectrometer, adjust for resonance, and then apply a microwave pulse at the resonant frequency. The pulse modifies the populations of the energy levels, and after it has ended the amplitude of the esr signal as a function of time indicates how the spin system returns to equilibrium. For details of such 'pulse saturation' experiments on rare earth salts, reference should be made to the work of Jefferies and co-workers [49, 50].

1.5.1 *Theory*

Much of the theory of the various relaxation processes is not particular to rare earth ions and hence it will only briefly be reviewed. Consider N identical ions, which have as ground state $|a\rangle$ and a first excited state $|b\rangle$ at energy δ. The change in populations N_a and N_b of these levels with time after saturation is characterized by the relaxation rate

$$\tau^{-1} = P_{ba} - P_{ab} \tag{1.60}$$

where P_{ba} and P_{ab} are the probabilities of single transitions $|b\rangle \rightleftharpoons |a\rangle$ respectively.

It is assumed that the transitions are stimulated by the vibrations of the ions in the crystal, and that they can be represented by the oscillating crystal field $V_c(t)$. The form of $V_c(t)$ is discussed in detail later. The simplest relaxation mechanism, known as the *direct process*, is dominant at low temperatures and involves transitions in which phonons with energy δ are emitted or absorbed. These phonons are said to be on 'speaking terms' with the spin system, and their number density at the crystal temperature T^0 is $p_0(\delta) = \{\exp(\delta/kT^0)-1\}^{-1}$. The probability of a downward transition being caused by this process is proportional to the number of phonons with energy δ, the population N_b, and the square of the matrix elements of $V_c(t)$, i.e. $P_{ba} \propto N_b p_0(\delta)|V_{ba}|^2$, where

$$V_{ba} = \langle a|V_c(t)|b\rangle. \tag{1.61}$$

For the reverse process $P_{ab} \propto N_a\{p_0(\delta)+1\}|V_{ab}|^2$. Putting in the appropriate factors it can be shown that within the Debye approximation the direct process relaxation rate τ_d is [53]

$$\tau_d^{-1} = \frac{3}{2\pi d v^5 \hbar}\left(\frac{\delta}{\hbar}\right)^3 |V_{ba}|^2 \coth\left(\frac{\delta}{2kT}\right), \tag{1.62}$$

where v = velocity of sound, and d is the density of the crystal. At low temperatures $\tau_d^{-1} \propto T$. While expression (1.62) is suitable for non-Kramers' levels it is actually zero for a Kramers' doublet. In this case it is necessary to consider admixtures of higher states, $|c\rangle$ etc., into $|a\rangle$ and $|b\rangle$ arising through the Zeeman interaction. As a result the matrix elements in (1.62) are replaced by

$$\sum_i \frac{2\mu_B g_J}{\Delta_i}\{\langle a|\mathbf{H}\cdot\mathbf{J}|i\rangle V_{bi} + V_{ia}\langle i|\mathbf{H}\cdot\mathbf{J}|b\rangle\}, \tag{1.63}$$

where Δ_i is the energy of the ith level.

In many rare earth salts the energy separation Δ_1 of the first excited state $|c\rangle$ from the ground state doublet is at an energy less than the maximum phonon energy $k\theta_{\text{Debye}}$ ($\theta_{\text{Debye}} \simeq 60°\text{K} \equiv 42\ \text{cm}^{-1}$ for the ethylsulphates), so that relaxation $|b\rangle \to |a\rangle$ may proceed by a two stage *Orbach process* [48, 51] via level $|c\rangle$. Firstly a phonon of energy Δ_1 is absorbed inducing the transition $|b\rangle \to |c\rangle$, and secondly relaxation $|c\rangle \to |a\rangle$ occurs with emission of a phonon of energy $\Delta_1 + \delta$. By solving the appropriate rate equations it may be shown that the relaxation time τ_0 for this process is given by

$$\tau_0^{-1} = \frac{3}{2\pi d v^5 \hbar}\left(\frac{\Delta_1}{\hbar}\right)^3 \frac{2|V_{ca}|^2|V_{bc}|^2}{|V_{ca}|^2+|V_{bc}|^2}\left\{\exp\left(\frac{\Delta_1}{kT}\right)-1\right\}^{-1}. \tag{1.64}$$

At sufficiently low temperatures the final factor may be approximated to give

$$\tau_0^{-1} = B\exp(-\Delta_1/kT). \quad (B = \text{constant}) \tag{1.65}$$

In the case of an excited doublet $|c\rangle$ and $|d\rangle$, expression (1.64) has to be multiplied by 2. Since the crystal field splitting Δ_1 is independent of magnetic field, so also is τ_0.

Finally, at higher temperatures the second order *Raman process* becomes important. It is a second order process involving the absorption of a phonon of energy Δ_i, thereby causing transition to a higher virtual state $|i\rangle$, and the simultaneous emission of another phonon of energy $\Delta_i + \delta$, with net result $|b\rangle \to |a\rangle$. Since $\delta \ll \theta_{\text{Debye}}$, almost all of the phonon spectrum is available to participate in the process, and consequently P_{ba} involves an integration over all phonon states. Compare the direct process, in which only those few phonons with correct energy δ are involved, and which is therefore only important at low temperatures. If there is only a singlet excited state $|c\rangle$ then the relaxation rate can be shown to be

$$\tau_r^{-1} = \frac{9\times 6!}{4d^2\pi^3 v^{10}}\left(\frac{k}{\hbar}\right)^7 \{|V_{ba}|^2 + \frac{1}{\Delta_1^2}|V_{ac}|^2|V_{bc}|^2\}T^7 \tag{1.66}$$

for a non-Kramers' doublet $|a\rangle, |b\rangle$. For a Kramers' doublet $|a\rangle$ and $|b\rangle$ and a higher doublet $|a'\rangle, |b'\rangle$ at $\Delta_{a'}, \Delta_{b'}$ we have:

$$\tau_r^{-1} = \frac{9!\,\hbar^2}{\pi^3 d^2 v^{10}}\left(\frac{k}{\hbar}\right)^9 \left\{ \sum_{i=a',b'} \frac{1}{\Delta_i^4} |V_{ia'}|^2 |V_{b'i}|^2 \right\} T^9. \tag{1.67}$$

The total spin lattice relaxation rate for a ground state doublet is the sum of the rates for all three processes. The temperature dependence for Kramers' (K) and non-Kramers' (NK) doublets is thus:

$$\tau_K^{-1} = A_K T + B \exp(-\Delta_1/kT) + C_K T^9, \tag{1.68}$$

$$\tau_{NK}^{-1} = A_{NK} T + B \exp(-\Delta_1/kT) + C_{NK} T^7, \tag{1.69}$$

where the constants A_K, C_K, A_{NK} and C_{NK} are given by comparison with equations (1.63), (1.67), (1.62), and (1.66) respectively.

It is usually assumed that the phonons created during a transition to the ground state pass instantaneously into the large helium bath surrounding the crystal, and experimentally the spin-bath relaxation rate τ_b is in fact measured. At low temperatures however, where the direct process dominates, there are so few phonons available with the correct energy δ that they may be unable to communicate energy from the spins to the bath immediately. In this case there is said to exist a *phonon bottleneck*, and there is a characteristic phonon-bath relaxation time τ_{ph} associated with it. The density of phonons $p(\delta)$ of energy δ is increased above its equilibrium value and is characterized by a 'hot phonon' temperature, T_p, defined by $p(\delta) = \{\exp(\delta/kT_p) - 1\}^{-1}$. In the absence of spins, $p(\delta)$ approaches p_0 exponentially with a time constant τ_{ph}. Re-working the theory for the direct process it is found that an important parameter is $\mathfrak{G} = C_s \tau_{ph}/C_p \tau_d$, where C_s/C_p is the ratio of spin and phonon specific heats. For $\mathfrak{G} \gg 1$ the bottleneck effect is much greater than that of the direct process, in which case it may be shown that the spins relax to the bath in a characteristic time $\tau_b \propto \tau_{ph}$. There are several possible mechanisms whereby the phonons can relax to the bath. At helium temperatures phonon–phonon scattering is negligible, as also is impurity scattering at the low frequencies used. Two possibilities are that the phonons are directly transmitted into the bath at the crystal boundary, or that they are inelastically scattered into other phonon states by the boundary. In both cases it would be expected that τ_{ph} is of the same order as the time for a phonon to cross the crystal, i.e. l/v where l is the linear crystal dimension. This size dependence has been observed [49] in Pr^{3+} in the lanthanum double nitrate where the bottleneck effect is large; see Fig. 1.11. Because of the large number of phonons available a bottleneck effect is unlikely to occur for the Raman process. It could conceivably occur for the Orbach process but it has not been observed.

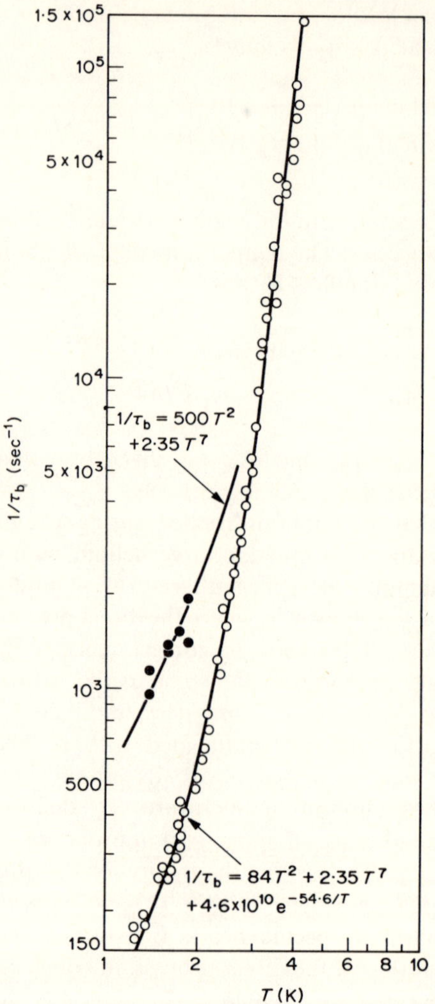

Fig. 1.11 Relaxation results for Pr^{3+} in lanthanum double nitrate, $La_2Mg_3(NO_3)_{12}\,24H_2O$, with an external field present of 6·88 KOe//z [49]. The circles are for a crystal 1·4 mm thick, containing 1% Pr, and the temperature dependence shows contributions from an Orbach process, a Raman process, and a direct phonon bottleneck. The dots are for a crystal 2·1 mm thick, containing 0·1% Pr, and illustrate the concentration and crystal size dependence of the phonon bottleneck effect.

1.5.2 *Comparison with experiment*

Most experimental determinations of spin-bath relaxation rates have been made on rare earth ions in lanthanum salts or in yttrium ethylsulphate using pulse saturation techniques. Experiments have also been reported on ions in other host lattices, e.g. lanthanum double nitrate [53] and CaF_2 [56], and using different techniques [55]. With few exceptions the temperature dependence can be fitted to either formula (1.68) or formula (1.69). In order to compare the coefficients in these expressions with theory it is necessary to have some estimate of the matrix elements V_{ab}, and hence some model of $V_c(t)$. Following Orbach [48] it is usually assumed that the average thermal lattice strain ε produces a deformation of the form

$$V_c(t) = \varepsilon \sum_{n,m} b_n^m r^n Y_n^m(\hat{\mathbf{r}}) \tag{1.70}$$

which perturbs the static crystal field, and in general has lower symmetry than the lattice. In evaluating V_{ba} it is convenient to use the operator equivalent form

$$V_c(t) = \varepsilon \sum_{n,m} a_n^m \langle r^n \rangle \theta_n O_n^m, \tag{1.71}$$

where the a_n^m are the *dynamic crystal field parameters*, which are related to the coefficients b_n^m by normalization constants g_n^m, and must be estimated empirically. For ions in La double nitrate and in La and Y ethylsulphates, Jefferies and co-workers [49, 50, 52, 53] have obtained good agreement with experiment by relating the a_n^m to the experimental static parameters A_n^0 through the empirical rule $|b_n^m| \simeq |B_n^0|$, which yields:

$$|a_n^m| = g_n^{|m|} |A_n^0|, \tag{1.72}$$

where the $g_n^{|m|}$ are factors given in Table 1.14. Explicitly the required matrix elements are then

$$|V_{ba}|^2 = \sum_{\substack{n=2,4,6, \\ |m| \leq n}} |\theta_n g_n^{|m|} |A_n^0 \langle r^n \rangle| \langle a|O_n^m|b \rangle|^2. \tag{1.73}$$

With one slight modification, namely using $g_6^{|6|} = A_6^6 \langle r^6 \rangle / A_6^0 \langle r^6 \rangle$, the same scheme also gives good agreement for ions in $LaCl_3$ [54]. However, for ions in CaF_2 [56] the experimental values for the relaxation times of the direct

Table 1.14 Values of the factors $g_n^{|m|}$ used in [53] to relate the dynamic crystal field parameters a_n^m to the static values A_n^m.

$\|m\|$	1	2	3	4	5	6
$n = 2$	4·90	2·45				
$n = 4$	8·95	6·32	23·6	8·37		
$n = 6$	12·9	10·2	20·2	11·2	52·7	15·2

process are two orders of magnitude shorter than theoretical estimates, and this may be due to the fact that the rare earth ion causes a relatively large disturbance of its host lattice, differing in both mass and charge from the Ca^{2+} ion which it replaces. At higher temperatures, theoretical predictions

Table 1.15 Contributions to the relaxation rates of rare earth ions in yttrium ethylsulphate [53]. Measurements were made at $1{\cdot}2 < T < 5°K$ on samples containing ~1% rare earth ions. For Pr^{3+} the direct process is bottlenecked so that only a lower limit for τ_d^{-1} can be obtained.

Ion	Field	τ_d^{-1} (sec^{-1})		τ_0^{-1} (sec^{-1})		τ_r^{-1} (sec^{-1})	
		Measured	Theory	Measured	Theory	Measured	Theory
Ce	4220 65° to z		$3{\cdot}39\,T$	$5{\cdot}1 \times 10^8 \exp(-25/T)$	$3{\cdot}2 \times 10^{10} \exp(-25/T)$		$1{\cdot}0\,T^9$
Pr	7010 $\parallel z$	$> 5 \times 10^4\,T$	$5 \times 10^5\,T$	$3{\cdot}8 \times 10^7 \exp(-19/T)$	$4{\cdot}2 \times 10^9 \exp(-20/T)$	$(30\,T^7)$	$1{\cdot}7\,T^7$
Nd	3240 $\perp z$	$1{\cdot}2\,T$	$0{\cdot}74\,T$			$1{\cdot}64 \times 10^{-4}\,T^9$	$3{\cdot}4 \times 10^{-5}\,T^9$
Sm	11 110 $\perp z$	$1{\cdot}3\,T$	$5{\cdot}0\,T$	$8{\cdot}0 \times 10^8 \exp(-51/T)$	$2{\cdot}4 \times 10^{10} \exp(-51/T)$	$4{\cdot}0 \times 10^{-4}\,T^9$	$8{\cdot}4 \times 10^{-4}\,T^9$
Er	759 $\perp z$	$5{\cdot}9\,T$	$4{\cdot}5\,T$	$5{\cdot}0 \times 10^{10} \exp(-63/T)$	$9{\cdot}2 \times 10^{10} \exp(-63/T)$	$3{\cdot}0 \times 10^{-3}\,T^9$	$1{\cdot}1 \times 10^{-3}\,T^9$

Table 1.16 Contributions to the relaxation rates of rare earth ions in $LaCl_3$ at concentrations < 2‰ and at temperatures $1{\cdot}65 < T < 4{\cdot}2$°K [54]. The T^2 behaviour of τ_d^{-1} for Pr^{3+} is ascribed to a phonon bottleneck.

Ion	Field (Oe)	τ_d^{-1}		τ_0^{-1}		τ_r^{-1}	
		Measured	Theory	Measured	Theory	Measured	Theory
Ce	1550 $\parallel z$	$0{\cdot}54\ T$	$7{\cdot}03 \times 10^{-3}\ T$	$2{\cdot}6 \times 10^9 \exp(-46/T)$	$2{\cdot}7 \times 10^{11} \exp(-46/T)$		
Pr (1%)	3750 $\parallel z$	$0{\cdot}272\ T^2$	$1{\cdot}79 \times 10^5\ T$	$2{\cdot}06 \times 10^9 \exp(48/T)$	$2{\cdot}81 \times 10^{10} \exp(-48/T)$		$0{\cdot}0733\ T^7$
Nd	2080 45° to z	$0{\cdot}581\ T$	$0{\cdot}295\ T$		$1{\cdot}22 \times 10^{12} \exp(-\Delta/T)$	$4{\cdot}85 \times 10^{-5}\ T^9$	$3{\cdot}75 \times 10^{-5}\ T^9$
Sm	112000 $\parallel z$	$1{\cdot}0\ T$	$0{\cdot}47\ T$	$3{\cdot}4 \times 10^{11} \exp(-58{\cdot}6/T)$	$2{\cdot}2 \times 10^{10} \exp(-58{\cdot}6/T)$	$0{\cdot}02\ T^9$	$0{\cdot}12 \times 10^{-3}\ T^9$
Tb	640 $\parallel z$	$4{\cdot}26\ T$	$1{\cdot}08\ T$	$4{\cdot}83 \times 10^{11} \exp(-82/T)$	$1{\cdot}18 \times 10^{11} \exp(-82/T)$	$0{\cdot}027\ T^7$	$0{\cdot}65 \times 10^{-3}\ T^7$
Er	725 $\perp z$	$2{\cdot}26\ T$	$2{\cdot}73\ T$	$1{\cdot}14 \times 10^{10} \exp(-55/T)$	$1{\cdot}66 \times 10^{10} \exp(-55/T)$	$3{\cdot}46\ T^9$	$0{\cdot}679 \times 10^{-3}\ T^9$

for the Raman and Orbach processes are in rough agreement with those obtained experimentally. Tables 1.15 and 1.16 contain results for some rare earth ions in yttrium ethylsulphate and lanthanum trichloride respectively.

The phonon bottleneck effect, which is characterized by a T^2 dependent inverse relaxation rate at low temperatures, has been observed for Pr^{3+} in all La and Y salts which have been studied, and for Nd^{3+}, Sm^{3+} and Ce^{3+} in the La double nitrates [52]. The effect is expected in Pr^{3+} since it is a non-Kramers' doublet ion with a fast direct relaxation rate. Results for Pr^{3+} in La double nitrate, shown in Fig. 1.11 for different crystal dimensions l, and concentrations d, indicate that the bottleneck rate is $\propto (ld)^{-1}$, in agreement with theory. For 1% Pr^{3+} in La double nitrate τ_b is smaller than τ_d by a factor $\mathfrak{G}(\simeq 10^3)$, and a saturating pulse of 1 W peak produces a hot phonon temperature $T_p \simeq 60°K$, which cools rapidly at a rate $\tau_b' \simeq 10^{-9}$ secs.

1.5.3 *Cross relaxation*

Measurements have also been made of the dependence of the relaxation times on the angle θ between the applied magnetic field **H** and the symmetry axes of the crystals. Theoretically one expects the Orbach and Raman relaxation rates to be independent of θ, and this is found to be so for the above-mentioned salts. However, for the direct process, for Kramers' ions, the direction θ enters the expression for τ_d^{-1} through the matrix elements like $\langle a|\mathbf{H}\cdot\mathbf{J}|i\rangle$ in (1.63).

Measured anisotropies have been as high as 15:1 for Sm in La double nitrate [53], and are mostly in reasonable agreement with the theory.

In experiments on the angular dependence of τ_d, anomalous peaks (rapid relaxation) have been observed for Nd^{3+} and Er^{3+} in La ethylsulphate, and Larson and Jefferies [53] have attributed them to a process involving Ce^{3+} impurity ions. The magnetic moment $\boldsymbol{\mu}_1$ of the rare earth ion can interact with the moment $\boldsymbol{\mu}_2$ of the impurity ion at $\mathbf{r}_{12}$ giving rise to a dipolar energy

$$\mathscr{H}_{\text{dip}} = \frac{\boldsymbol{\mu}_1\cdot\boldsymbol{\mu}_2}{r_{12}^3} - \frac{3(\boldsymbol{\mu}_1\cdot\mathbf{r}_{12})(\boldsymbol{\mu}_2\cdot\mathbf{r}_{12})}{r_{12}^5}. \tag{1.74}$$

If under certain conditions the separations between the lowest pairs of levels of the rare earth and impurity ions are equal, then it is possible through $\mathscr{H}_{\text{dip}}$ for the rare earth spin to flip 'down' and simultaneously an impurity to flip 'up'. The impurity ion may then be able to relax to the bath in a shorter time than the rare earth ion. The process is known as *cross-relaxation.* For two spin systems α, (Er^{3+} say), and β, (Ce^{3+} say), with equal energy separations, one can define a cross-relaxation rate τ_{12}^{-1} in terms of the probability ω of an α spin undergoing a mutual spin flip with a β spin. Assuming spin lattice rates are negligible:

$$\tau_{12}^{-1} = \omega(N_\alpha + N_\beta),$$

where N_α and N_β are the numbers of α and β spins respectively. The return to equilibrium involves two time constants, the slower of which τ_- usually dominates the faster one τ_+.

For Ce^{3+} ions in the ground state it can be shown that τ_{12} is insensitive to θ, and in order to explain the anomalous peaks it is necessary to assume that cross-relaxation also occurs with excited Ce^{3+} ions. In this case the energy level separations of Er^{3+} and Ce^{3+} ions become equal when the magnetic field is at angle $\theta = 21°$, which is precisely when the anomalous relaxation peak occurs; see Fig. 1.12. The same workers have also shown that

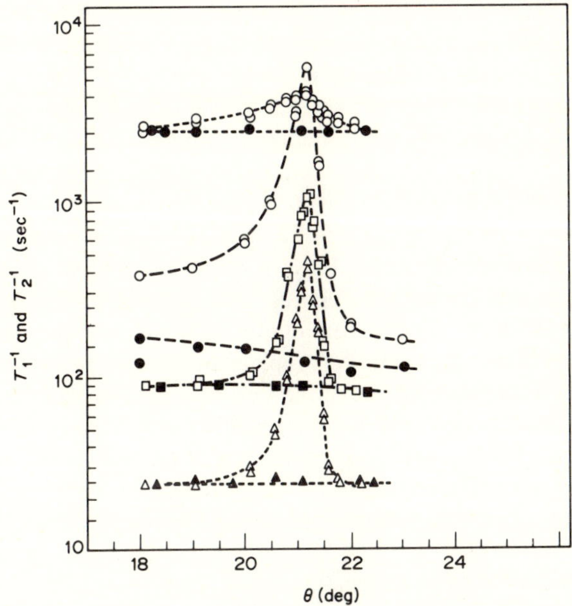

Fig. 1.12 Angular variation of the two observed relaxation rates, open and solidus symbols, T_1 and T_2 respectively, for 1% Er in lanthanum ethylsulphate containing Ce^{3+} impurities [53]. θ is the angle between applied magnetic field and c-axis, and the peaks at $\theta = 21°$ are due to cross-relaxation. The curves are: upper circles for 0·01% Ce at 3·5°K; squares for 0·01% Ce at 2·6°K; triangles for 0·01% Ce at 1·8°K; and lower circles for 0·1% Ce at 2·16°K.

the relaxation time T_1 ($\simeq \tau_+$) depends markedly upon the concentration of Ce^{3+}, which is consistent with cross-relaxation processes.

1.5.4 *Other ions*

Most ions considered in relaxation experiments have been chosen with $I = 0$ in order to avoid hyperfine splittings of the Zeeman levels. For ions with non-zero I, e.g. Pr^{3+}, the allowed paramagnetic transitions are those between levels such that $\Delta M_I = 0$, but forbidden transitions with $\Delta M_I = \pm 1$ are sometimes observed. In general, with so many transitions possible, the exponential recovery will depend upon many time constants.

The case of Gd^{3+} is exceptional since the crystal field splittings are small and many levels may take part in the relaxation. It has been shown [57] that the Raman process rate for such systems is $\propto T^{-5}$, and this is observed [56] for Gd^{3+} in CaF_2. At low temperatures a $T^{-\frac{1}{2}}$ temperature dependence is observed instead of the direct process T^{-1} dependence, and this is not understood.

1.6 Spin-spin interactions

At very low temperatures (< 2°K) there usually exist interactions between neighbouring ions in crystals. For example, Ce ethylsulphate exhibits anti-ferromagnetic ordering below 0·08°K [78], while $GdCl_3$ [74] becomes ferro-magnetic below 2·2°K. There are several mechanisms by which this coupling may occur, the most important being (i) dipole–dipole, (ii) quadrupole–quadrupole, (iii) isotropic superexchange and (iv) anisotropic exchange interactions. Experimentally, these interactions are most easily studied by obtaining the esr spectra of pairs of rare earth ions introduced into non-magnetic host lattices. The host lattices which have been most commonly used are $LaCl_3$ and the ethylsulphates, and of these $LaCl_3$ is the simpler system in that there is only one kind of anion. In this case the most desirable situation is when the rare earth ions replace neighbouring La ions along the crystal *c*-axis, but the case of one of the ions on a next-nearest neighbour site (Fig. 1.13) has also been studied. In practice, to reduce the number of larger clusters of rare earth ions, the concentration of magnetic ions is kept small (~ 1%). Other ambiguities, such as the orientation of the pair, may be resolved by varying the direction of the applied magnetic field.

1.6.1 *Dipole interactions*

The dipolar interaction between two ions $\mathcal{H}_{\text{dip}}$, has been discussed previously in connection with cross-relaxation. It is an interaction between magnetic moments $\boldsymbol{\mu} = -\mu_B(\mathbf{L}+2\mathbf{S}) = -g\mu_B\mathbf{J}$, where g is a tensor and in general is anisotropic. In the isotropic case $\mathcal{H}_{\text{dip}}$ may be written in the spin Hamiltonian form,

$$\mathcal{H}^{ij}_{\text{dip}} = \frac{g^2\mu_B^2}{|\mathbf{r}_i-\mathbf{r}_j|^3}\{\mathbf{S}_i\cdot\mathbf{S}_j-3S_{iz}S_{jz}\}, \tag{1.75}$$

and more generally, for identical nearest-neighbour ions along the *c*-axis:

$$\mathcal{H}^{ij}_{\text{dip}} = A_{ij}S_{iz}S_{jz}+\tfrac{1}{2}B_{ij}(S_{i+}S_{j-}+S_{i-}S_{j+}), \tag{1.76}$$

where

$$A_{ij} = -2g_{\parallel}^2\mu_B^2|\mathbf{r}_i-\mathbf{r}_j|^{-3} \quad \text{and} \quad B_{ij} = g_{\perp}^2\mu_B^2|\mathbf{r}_i-\mathbf{r}_j|^{-3}.$$

If the interaction is comparable with the energies of the individual ionic

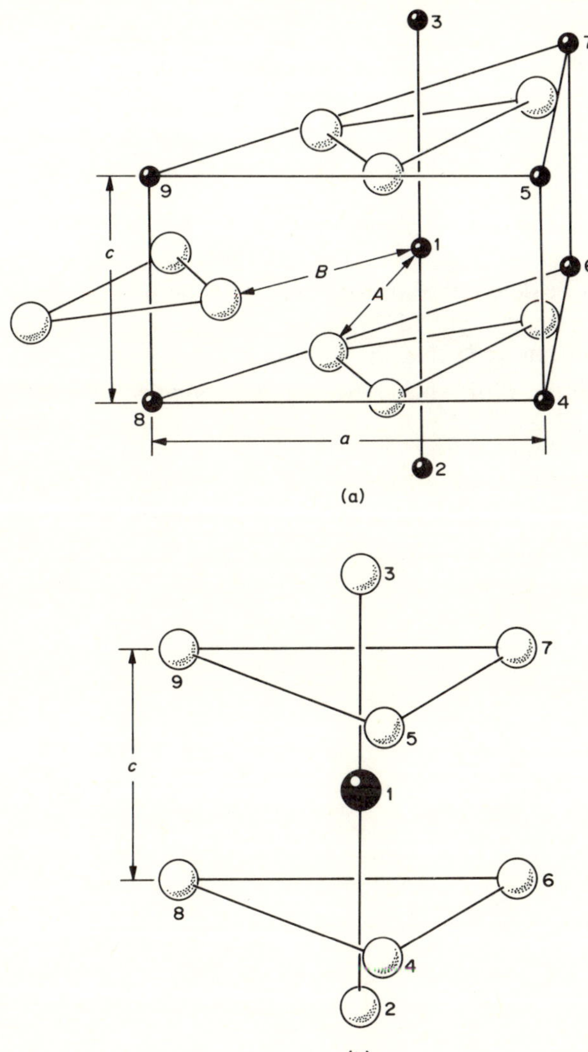

Fig. 1.13 (a) The crystal structure of the rare earth trichlorides lanthanum to gadolinium. • = rare earth ion. O = chlorine ion. a is the lattice spacing quoted in Table 1.17, and for $LaCl_3$, $A = 2{\cdot}91$Å and $B = 2{\cdot}96$Å. (b) The positions of rare earth ions in the crystal structure of the ethylsulphates. Ions 2 and 3 are nearest-neighbours to ion 1 ($\sim$7Å), and ions 4 to 9 are next-nearest-neighbours ($\sim$8·8Å).

spins, the spins are coupled, and the pair may be considered as a single entity with energy levels depending upon the total spin $\mathbf{T} = \mathbf{S}_i + \mathbf{S}_j$, say. For a ground state doublet, e.g. Nd^{3+}, with effective spin $S = \frac{1}{2}$, the interaction produces a triplet state $\mathbf{T} = 1$, and a singlet state $\mathbf{T} = 0$. In the presence of a magnetic field, splittings of these energy levels are produced by the Zeeman interactions of the individual ions:

$$\mathscr{H}_{H}^{ij} = \sum_{i=i,j} \{g_{\parallel}\mu_{B}H_{z}S_{iz} + g_{\perp}\mu_{B}(H_{x}S_{ix} + H_{y}S_{iy})\}. \tag{1.77}$$

The resulting energy level structure for the case **H** parallel to the *c*-axis is shown in Fig. 1.14, and the allowed transitions $\Delta S_{iz} = \pm 1$, $\Delta S_{jz} = 0$, or vice versa, are given by

$$h\nu = g_{\parallel}\mu_{B}H \pm \tfrac{1}{2}(A_{ij} - B_{ij}). \tag{1.78}$$

The sign can only be determined from the relative intensities of the observed lines. For nearest-neighbour Nd^{3+} ions in La and Y ethylsulphates esr measurements indicate [67] that the dipolar interaction is dominant. The situation is slightly complicated by hyperfine interactions and lattice distortions but the values of $A_{ij} = 277 \pm 2 \times 10^{-4}$ cm^{-1} and $B_{ij} = +49 \pm 4 \times 10^{-4}$ cm^{-1} obtained from experiments are in good agreement with calculated dipolar values of $-309 \pm 4 \times 10^{-4}$ cm^{-1} and $+52 \pm 0{\cdot}6 \times 10^{-4}$ cm^{-1} respectively. It also appears that dipolar coupling is important in Dy

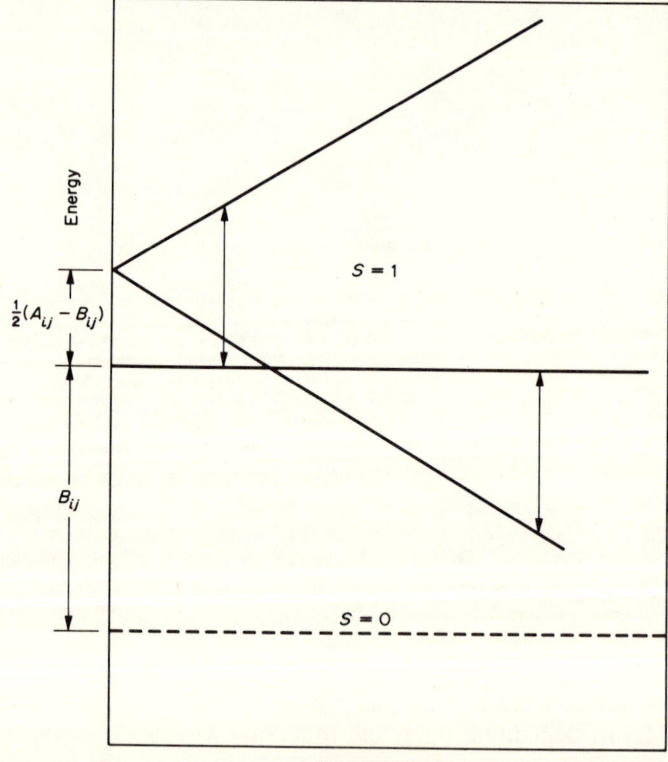

Fig. 1.14 A schematic diagram of the energy levels of a pair of Nd^{3+} ions in $LaCl_3$ as a function of applied field (parallel to the crystallographic *c*-axis) [83]. The allowed transitions are shown.

ethylsulphate, for $Gd^{3+}-Dy^{3+}$ interactions in that salt [84], and for $Gd^{3+}-Ce^{3+}$ interactions in Ce ethylsulphate [86].

1.6.2 *Quadrupole interactions*

In many other rare earth salts it is clear that dipolar interactions are unable to account for the magnitude of the spin–spin coupling. For example, in Ce ethylsulphate the ordered state at low temperatures requires an interaction which is six times larger than $\mathscr{H}^{ij}_{\text{dip}}$ and of opposite sign [78, 69]. It has been suggested [77] that the quadrupole–quadrupole interaction is dominant in this salt. The quadrupole–quadrupole interaction is most conveniently expressed in terms of spherical harmonics:

$$\mathscr{H}^{ij}_{\text{quad}} = \frac{4\pi e^2 \langle r_i^2 \rangle \langle r_j^2 \rangle}{5\varepsilon_D |\mathbf{r}_i - \mathbf{r}_j|^5} \sum_{m=-2}^{2} c_m Y_2^m(\hat{\mathbf{r}}_i) Y_2^{-m}(\hat{\mathbf{r}}_j), \tag{1.79}$$

where ε_D is an effective dielectric constant and $c_m = c_{-m}$, $c_0 = 6$, $c_1 = 4$ and $c_2 = 1$. Because of the $|\mathbf{r}_i - \mathbf{r}_j|^{-5}$ factor it is only appreciable for nearest and next-nearest ions, and it is largest for ions which have large $\langle r_{4f}^2 \rangle$ values, such as Ce^{3+}. For cases in which the *c*-axis is the axis of quantization, equation (1.79) has the operator equivalent form:

$$\mathscr{H}^{ij}_{\text{quad}} = A\{4O_{2i}^0 O_{2j}^0 - 16(O_{2i}^{+1} O_{2j}^{-1} + O_{2j}^{-1} O_{2j}^{+1}) + (O_{2i}^{+2} O_{2j}^{-2} + O_{2i}^{-2} O_{2j}^{+2})\}, \tag{1.80}$$

where for like ions:

$$A = \frac{3e^2 |\langle r_{4f}^2 \rangle|^2 \alpha_J}{8\varepsilon_D |\mathbf{r}_i - \mathbf{r}_j|^5 hc}. \tag{1.81}$$

In first order perturbation theory the quadrupole interaction has no effect on a Kramers' doublet, as is appropriate for Ce^{3+}, so that it can only produce interactions through second order processes involving higher states. If $|+\rangle$ and $|-\rangle$ are the ground states of a single ion then the basis states of an ion pair may be constructed as products of them, and labelled, say, $|++\rangle$, $|--\rangle$ and $|+-\rangle$. Since the latter is degenerate with $|-+\rangle$ it is necessary, in order to use non-degenerate perturbation theory, to replace both by the symmetric and antisymmetric combinations $\frac{1}{2}\{|+-\rangle + |-+\rangle\}$ and $\frac{1}{2}\{|+-\rangle - |-+\rangle\}$. The matrix elements of interest are then:

$$\langle ab|\mathscr{H}_{\text{quad}}|cd\rangle = \sum_{kl} \frac{\langle ab|\mathscr{H}_{\text{quad}}|kl\rangle\langle kl|\mathscr{H}_{\text{quad}}|cd\rangle}{E_{kl} - E_{ab}}, \tag{1.82}$$

where the sum is over excited states $|kl\rangle$. It is possible to express the resulting energy level splittings in terms of a spin Hamiltonian identical in form to (1.76). For Ce^{3+} in the concentrated ethylsulphate $\mathscr{H}^{ij}_{\text{quad}}$ produces mixing of the three doublet levels. Evaluation of (1.82) and comparison with (1.76) yields [77] expressions for A_{ij} and B_{ij}. It is found that the quantity

$$-\sum_{i,j}(A_{ij}-B_{ij}),$$

which determines the esr spectrum, shows a negative contribution from nearest neighbours, indicating antiferromagnetic coupling, and a larger positive ferromagnetic contribution from next-nearest-neighbours. The net result is of the right order of magnitude, predicting a frequency shift of $+0{\cdot}078\ \mathrm{cm}^{-1}/\mathrm{h}$, compared with the experimental non-dipolar shift of $-0{\cdot}113\ \mathrm{cm}^{-1}/\mathrm{h}$, but it has the wrong sign.

1.6.3 *Exchange coupling*

For rare earth ions in $LaCl_3$ there is evidence that the spins are coupled by an exchange interaction. The basic mechanism is isotropic superexchange between the real spins of the pair, i.e. $\mathcal{J}_{ij}\ \mathbf{S}_i\cdot\mathbf{S}_j$, but there may be a small anisotropic interaction also. The isotropic interaction between spins may be written in terms of the total momentum $\mathbf{J}_i$ by combining $\mathbf{J} = \mathbf{L}+\mathbf{S}$ with $\mathbf{L}+2\mathbf{S} = g_J\mathbf{J}$, so that

$$\mathbf{S} = (g_J-1)\mathbf{J}. \tag{1.83}$$

At low temperatures, for a ground state doublet, it is convenient to introduce an effective spin S' by defining a splitting factor g such that the matrix elements of $g\mathbf{S}'$ are the same as those of $g_J\mathbf{J}$. The spin Hamiltonian is then anisotropic, i.e. for symmetry about the c-axis:

$$\mathcal{H}^{ij}_{ex} = \mathcal{J}_{ij}\left(\frac{g_J-1}{g_J}\right)^2\{g_{\parallel}^2S_{iz}S_{jz}+\tfrac{1}{2}g_{\perp}^2(\mathrm{S}_{i+}\mathrm{S}_{j-}+\mathrm{S}_{i-}\mathrm{S}_{j+})\}, \tag{1.84}$$

where the primes on the fictitious spins have been dropped. This has the same form as (1.76).

No resonances attributable to pairs have been detected for Er^{3+} in $LaCl_3$ [68]. Measurements of pair spectra have been made on Nd^{3+} in $LaCl_3$ [68, 83] and in $LaBr_3$ [68] and, in the notation of equation (1.76), it is found that $(A_{ij}-B_{ij}) > 0$ for a nearest-neighbour (*nn*) interacting pair. Since $g_{\parallel} > g_{\perp}$ this indicates that $\mathcal{J}_{nn} > 0$, so that the coupling is antiferromagnetic. For pairs such that one of the ions is on a next-nearest-neighbour (*nnn*) site, and choosing the crystallographic c-axis as z-axis, the spin Hamiltonian takes the more complex form:

$$\mathcal{H}^{ij}_{ex} = a_{zz}S_{iz}S_{jz}+a_{xx}S_{ix}S_{jx}+a_{yy}S_{iy}S_{jy}+a_{xz}(S_{ix}S_{jz}+S_{iz}S_{jx}). \tag{1.85}$$

If the exchange interaction between real spins is isotropic then its contributions to a_{xx} and a_{yy} are equal. However, experiment shows that $(a_{xx}-a_{yy})>0$, so that it has opposite sign to that due to the dipolar interaction, and hence indicates that the non-dipolar contributions to a_{xx} and a_{yy} are unequal. It appears [83], therefore, that the superexchange between real spins must be anisotropic, though there may be a small contribution from quadrupole–quadrupole interactions.

Pair spectra have also been investigated for Gd^{3+} in $LaCl_3$ [66], $EuCl_3$ [85], and in $GdCl_3$ [85], and for Ce^{3+} in $LaCl_3$ [86]. It would be expected that since Gd^{3+} is an S-state ion its interactions would arise predominantly from dipolar and isotropic exchange, and this appears to be the case when **H** is along the c-axis. The total Hamiltonian then has the simple form

$$\mathscr{H}^{ij} = g\mu_{\rm B}H(S_{iz}+S_{jz})+\alpha(\mathbf{S}_i\cdot\mathbf{S}_j-3S_{iz}S_{jz})+\mathscr{J}_{ij}\mathbf{S}_i\cdot\mathbf{S}_j, \tag{1.86}$$

where $g = 1{\cdot}991$ is the single ion, isotropic, splitting factor, and $\alpha = g^2\mu_{\rm B}^2/|\mathbf{r}_i-\mathbf{r}_j|^3$. The energy levels given by (1.86) depend only upon the ratios α/g and $\mathscr{J}_{ij}/\alpha$, and hence effectively upon $\mathscr{J}_{ij}$ alone. It is also possible to include crystal field effects by adding to (1.86) the crystal fields for the individual ions,

$$\sum_{i=1,2} V_c(i),$$

which have C_{3h} symmetry, and are therefore characterized by the crystal field parameters B_2^0, B_4^0 and B_6^0. When the crystal field is included it is convenient to evaluate the energy matrices in the $|S_i, S_j, M_i, M_j\rangle$ representation and then transform to the $|S_i, S_j, T, M_T\rangle$ representation.

In Fig. 1.15 the experimental and best fit calculations are shown for the nearest-neighbour and next-nearest-neighbour pair spectra of Gd^{3+} in $LaCl_3$ [66]. In Table 1.17 are given the calculated, best fit parameters in the Hamiltonian. It is found that nearest-neighbours are antiferromagnetically coupled, but this coupling is weaker than next-nearest-neighbour interactions which are ferromagnetic. This is also the case for Gd^{3+} pairs in other chlorides [85] and for $Eu^{3+}-Gd^{3+}$ pairs in $EuCl_3$ [85]. For nearest-neighbouring Gd^{3+} in $LaCl_3$, $\mathscr{J}_{nn}$ decreases with temperature from 0·01330 cm^{-1} at 20°K to 0·01254 cm^{-1} at 360°K, while $\mathscr{J}_{nnn}$ increases from $-0{\cdot}0602$ cm^{-1} at 20°K to $-0{\cdot}0532$ cm^{-1} at 360°K. This temperature dependence of the exchange interaction is thought to arise mainly from thermal expansion effects, and hence it reflects the dependence upon the ionic separation. The variation of separation with temperature may be estimated from the variation of the (best fit) dipole interaction constant α with temperature. In this way it is found that $\mathscr{J}_{nn} \propto r^{13}$ and $\mathscr{J}_{nnn} \propto r^{22}$, neither of which can be explained qualitatively by superexchange. No quantitative calculations of superexchange involving rare earth ions have yet been made. They will not be simple because of the presence of the $5s^25p^6$ shells outside the $4f$ electrons.

The use of $EuCl_3$ as a host lattice leads to complications because it is weakly paramagnetic. Single foreign ions in it interact with the neighbouring Eu^{3+} ions, and their g-values are shifted. For Gd^{3+} in $EuCl_3$ it may be assumed that the interactions are isotropic exchange and dipolar, and if only

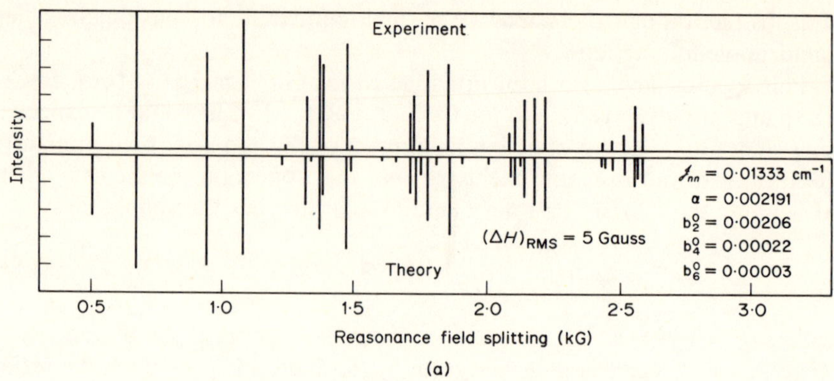

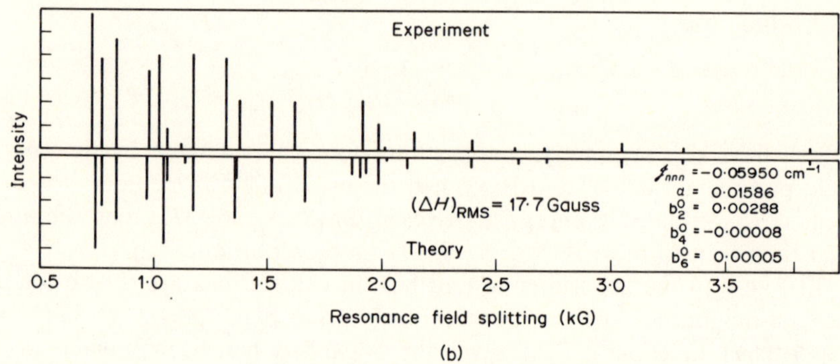

Fig. 1.15 Experimental and best fit theoretical pair spectra for Gd^{3+} ions in $LaCl_3$. (a) nearest-neighbour pairs, (b) nest-nearest-neighbour pairs, [66]. $b_2^0 = 3B_4^0$, $b_4^0 = 60B_4^0$ and $b_6^0 = 1260B_6^0$.

Table 1.17 Results for ion pairs in the trichlorides at 77°K [85]. Here r_{nn}, $\mathcal{J}_{nn}$ and α_{nn} are respectively the separation, exchange constant and dipolar interaction constant for nearest-neighbour ions. Subscript *nnn* refers to next-nearest-neighbour ions. (*) Values of $\mathcal{J}_{nn}$ and $\mathcal{J}_{nnn}$ for $GdCl_3$ are extrapolations from those of $LaCl_3$ and $EuCl_3$, and α_{nn} and α_{nnn} are values calculated taking $g = 1{\cdot}9915$ and assuming a 3% thermal contraction in lattice parameters.

Ion pairs	Salt	Lattice spacing (Å)	r_{nn} (Å)	r_{nnn} (Å)	$\mathcal{J}_{nn}$ (cm⁻¹)	$\mathcal{J}_{nnn}$ (cm⁻¹)	α_{nn} (cm⁻¹)	α_{nnn} (cm⁻¹)
Gd^{3+}–Gd^{3+}	$LaCl_3$	7·483	4·375	4·843	0·0133	−0·0595	0·0219	0·0159
Gd^{3+}–Gd^{3+}	$EuCl_3$	7·369	4·133	4·730	0·0488	−0·0637	0·0246	0·0166
Gd^{3+}–Gd^{3+}	$GdCl_3$(*)	7·363	4·105	4·721	0·056	−0·064	0·025	0·0165
Gd^{3+}–Eu^{3+}	$EuCl_3$	7·369	4·133	4·730	0·052	−0·061		

nearest and next-nearest-neighbour interactions are important, the components of the g-shift can be shown [85] to satisfy:

$$\begin{aligned} \Delta g_{\parallel}+2\Delta g_{\perp} &= -(48/\delta)(\mathscr{J}_{nn}+3\mathscr{J}_{nnn}), \\ \Delta g_{\parallel}-\Delta g_{\perp} &= 0{\cdot}0013, \end{aligned} \tag{1.87}$$

where $\delta = 370\ \mathrm{cm}^{-1}$ is an average of the first excited energy separation ($^7F_1$) for Eu^{3+}. For a pair of nearest neighbour Gd^{3+} ions there is a g-shift for each Gd^{3+} ion, and in this case, for the interactions assumed in obtaining (1.87), it can be shown that:

$$\Delta g_{\parallel}(\text{single ion})-\Delta g_{\parallel}(nn\ \text{pair}) = (-8/\delta)(\mathscr{J}_{nn}-2\alpha'_{nn}), \tag{1.88}$$

where $\alpha'_{nn} = +0{\cdot}0123\ \mathrm{cm}^{-1}$ for $EuCl_3$. A measurement of $\Delta g_{\parallel}$ for both a single ion and a pair of ions thus enables a determination of $\mathscr{J}_{nn}$ and $\mathscr{J}_{nnn}$ for $Gd^{3+}-Eu^{3+}$ pairs, and some results are shown in Table 1.17. They are very similar to those for $Gd^{3+}-Gd^{3+}$ in $EuCl_3$.

$GdCl_3$ has a lattice spacing intermediate between $LaCl_3$ and $EuCl_3$ so that estimates of the $Gd^{3+}-Gd^{3+}$ interactions in this salt obtained by interpolation may be expected to be reasonably good.

1.7 Magnetic susceptibilities

For a truly paramagnetic salt the static susceptibility χ is given by formula (1.41), in which the energy levels depend upon the crystal field, so that in general χ is anisotropic. As remarked earlier, at high temperatures this expression reduces to Hund's formula (1.37). At low temperatures it is necessary to use the complete formula, and in general this gives a good fit to experimental data, as illustrated in Fig. 1.16 for Er ethylsulphate [87]. In the case of a non-Kramers' salt the situation can arise where only the lowest singlet state is occupied, and then the application of a magnetic field is unable to orientate the spins, so that the salt would be diamagnetic were it not for the Van Vleck term which contributes a temperature *independent* susceptibility $N\alpha$. Pr ethylsulphate, Fig. 1.17, provides an example of this behaviour [88].

At very low temperatures the interactions between ions become important. Their effects on the susceptibility at high temperatures are usually included by employing a Curie–Weiss law, or more generally by expanding χ^{-1} in powers of T^{-1}, i.e.

$$\chi^{-1} = \frac{T-\theta}{C} = \frac{T}{C}\left\{1+\sum_{n=1}^{\infty}\frac{B_n}{(kT)^n}\right\}, \tag{1.89}$$

where θ, C and the B_n's are constants.

For many rare earth ethylsulphates [87, 91, 97, 98] the interactions are predominantly dipolar, in which case for a ground state doublet it may be

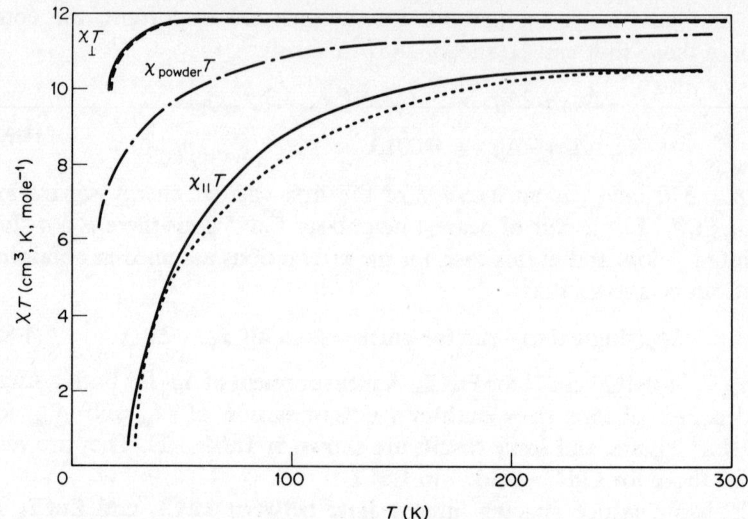

Fig. 1.16 The magnetic susceptibility of Er ethylsulphate as a function of temperature [87]. The broken curves are experimental results, and the full curves show the calculated susceptibilities. The free ion value of χT is 11·48.

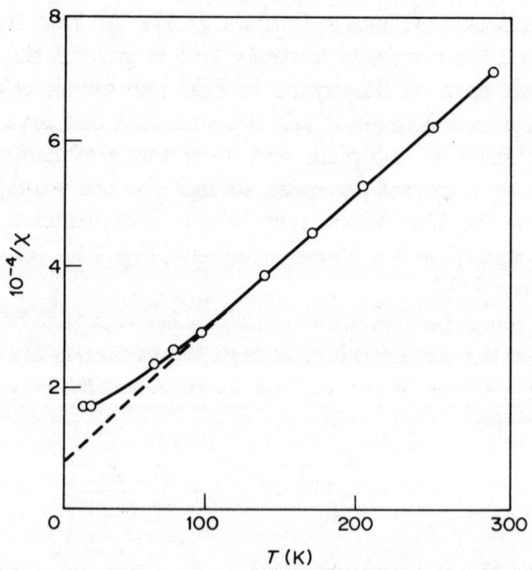

Fig. 1.17 The magnetic susceptibility of Pr ethylsulphate at low temperatures [88]. (cgs units)

shown [89] that for $\chi_{\|}(\text{mol}^{-1})$, $C = Ng_{\|}^2\mu_B^2/4k$, and

$$B_1^d \equiv -\theta = \frac{g_{\|}^2\mu_B^2}{4k}\sum_j \frac{3z_{ij}-r_{ij}}{r_{ij}^5} \qquad (°\text{K}), \tag{1.90}$$

where the sum is over all lattice sites excluding *i*. For some salts at temperatures less than 5°K, e.g. Ho^{3+} [87], it is necessary to consider B_2, and also contributions from off-diagonal terms. Below 0·13±0·01°K it is found [98] that Dy ethylsulphate becomes ferromagnetically ordered with the spins parallel to the *c*-axis. Ce ethylsulphate is exceptional in that quadrupole–quadrupole interactions are thought to be responsible for its ordering at 0·89°K [92].

At very low temperatures several of the rare earth trichlorides exhibit magnetic transitions. These can be explained by assuming that besides dipolar interactions there are also ferromagnetic exchange interactions between nearest-neighbour ions, and antiferromagnetic exchange coupling between next-nearest-neighbour ions. For temperatures much greater than the ordering temperatures, the effects of exchange upon the susceptibility $\chi_{\|}$ can be represented [93] by a contribution to the Curie–Weiss constant of

$$B_1^{ex} = -21(\mathcal{J}_{nn}+3\mathcal{J}_{nnn}) \qquad (°\text{K}), \tag{1.91}$$

where $\mathcal{J}_{nn}$ and $\mathcal{J}_{nnn}$ are in °K. For $GdCl_3$, which exhibits ferromagnetic ordering at 2·2°K [74], the dipolar interaction contributes approximately one third of the total B_1 value [93], and the theoretical estimate of B_1 (−3·0°K) using $\mathcal{J}_{nn}$ and $\mathcal{J}_{nnn}$ of Table 1.17 is in good agreement with experiment ($B_1 = -3{\cdot}2$°K) [85]. The susceptibilities of $NdCl_3$ and $CeCl_3$ show anomalies below 2°K and these have been associated with magnetic transitions [94].

1.8 Specific heats

The contributions of the spins and phonons to the total specific heat of a crystalline salt may be considered to be independent. In rare earth salts it is the magnetic specific heat C_m at low temperature which is primarily of interest. At such temperatures the lattice contribution is usually negligibly small, being $\propto T^3$, but if it is not it may be estimated experimentally from measurements on similar salts which show no magnetic contribution.

At absolute zero only the ground state is populated and as T is increased energy must be provided to populate the higher energy levels. At very high temperatures all levels are equally populated so that no further energy is required for repopulation. C_m would therefore be expected to show a maximum at a temperature such that the thermal energy is the order of the separation of the levels. This phenomenon is known as the *Schottky anomaly*.

If an ion has n energy levels E_i which have degeneracies w_i, its internal energy at temperature T is

$$\sum_{i=0}^{n-1} w_i E_i \exp(-E_i/kT)/Z,$$

where

$$Z = \sum_{i=0}^{n-1} w_i \exp(-E_i/kT)$$

is the partition function.

The temperature derivative of this expression gives C_m:

$$C_m = RT\frac{\partial^2}{\partial T^2}(T \ln Z). \tag{1.92}$$

If only the lowest two levels, E_0 and E_1 are involved, as for a Kramers' doublet in a magnetic field, this reduces to

$$C_m = \frac{R\Delta^2 w_0}{kT^2 w_1} \frac{\exp(\Delta/kT)}{[1+(w_0/w_1)\exp(\Delta/kT)]}, \tag{1.93}$$

where $\Delta = E_1 - E_0$. Accompanying the repopulation of the energy levels as T is increased is a change in entropy given by $S = k \ln \Omega$, where Ω is the number of ways in which the levels can be populated. For a single ion $\Omega = \Sigma_i(w_i/w_0)$, and for N identical ions this quantity is raised to the power of N, so that:

$$S = R \ln(\Sigma_i w_i/w_0) \quad \text{mole}^{-1} \tag{1.94}$$

For the case of just two singlet levels the change in entropy is $S = R \ln 2$.

Experimental results are available [61, 62, 69, 91] for the specific heats of the concentrated ethylsulphates of La, Ce, Pr, Nd, Dy, Yb and Y. Of these salts, Ce, Pr and Dy exhibit Schottky anomalies, and that for Ce is shown in Fig. 1.18. In that figure it can be seen that the theoretical curve is much less sharp than that observed, and the interactions between neighbouring ions are generally thought to be responsible for the discrepancy. The effect of these interactions upon C_m at temperatures well above the Schottky anomaly can be written into the constants C_n in the general expression:

$$C_m = R\sum_{n=2}^{\infty}\frac{C_n}{(kT)^n}. \tag{1.95}$$

In the ethylsulphates the dipole–dipole interaction dominates, and for two singlet states, assuming an isotropic g-value, Van Vleck [90] has shown that

$$C_2^d = -\frac{3g^4\mu_B^4}{16}\sum_j \frac{1}{r_{ij}^6}. \tag{1.96}$$

This has been generalized by Daniels [89] for ions with anisotropic g-values. In the Ce salt the quadrupole–quadrupole interaction is also important. All

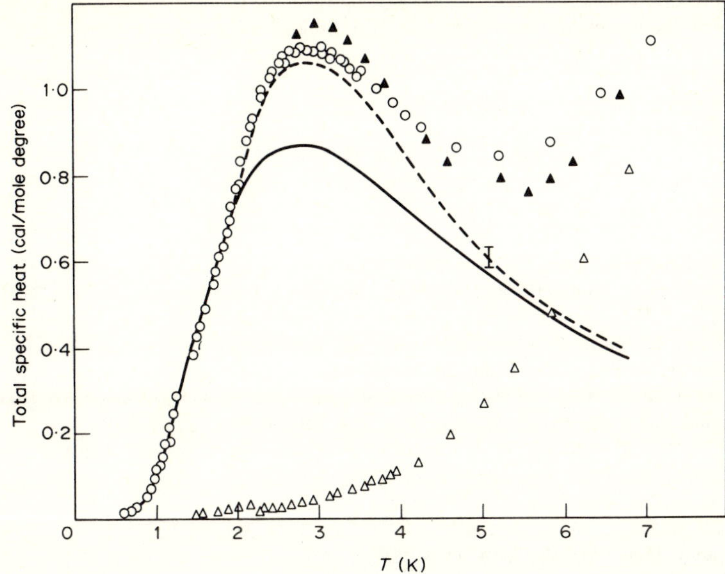

Fig. 1.18 Experimental values of the total specific heats of cerium- (○ and ▲) and lanthanum (△) ethylsulphates [61]. The broken curve is the experimental Schottky anomaly of the cerium salt obtained by subtracting the total (lattice) specific heat of Nd ethylsulphate. The solid curve is a calculation of the anomaly assuming a level splitting of 4·2 cm^{-1}.

attempts [63, 69] to account for the discrepancy between theory and experiment for this salt have been unsuccessful.

At very low temperatures the interactions between Gd ions in $GdCl_3$ arise due to exchange. Marquard [93] has shown that

$$C_2^{ex} = 330{\cdot}76\,(\mathscr{J}_{nn}^2 + 3\mathscr{J}_{nnn}^2). \tag{1.97}$$

This shows good agreement with the experimental results $C_m T^2/R^2 = 3{\cdot}25\ (°K)^2$ at 77°K obtained by Clover and Wolf [70], using an rf relaxation method, and taking values of $\mathscr{J}_{nn}$ and $\mathscr{J}_{nnn}$ given in Table 1.17.

Another mechanism has been proposed to account for the discrepancy between theory and experiment for salts like Ce ethylsulphate. This assumes that the spins and lattice are not independent, but on the contrary are coupled sufficiently strongly to distort the phonon spectrum, and thereby produce a significant contribution to the total specific heat [64, 65]. This process is also of interest in connection with the theory of thermal conductivity [71]. Suffice to say here that it is unable to account for the disagreement.

References

[1] B. G. Wybourne, *Spectroscopic Properties of Rare Earths*, John Wiley and Sons, Inc., New York (1965).

[2] G. H. Dieke, *Spectra and Energy Levels of Rare Earth Ions in Crystals*, John Wiley and Sons, Inc., New York (1969).

[3] D. S. McClure, *Solid State Phys.*, **9**, 399 (1959).
[4] L. I. Schiff, *Quantum Mechanics*, McGraw Hill, 3rd ed. (1968).
[5] F. Herman and S. Skillman, *Atomic Structure Calculations*, Prentice Hall Inc., New Jersey (1963).
[6] M. G. Mayer, *Phys. Rev.*, **60**, 184 (1941).
[7] R. Latter, *Phys. Rev.*, **99**, 510 (1955).
[8] C. A. Coulson and C. S. Sharma, *Proc. Phys. Soc.*, **79**, 920 (1962).
[9] S. Arajs and R. V. Colvin, *Rare Earth Research*, MacMillan (1961) p. 178.
[10] J. C. Slater, *Phys. Rev.*, **36**, 51 (1930).
[11] G. Burns, *J. Chem. Phys.*, **41**, 1521 (1964).
[12] E. C. Ridley, *Proc. Camb. Phil. Soc.*, **56**, 41 (1960).
[13] B. R. Judd and I. Lindgren, *Phys. Rev.*, **122**, 1802 (1961); I. Lindgren, *Nucl. Phys.*, **32**, 151 (1962).
[14] A. J. Freeman and R. E. Watson, *Phys. Rev.*, **127**, 2058 (1962).
[15] O. J. Sovers, *J. Phys. Chem. Solids*, **28**, 1073 (1967).
[16] E. U. Condon and G. H. Shortley, *Theory of Atomic Spectra*, C.U.P., New York (1953).
[17] K. Rajnak and B. G. Wybourne, *Phys. Rev.*, **132**, 280 (1963) (see also ref. 1).
[18] J. P. Elliott, B. R. Judd and W. A. Runciman, *Proc. Roy. Soc.*, **A240**, 509 (1957).
[19] B. R. Judd, *Proc. Phys. Soc.*, **A69**, 157 (1956).
[20] B. R. Judd, *Proc. Phys. Soc.*, **82**, 874 (1963).
[21] R. J. Elliott and K. W. H. Stevens, *Proc. Roy. Soc.*, **A218**, 553 (1953).
[22] B. Bleaney, *Proc. Phys. Soc.*, **68A**, 937 (1955).
[23] A. Abragam, *The Principles of Nuclear Magnetism*, Oxford (1961) pp. 170–173.
[24] B. Bleaney, *Hyperfine Interactions*, eds. A. J. Freeman and R. B. Frankel, Academic Press, London (1967) pp. 1–50.
[25] R. E. Watson, A. J. Freeman, *ibid*, p. 53.
[26] B. Bleaney and K. W. H. Stevens, *Repts. Progr. Phys.*, **16**, 108 (1953); K. D. Bowers and J. Owen, *Repts. Progr. Phys.*, **18**, 304 (1955); W. Low, *Solid State Phys.*, *Suppl.* **2**, 329 (1960).
[27] K. W. H. Stevens, *Proc. Phys. Soc.*, **A65**, 209 (1952).
[28] M. T. Hutchings, *Solid State Phys.*, **16**, 227 (1964).
[29] K. R. Lea, M. J. M. Leask and W. P. Wolf, *J. Phys. Chem. Solids*, **23**, 1381 (1962).
[30] R. W. Bierig and M. J. Weber, *Phys. Rev.*, **134**, A1492 (1964).
[31] B. Bleaney, *Proc. Phys. Soc.*, **A73**, 937 and 939 (1959).
[32] D. P. Schumacher and C. A. Hollingsworth, *J. Phys. Chem. Solids*, **27**, 749 (1966).
[33] J. M. Baker, B. Bleaney and W. Hayes, *Proc. Roy. Soc.*, **A247**, 141 (1958).
[34] B. G. Wybourne, *Phys. Rev.*, **148**, 317 (1966).
[35] B. Bleaney, R. J. Elliott, H. E. D. Scovil and R. S. Trenam, *Phil. Mag.*, **42**, 1062 (1951).
[36] M. T. Hutchings and D. K. Ray, *Proc. Phys. Soc.*, **81**, 663 (1963).
[37] D. K. Ray, *Proc. Phys. Soc.*, **82**, 47 (1963).
[38] M. N. Ghatikar, A. K. Raychaudhuri and D. K. Ray, *Proc. Phys. Soc.*, **86** 1235 (1965).
[39] M. N. Ghatikar, A. K. Raychaudhuri and D. K. Ray, *Proc. Phys. Soc.*, **86**, 1239 (1965).
[40] A. K. Raychaudhuri and D. K. Ray, *Proc. Phys. Soc.*, **90**, 839 (1967).
[41] U. Ranon, *J. Phys. Chem. Solids*, **25**, 1205 (1964).
[42] M. M. Ellis and D. J. Newman, *J. Chem. Phys.*, **49**, 4037 (1968).
[43] C. K. Jorgensen, R. Pappalardo and H. H. Smidtke, *J. Chem. Phys.*, **39**, 1422 (1963).
[44] J. D. Axe and E. Burns, *Phys. Rev.*, **152**, 331 (1966).
[45] R. E. Watson and A. J. Freeman, *Phys. Rev.*, **156**, 251 (1967).
[46] M. M. Curtis, D. J. Newman and G. E. Stedman, *J. Chem. Phys.*, **50**, 1077 (1969).
[47] M. M. Ellis and D. J. Newman, *J. Chem. Phys.*, **47**, 1086 (1967).
[48] R. Orbach, *Proc. Roy. Soc.*, **A264**, 458 (1961).
[49] P. L. Scott and C. D. Jeffries, *Phys. Rev.*, **127**, 32 (1962).
[50] G. H. Larson and C. D. Jeffries, *Phys. Rev.*, **141**, 461 (1966).
[51] C. B. P. Finn, R. Orbach and W. P. Wolf, *Proc. Phys. Soc.*, **77**, 261 (1961).

[52] R. H. Ruby, H. Benoit and C. D. Jeffries, *Phys. Rev.*, **127**, 51 (1962).
[53] G. H. Larson and C. D. Jeffries, *Phys. Rev.*, **145**, 311 (1966).
[54] R. C. Mikkelson and H. J. Stapleton, *Phys. Rev.*, **140**, A1968 (1965).
[55] A. Kiel and W. B. Mims, *Phys. Rev.*, **161**, 386 (1967).
[56] R. W. Bierig, M. J. Weber and S. I. Warshaw, *Phys. Rev.*, **134**, A1504 (1964).
[57] R. Orbach and M. Blume, *Phys. Rev. Lett.*, **8**, 478 (1962).
[58] R. J. Elliott, *Proc. Phys. Soc.*, **B70**, 119 (1957).
[59] B. R. Judd, C. A. Lovejoy and D. A. Shirley, *Phys. Rev.*, **128**, 1733 (1962).
[60] D. T. Edmonds, *Phys. Rev. Letts.*, **10**, 129 (1963).
[61] H. Meyer and P. L. Smith, *J. Phys. Chem. Solids*, **9**, 285 (1959).
[62] H. Meyer, *J. Phys. Chem. Solids*, **9**, 296 (1959).
[63] R. Finkelstein and A. Mencher, *J. Chem Phys.*, **21**, 472 (1953).
[64] F. Perisco and K. W. H. Stevens, *Proc. Phys. Soc.*, **82**, 855 (1963).
[65] J. W. Tucker, *Proc. Phys. Soc.*, **85**, 559 (1965).
[66] M. T. Hutchings, R. J. Birgeneau and W. P. Wolf, *Phys. Rev.*, **168**, 1026 (1968).
[67] J. M. Baker, *Phys. Rev.*, **136**, A1341 (1964).
[68] K. L. Brower, H. J. Stapleton and E. O. Brower, *Phys. Rev.*, **146**, 233 (1966).
[69] C. E. Johnson and H. Meyer, *Proc. Roy. Soc.*, **A253**, 199 (1959).
[70] R. B. Clover and W. P. Wolf, *Sol. Stat. Comm.*, **6**, 331 (1968).
[71] P. V. E. McClintock, I. P. Morton, R. Orbach and H. M. Rosenberg, *Proc. Roy. Soc.*, **A298**, 359 (1967).
[72] M. Blume, A. J. Freeman and R. E. Watson, *Phys. Rev.*, **134**, A320 (1964).
[73] C. A. Hutchison, Jr. and E. Wong, *J. Chem. Phys.*, **29**, 754 (1958).
[74] W. P. Wolf, M. J. M. Leask, B. Mangum and A. F. G. Wyatt, *J. Phys. Soc. Japan*, **17**, B–1, 487 (1961).
[75] M. M. Abraham, L. A. Boatner, C. B. Finch, E. J. Lee and R. A. Weeks, *J. Phys. Chem. Solid*, **28**, 81 (1967).
[76] B. Bleaney, R. J. Elliott and H. E. D. Scovil, *Proc. Phys. Soc.*, **A64**, 933 (1951).
[77] J. Dweck and G. Seidel, *Phys. Rev.*, **146**, 359 (1966).
[78] A. H. Cooke, S. Whitley and W. P. Wolf, *Proc. Phys. Soc.*, **B68**, 415 (1955).
[79] J. M. Baker, G. M. Copland and B. M. Wanklyn, *J. Phys. C.*, **2**, 862 (1969).
[80] R. J. Elliott and K. W. H. Stevens, *Proc. Roy. Soc.*, **A215**, 437 (1952).
[81] C. L. Herzenberg, L. Meyer-Schultzmeister, L. L. Lee and S. S. Hanna, *Bull. Am. Phys. Soc.*, **7**, 39 (1962).
[82] D. A. Shirley, R. B. Frankel and H. H. Wickman, 3rd Mössbauer Conf., p. 392.
[83] J. M. Baker, J. D. Riley and R. G. Shore, *Phys. Rev.*, **150**, 198 (1966).
[84] J. Dweck and G. Seidel, *Phys. Rev.*, **151**, 289 (1966).
[85] R. J. Birgeneau, M. T. Hutchings and W. P. Wolf, *Phys. Rev.*, **179**, 275 (1969).
[86] R. J. Birgeneau, M. T. Hutchings and R. N. Rogers, *Phys. Rev.*, **175**, 1116 (1968).
[87] A. H. Cooke, R. Lazenby and M. J. M. Leask, *Proc. Phys. Soc.*, **85**, 767 (1965).
[88] C. J. Gorter and W. J. de Haas, *Leiden Comm.*, **218b**, 303 (1931).
[89] J. M. Daniels, *Proc. Phys. Soc.*, **A66**, 673 (1953).
[90] J. H. Van Vleck, *J. Chem. Phys.*, **5**, 320 (1937).
[91] L. D. Roberts, C. C. Sartain and B. Barie, *Phys. Rev.*, **87**, 192 (1952); *Rev. Mod. Phys.*, **25**, 170 (1953).
[92] A. H. Cooke, S. Whitley and W. P. Wolf, *Proc. Phys. Soc.*, **B68**, 415 (1955).
[93] C. D. Marquard, *Proc. Phys. Soc.*, **92**, 650 (1967).
[94] J. C. Eisenstein, R. R. Hudson and B. W. Mangum, *Phys. Rev.*, **137**, A1886 (1965).
[95] M. E. Rose, *Relativistic Electron Theory*, John Wiley and Sons, London (1961).
[96] D. Liberman, J. T. Waber and D. T. Cromer, *Phys. Rev.*, **137**, A27 (1965).
[97] J. M. Baker and F. I. B. Williams, *Proc. Roy. Soc.*, **A267**, 283 (1962).
[98] W. Low, *Progress in Science and Technology of the Rare Earths*, Pergamon Press, London, **2**, 123 (1966).

2

Structural Behaviour of the Rare Earth Metals and Alloys

Since the original investigations of Klemm and Bommer in 1937 [1] numerous investigations have served to complete the classification of the room temperature crystal structures. Relatively few observations have been made at high temperatures where structure changes exist for most of the pure metals. Rather, low temperature measurements have been more common because of the important effects associated with the magnetically ordered phases of these materials. It is not the purpose of the present chapter to discuss such changes, and a detailed account is given in Chapter 4, where their importance can be considered directly in relation to the magnetic behaviour.

The most significant contributions to an understanding of the rare earth structures have come from an examination of rare earth alloying behaviour and from the observation of several polymorphic transitions induced by the application of high pressures to the metallic elements. These effects will be considered independently in the following. First however, it is appropriate to discuss the room temperature structures of the metals themselves at atmospheric pressure.

2.1 Crystal structures of the pure metals

The sequence of crystal structures across the rare earth series from lanthanum to lutetium, provides many interesting features. In the solid phase the metals can be observed in all three of the normal metallic configurations (body centred cubic, face centre cubic and hexagonal close packed), in addition to the double-hexagonal close packed and samarium structures.

The room temperature structures and the corresponding lattice parameters are given in Table 2.1a while those for the other phases observed at atmospheric pressure at other temperatures are listed in Table 2.1b.

The atomic volume and mean ionic radius* decrease continuously (with

* By this is implied a radius derived from the close packed structures, although as discussed in the previous chapter this is also seen, as may be anticipated, in the theoretical ionic radius.

Table 2.1(a) The room temperature structural properties and lattice parameters of the rare earth metals. Distances are measured in Ångstroms, density is in gm cm^{-3}, and atomic volume is in cm^3 mole^{-1}.

*The samarium structure is given in terms of a hexagonal unit cell.

†The c/a ratios of the d-hex and Sm structures are given as $c/2a$ and $c/4{\cdot}5a$ respectively.

	Structure	Space group	*a*-spacing	*c*-spacing	c/a-ratio†	Metallic radius	Density	Atomic volume
Lanthanum	d-hex (A3′)	$P6_6/mmc$	3·769	12·159	1·612	1·877	6·162	22·5
Cerium	d-hex (A3′)	$P6_3/mmc$	3·673	11·802	1·607	1·82	6·678	20·70
Praseodymium	d-hex (A3′)	$P6_3/mmc$	3·671	11·831	1·611	1·828	6·769	20·78
Neodymium	d-hex (A3′)	$P6_3/mmc$	3·656	11·795	1·613	1·821	7·016	20·60
Samarium	rhomb (Sm)	R3m	3·628	26·231*	1·607	1·802	7·536	19·0
Europium	bcc (A2)	Im3m	4·581			2·042	5·245	28·91
Gadolinium	hcp (A3)	$P6_3/mmc$	3·634	5·781	1·5910	1·802	7·886	19·88
Terbium	hcp (A3)	$P6_3/mmc$	3·606	5·697	1·5800	1·782	8·253	19·25
Dysprosium	hcp (A3)	$P6_3/mmc$	3·593	5·654	1·5735	1·773	8·559	19·03
Holmium	hcp (A3)	$P6_3/mmc$	3·580	5·627	1·5720	1·766	8·799	18·74
Erbium	hcp (A3)	$P6_3/mmc$	3·561	5·593	1·5706	1·757	9·062	18·47
Thulium	hcp (A3)	$P6_3/mmc$	3·543	5·571	1·5725	1·746	9·318	18·15
Ytterbium	fcc (A1)	Fm3m	5·463			1·940	6·959	24·80
Lutetium	hcp (A3)	$P6_3/mmc$	3·510	5·567	1·5797	1·734	9·849	17·78

Table 2.1(b) The structure changes with temperature, and transition temperatures, of the rare earth metals.

	Transitions and transition temperatures (°K)	Melting point	Boiling point
Lanthanum	d-hex(α) $\xrightarrow[583]{}$ fcc(β) $\xrightarrow[1151]{}$ bcc(δ)	1193	3742
Cerium	fcc(α) $\longrightarrow$ d-hex(β) $\longrightarrow$ fcc(γ) $\xrightarrow[1003]{}$ bcc(δ)	1068	3740
Praseodymium	d-hex $\xrightarrow[1071]{}$ bcc	1208	3300
Neodymium	d-hex $\xrightarrow[1141]{}$ bcc	1297	3300
Samarium	rhomb $\xrightarrow[1190]{}$ bcc	1345	2077
Europium	No transitions	1099	1712
Gadolinium	hcp $\xrightarrow[1535]{}$ bcc	1585	3270
Terbium	hcp $\longrightarrow$ bcc	1629	3070
Dysprosium	hcp $\longrightarrow$ bcc	1680	2870
Holmium	hcp $\longrightarrow$ bcc	1734	2870
Erbium	No transitions	1823	2923
Thulium	No transitions	1818	2000
Ytterbium	fcc $\xrightarrow[1071]{}$ bcc	1101	1700
Lutetium	No transitions	1925	3600

the exception of europium and ytterbium) with increasing atomic number. This behaviour is the well-known lanthanide contraction. It arises because of the incomplete screening of one 4*f* electron by another as discussed in the previous chapter. It is consequently an ionic property whose effect is seen

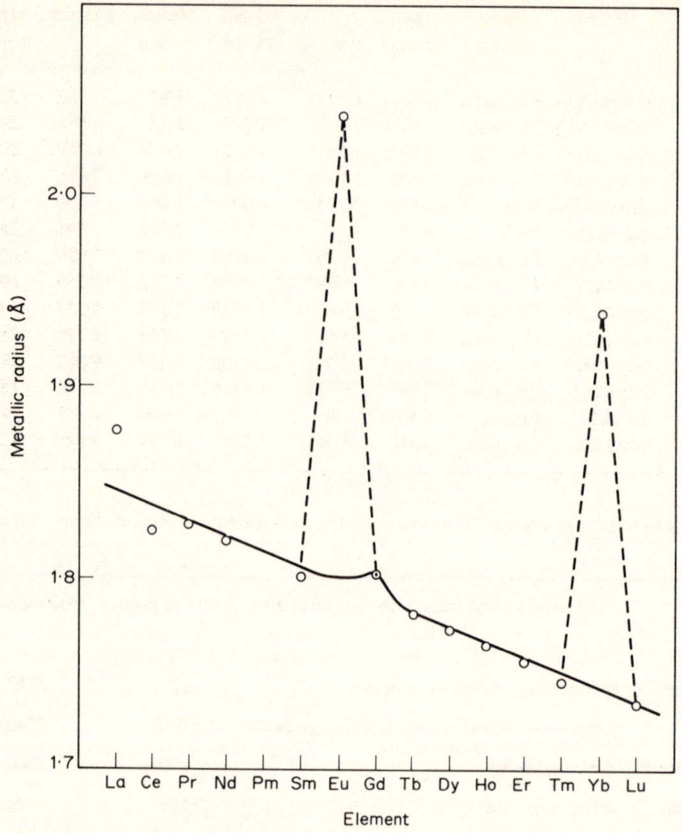

Fig. 2.1 Variation of the metallic radius of the rare earth elements with atomic number, illustrating the lathanide contraction.

throughout the structural properties of solids involving these ions. Fig. 2.1 shows the derived value of the ionic radii of the pure metals, from which the contraction is evident.

The structures of all the elements at normal temperatures, with the exception of europium, are of a close packed nature, and may be described in terms of stacking sequences involving three types of layer. These may be defined as A, B and C and are shown in Fig. 2.2a for the hexagonal representation of the fcc structure. It is evident that any one of these layers is related by a translational movement to any other layer. In the heavy rare earths all the elements but ytterbium, which is not a typical member of the series,

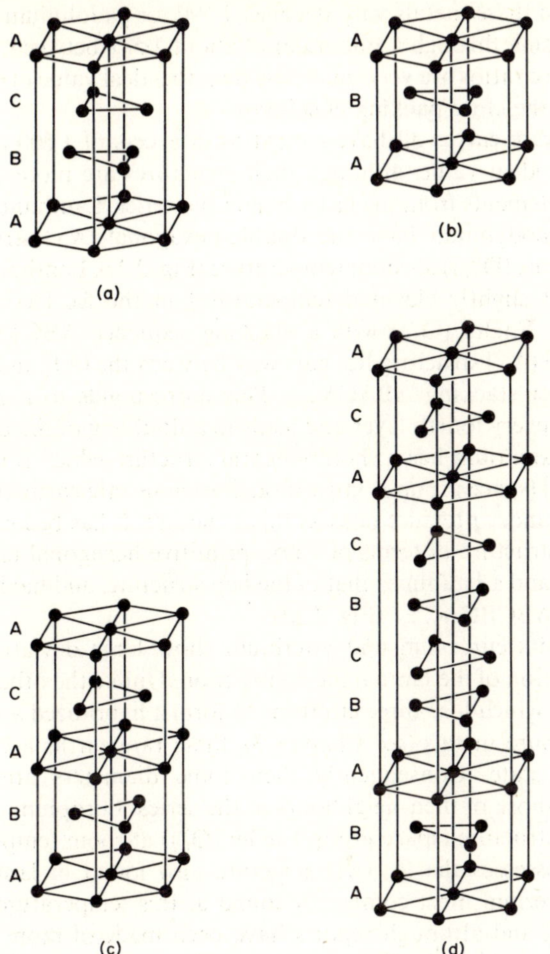

Fig. 22 The four close packed structures observed for the rare earth metals (1) fcc; (b) hcp; (c) d-hex; (d) Sm-type.

have a room temperature structure which is the hexagonal close packed (A3) type with space group $P6_3/mmc$ (D^4_{6h}). This structure has a stacking sequence ABAB ... and is shown in Fig. 2.2b along with the other modifications to be discussed below. The lattice spacings of the heavy elements, with the exception of gadolinium and ytterbium, follow the lanthanide contraction very closely (see Fig. 2.1 and Table 2.1). The anomalous values for gadolinium are associated with the fact that unlike the other elements, this metal is ferromagnetically ordered at room temperature [2]. If the thermal expansion measurements of Bozorth *et al* [3] are used to obtain pseudo-lattice parameters for hypothetical non-magnetic gadolinium, by extrapolation from the high temperature measurements, then this anomaly largely disappears. The axial ratios of these metals vary between 1·590 for gadolinium and 1·586 for lutetium passing through a minimum of about 1·571 between holmium and thulium. These ratios are very much less than the ideal value of $(8/3)^{\frac{1}{2}} = 1{\cdot}633$ obtained for the close packing of spheres.

The light rare earths all have c/a ratios in excess of 1·600 and are much closer to the ideal value, although their structures are more complex than those of the elements from the latter half of the series. Lanthanum, praseodymium and neodymium have the double-hexagonal (A3′) structure (space group $P6_3/mmc$ (D_{6h}^4)) at room temperature (Fig. 2.2c). Lanthanum however, also exists at slightly elevated temperatures in the fcc (A1) modification (space group Fm3m (O_h^5)) with a stacking sequence ABCABC shown in Fig. 2.2a. The (A3′) structure lies part way between the (A1) and (A3) lattices, having a layer stacking ABACA.... This corresponds to a stacking fault appearing in every fourth layer and leads to a doubling of the unit cell c-axis parameter. Samarium has a rhombohedral structure which is unique to this element [4, 5] (space group R3m) although various rare earth alloys and rare earth metals under pressure possess this structure. It has become customary to view this structure in terms of a non-primitive hexagonal unit cell whose c-axis is four and a half times that of the hcp structure, and having a stacking sequence ABABCBCAC ... (Fig. 2.2d).

The elements europium and ytterbium show large deviations from the smooth variation of the lanthanide contraction. Unlike the other elements so far discussed, which lose three electrons to form a hybridized s–d conduction band in the pure metals (see Chapter 3), these rare earths are ‘divalent’ in the metallic state. Consequently their ionic diameters are appreciably greater than those of their neighbours in the series. Europium crystallizes in the bcc (A2) structure (space group I m3m (O_h^9)) at room temperature while ytterbium possesses the fcc (A1) structure also found in lanthanum. The structure of cerium most commonly found at this temperature is again the fcc (A1) form, and although reports have been made of more than one fcc allotrope existing here [6, 7] these appear to be associated with oxide phases also present in the specimen [8]. The effective valence of cerium at this temperature appears to be in excess of three and the many complex structural and electronic properties which occur in this metal are related to its anomalous value and changes in this value caused by changes in specimen temperature, pressure, etc.

2.2 Structural stability

The close-packed structures of most of the rare earths are intimately related and, as may be expected, structural transformations occur as a result of varying the conditions of the specimens under investigation. Similar transformations are also observed in the alloying behaviour of the metals amongst themselves. Since the early investigations of the structural stability by

Spedding and co-workers [9] with high temperature X-ray methods, the principal work has been carried out using expansion measurements, alloy structure observations and high pressure techniques, in an effort to understand this structural behaviour. This has resulted in some appreciation of the factors which affect the structures and their relation to other physical properties of the pure metals and intra-rare earth alloys.

In the following, the effects of temperature, pressure and alloying on the crystal structures are discussed in detail.

2.2.1 *Alloying behaviour*

The general rules relating to the existence of metal alloys were set out by Hume–Rothery [10] in 1936 and relate to ionic size, electronegativity and valence. These rules have been shown to apply to the rare earth elements and their relation to these metals has been discussed at length by Gschneidner [11–13] and will not be dealt with further.

A graphical interpretation of the simultaneous effect of the size factor and electronegativity is available in the form of the well-known Darken–Gurry maps [14]. In these, the elements are represented by points in a graph of electronegativity versus ionic radius and the alloying behaviour of a

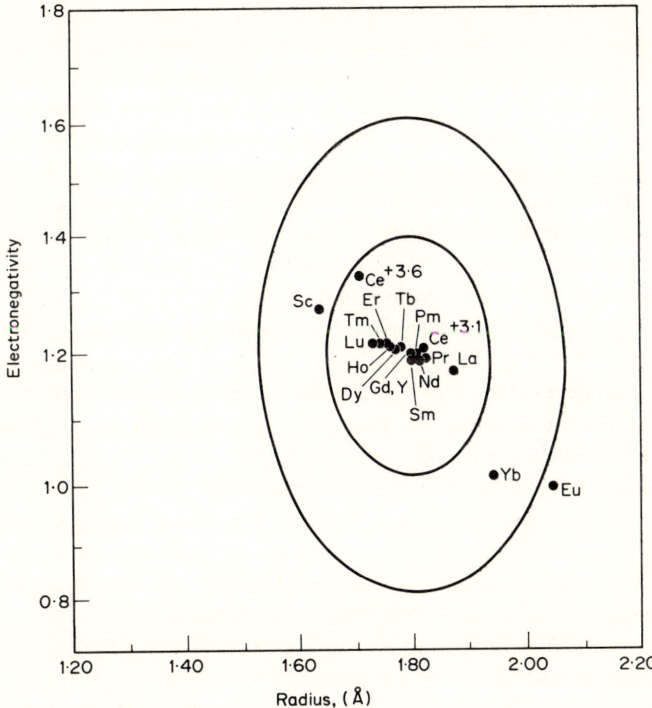

Fig. 2.3 A Darken-Gurry plot of the rare earths centred on gadolinium [14].

particular element is obtained by superimposing an ellipse on this map, centred at the element in question and with its semi-axes of length corresponding to the criteria of the Hume–Rothery rules (±15% in size and ±0·4 in electronegativity). Elements within this ellipse can be expected to show extensive solid solubility in the reference element while the solubility of those outside will be negligible. A section of such a map, obtained in the vicinity of the rare earth elements is shown in Fig. 2.3. In this two ellipses are shown, the inner with dimensions (±8% ±0·2) and the outer (±15% ±0·4). This is now standard practise and gives the regions of high, medium and negligible solubility respectively for the inner, intermediate and outer zones of the ellipses.

The ellipses shown are centred at gadolinium and it is evident that one will obtain extensive solid solubility of almost all the rare earths in this element (ytterbium and europium excepted). Since the spread of the points corresponding to the other elements about gadolinium is small with respect to the dimensions of the ellipse, this extensive solubility is also to be expected for all intra-rare earth alloys (again excluding Yb and Eu). It should of course be realized that these maps give only an indication of likely extensive solubility based on two parameters, and in practise it is necessary to take other factors into account, for example the structures for the terminal elements and the ionic valence, before reliable predictions are possible.

An extensive, although by no means exhaustive, investigation of intra-rare earth alloy systems has been performed by many workers (see for example [15–20]) with many objectives in view, and in the following those results which relate to the stability of the various structures observed in the rare earth series will be considered in detail.

The structures of the pure rare earth metals show a systematic variation through the series from lanthanum to lutetium. By suitably alloying light rare earths with heavy rare earths, structures are obtained which are intermediate between those of the component elements. In the extreme case, alloys formed over the whole concentration range between cerium and terbium show the structures: hcp, Sm-type, d-hex, and fcc, with increasing cerium content [19]. Some evidence for the formation of these intermediate structures was obtained by the early workers [21–23]. No attempt at a systematic study was made however, until the work of Spedding *et al* [15], and Nachman *et al* [16]. The former investigated, among other systems, the alloying behaviour of lanthanum–gadolinium metals. The phase diagram, which is typical of alloys formed between d-hex and hcp rare earth metals, is shown in Fig. 2.4. This illustrates clearly that with increasing gadolinium (hcp component) content the d-hex structure becomes unstable with respect to the Sm-type structure. This samarium-like phase exists over a relatively narrow composition range and in turn is replaced by the hcp structure. At the same time the high temperature fcc phase becomes less stable with respect to the d-hex lattice, higher thermal energies being necessary to cause the transition,

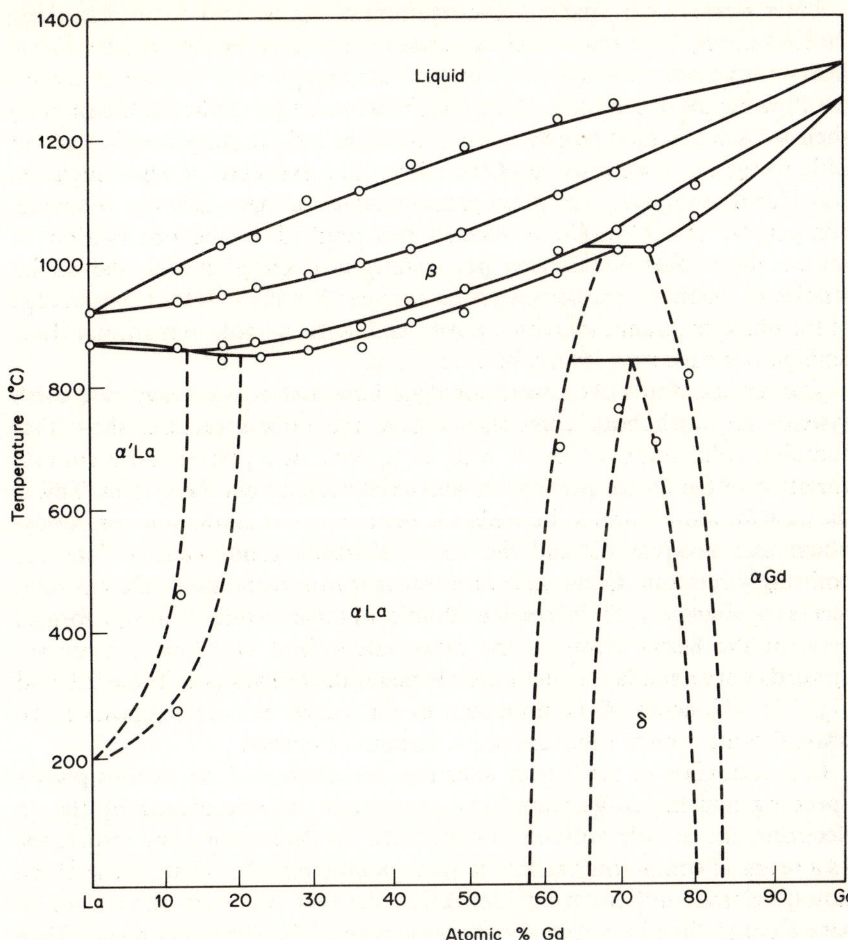

Fig. 2.4 The phase diagram of the La–Gd system showing the structural sequence hcp-Sm-type-d-hex-fcc with increasing lanthanum content [15].

and disappears completely for gadolinium concentrations greater than about 20 at.%. The bcc phase observed in the vicinity of the melting point persists throughout however, and at a fixed composition the Sm-type structure develops as the result of a solid–solid transformation from the bcc phase. These high temperature phases will be discussed in more detail in Section 2.2.2.

The room temperature structure changes are accompanied by a variation in the axial-ratio. In the d-hex phases this ratio appears to be reasonably constant at $c'/a = 1{\cdot}612$ (reduced value $c' = c/2$), but changes rapidly to a value of approximately $c''/a = 1{\cdot}60$ for the Sm-type structure ($c'' = c/4{\cdot}5$) compared with a value of 1·607 for samarium metal and again shows a rapid decrease to about 1·59 when the hcp structure is most stable.

Later workers, carrying out investigations along the lines dictated by these investigations, have made a more extensive coverage of the possible light–heavy rare earth alloy systems but with the exception of the use of cerium (so allowing the observation of the fourth, fcc, structure as already mentioned) their work has served largely to confirm these early findings and has added little to the basic knowledge of the alloys. The existence of what might be considered as impurity stabilized phases (and as we shall see in the following, comparable stress stabilized phases) has resulted in the observation of several misleading phenomena particularly in connection with the earlier studies of magnetic properties of the rare earth metals. With a knowledge of the phase diagrams, increased purity and better sample preparation these ambiguities have now largely been resolved.

The results of this alloy work for light–light and heavy–heavy rare earth systems in which both components have the same structure, show that complete solid solubility exists with no intermediate phases and a smooth variation of the lattice parameters and axial ratio across the system. This is the case for alloys formed between the light elements lanthanum, praseodymium and neodymium and the heavy elements gadolinium to lutetium omitting ytterbium. In the case of intra-light-rare earth alloys the c/a ratio increases slightly with increasing atomic number, while in alloys formed between the heavy elements the ratio values tend to follow a concave-upwards curve similar to the pure elements themselves (see Table 2.1 and Fig. 2.5). The depth of the minimum in this curve, however, appears to be related to the atomic numbers of the terminal elements.

In a discussion of the factors affecting the stability of the various phases Spedding and his co-workers [15] considered the role played by the $4f$ electrons, the atomic volume, and c/a ratio in determining the structures. As a result of comparing the lanthanum–gadolinium alloys with those of the closely-related lanthanum-yttrium series, the first two were shown not to have a controlling influence in the appearance of the samarium phase. They did however, for the first time, successfully correlate the c/a ratio with the structure, as has previously been given and the subsequent observations of other systems have served to support and emphasize this relation. Using the results obtained in these alloy studies Fig. 2.5 has been drawn (as indicated by Spedding *et al* [15]) to show the variation of the axial-ratio across the rare earth series and to compare its value with the existence of the d-hex, Sm-type and hcp phases. We have attempted to cover continuously as much of the range from lanthanum to lutetium as published results allow.

The correlation between the c/a ratio and structure type is clearly visible in this figure as is also the evidence to support the increased c/a ratio for lutetium. In the d-hex region of this graph it is uncertain whether the positive slope of c/a with atomic number is real, as many of the systems show little or no variation of the axial ratio in this range. It is only in alloys containing cerium as one component that any volume of evidence exists for the upward

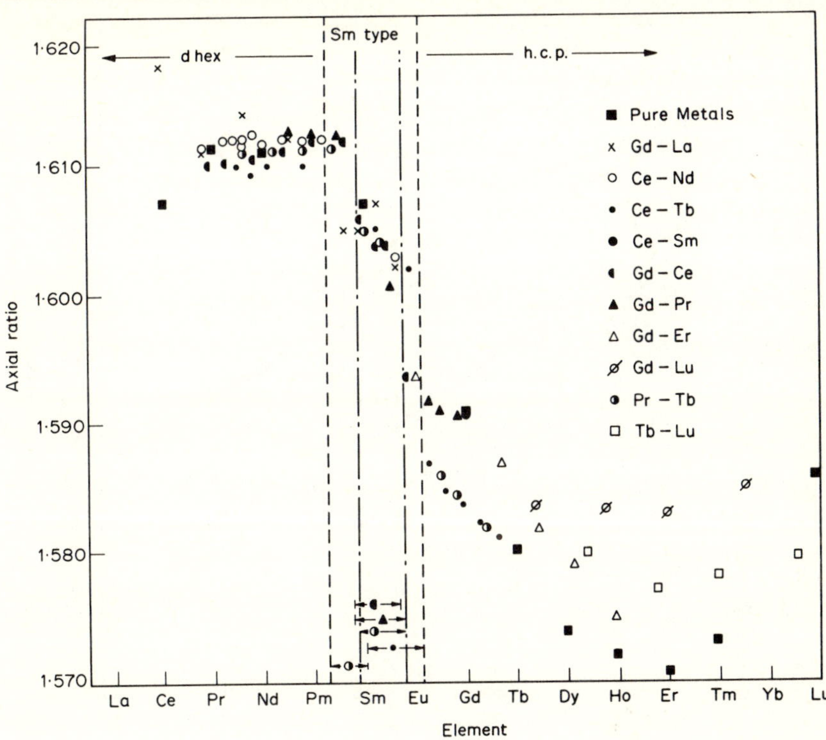

Fig. 2.5 Variation of the c/a ratios for the pure rare earth metals and intra-rare earth binary alloys.

trend in c/a. The onset of the Sm-type structure with increasing atomic number is very sudden and does not appear to be related to a critical c/a ratio in the d-hex phase. The structure exists over a narrow but rather diffuse range of atomic numbers centred slightly to the right of samarium metal and has a spread of reduced c/a values given by $c/a = 1{\cdot}603 \pm 0{\cdot}003$. These show some tendency to decrease with increasing atomic number and as such probably represent a continuous rapid decrease of the axial ratio. The range of c/a values obtained in the hcp region of the graph is appreciably wider at a given atomic number than those for either the d-hex or Sm-type phases. As mentioned previously, the detailed variation of the ratio seems to depend to a considerable extent on the atomic number of the heavy rare earth involved in light–heavy alloys and on the position of both elements in the case of heavy–heavy systems.

The variation of the lattice parameters of the metals shows a general trend as an alloy system goes from the d-hex to the hcp structure. The behaviour of the Pr–Tb system is shown in Fig. 2.6 from which it can be seen that with increasing Tb content both the a- and c-spacings contract linearly with negative and positive deviations respectively from Vegard's law. The scale

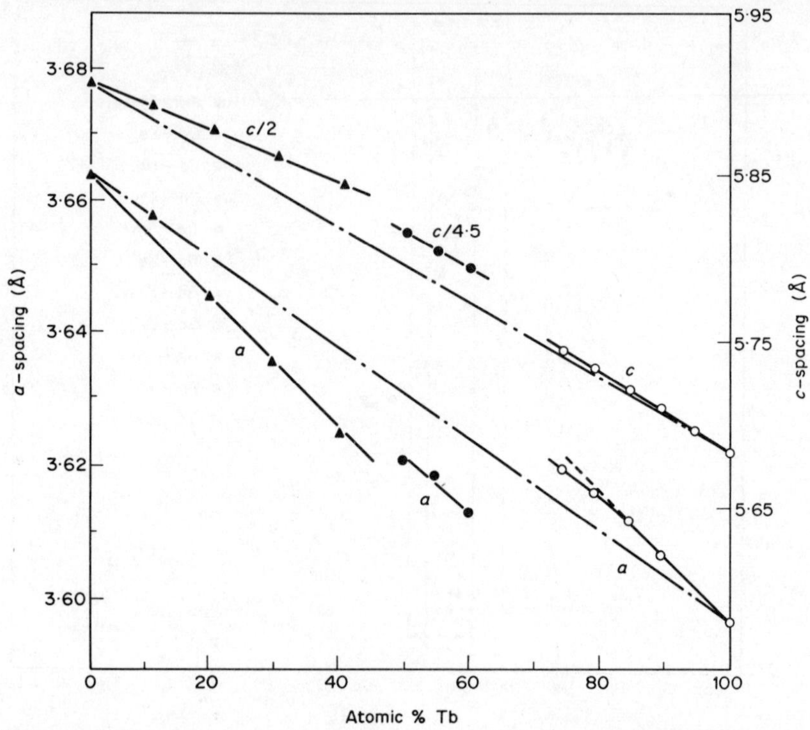

Fig. 2.6 Variation of the lattice parameters across the series Pr–Tb. ▲ hcp; ● Sm-type; ○ d-hex.

of these changes is such as to keep c/a nearly constant. Over the concentration range in which the samarium structure exists there is insufficient data to give a clear description of the parameter changes. In the region of the hcp phase (high Tb content) the c-parameter again decreases linearly but with a much smaller deviation from ideality, while the a-spacing variation appears to be at all times concave downwards. The combination of these two variations results in the rapid fall of the axial ratio from the Sm- to the hcp structures which has been illustrated in Fig. 2.5. Much more detailed work is necessary in the intermediate composition regions in order to completely specify the spacing variation as one phase becomes unstable with respect to another. Unfortunately these regions show the coexistence of both structures, due presumably to the increase in stacking errors which enables the transition to occur and the precise structure details must be somewhat ambiguous.

Yttrium appears to behave in a similar manner to the heavy rare earth metals in its alloying behaviour with the light rare earths and consequently is frequently used as a diluent in studies of rare earth systems.

In cerium alloys there is some evidence [24] that cerium may vary its effective valence in solution towards the valence of the host metal. Thus, for example, the structure details and room temperature susceptibility of a series of Ce–Tb alloys are more satisfactorily accounted for in terms of $Ce^{3 \cdot 4+}$ ions rather than Ce^{3+} ions as in the case for the pure metal.

In addition to the correlation of structure with axial ratio observed by Spedding *et al* [15], both Beaudry *et al* [25] and Raynor *et al* [26] have derived elementary models which allow the composition of the samarium phase in any given alloy system to be predicted. Both of these empirical methods appear to work reasonably well indicating that ultimately it may be possible to derive a single picture to explain the structure sequence that occurs during alloying in intra-rare earth systems and the comparable structure sequence that occurs in the pure elements. Unfortunately neither of these models provides any understanding of the mechanisms of the phase changes or of the relative values of the free energies of the various structures.

2.2.2 *The effects of temperature on the structure*

The room temperature structures of most of the rare earth metals undergo solid–solid transitions as the temperature is varied. In cerium, transitions are observed below room temperature, the lowest occurring at 100°K, and as such this metal has the lowest temperature solid–solid phase change of any of the pure metals.

Fig. 2.7 shows the existence regions, at atmospheric pressure, for the different phases of all the elements. Most of the elements are seen to undergo a transition to a bcc (A2) structure at temperatures in the vicinity of the melting point. The range of temperatures over which this exists is relatively small except for divalent europium. The other divalent metal, ytterbium, also crystallizes in a cubic modification (in this case fcc) as has previously been mentioned. It does however, show the transition to the bcc polymorph. The elements lanthanum and cerium, while transforming into this cubic structure before melting, undergo other solid–solid phase changes at lower temperatures. The d-hex structure of the former changes to the fcc (A1) phase at 583°K before transforming to the body centred phase at 1141°K.

The behaviour of cerium is much more complex. This metal exists in several allotropic forms, the details of which have been studied by both metallurgical and physical techniques [27–29]. On cooling below room temperature, the fcc (γ) phase partly transforms to a d-hex phase (β) at 263°K and at still lower temperatures (116°K) a new fcc phase (α) forms directly from the γ phase with a much reduced lattice spacing. Although other allotropes have been reported it now seems certain that these have their origin in impurities and we shall consider only the three phases mentioned, in addition to the body-centred cubic phase observed near to the melting point. The low temperature phase transformation temperatures are not well established as there is an extensive thermal hysteresis behaviour between 100°K and room temperature, in addition to which appreciable overlapping of the three phases occurs. Rashid and Altstetter [29] have recently investigated this behaviour and have shown that for their specimens

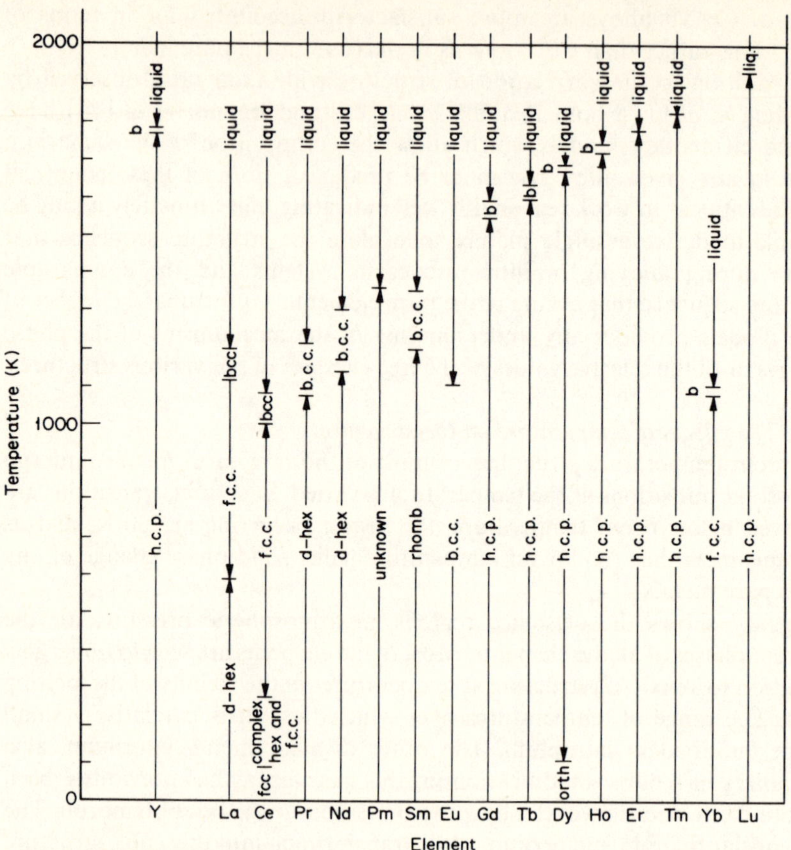

Fig. 2.7 Existence temperature ranges of the various structures observed for the rare earth elements at atmospheric pressure.

the thermodynamically stable phase at room temperature is the d-hex structure, the γ phase being metastable on cooling to this temperature. The amount of β phase present at room temperature could be increased by cycling through the α–γ transformation. These findings are somewhat at variance with those of other workers (see [13]) but the detailed behaviour of cerium is so highly dependent on its past history and purity that a full understanding of the observed structures may have to wait for higher specimen perfection than is presently available.

The transformation to the collapsed fcc structure (either β–α or γ–α transitions) has been associated with a change in the valence state of the ions at the transition. In the γ and β modification this appears to be $3{\cdot}06 \pm 0{\cdot}06$ whereas in the α phase it is increased to $3{\cdot}67 \pm 0{\cdot}09$ [28]. This latter value is extremely sensitive to both temperature and pressure. A tetravalent state has been proposed for α-cerium [30, 31], but much of the evidence suggests that this does not exist at atmospheric pressure.

It has been emphasized in the past that there is an indication that the structures again fall into a sequence with increasing temperature of the form hcp–Sm–d–hex–fcc–bcc, which is similar to that obtained with decreasing atomic number both in the pure metals and their alloys. No element shows this complete series however, and only lanthanum passes through more than two phases; indeed all except lanthanum, cerium, europium and ytterbium transform directly to the bcc phase from the room temperature structure. Erbium, thulium and lutetium do not form the cubic phase at all and remain hexagonal close packed to the melting point.

The existence of the high temperature transformation in praseodymium, neodymium and ytterbium was originally observed in resistivity and X-ray measurements at temperatures up to 1170°K [32]. The allotropic form of this new phase was not identified however. Spedding and his co-workers [9] later showed that the high temperature structure of the closely-related element yttrium was the same as that of lanthanum (i.e. bcc) by the observation of continuous solid solubility at high temperatures in the yttrium–lanthanum system. Similarly, other alloy studies [15, 33–35] suggested that the bcc structure should exist for these also. These latter observations relied on the retention of the high temperature allotrope after quenching the alloy and as such are not confirmatory. The more recent phase diagram studies of various systems however, have verified the existence of the cubic structure for many of the metals using arguments similar to those of reference [28]. During the course of his investigations at high pressures using differential thermal analysis measurements, Jayaraman [36] was unable to detect a thermal arrest for Er, Tm and Lu and concluded that these metals did not undergo the hcp–bcc transition. As may be seen from Fig. 2.8 the temperature range of the bcc phase on the second half of the rare earth series decreases rapidly with increasing atomic number and would be extremely small or

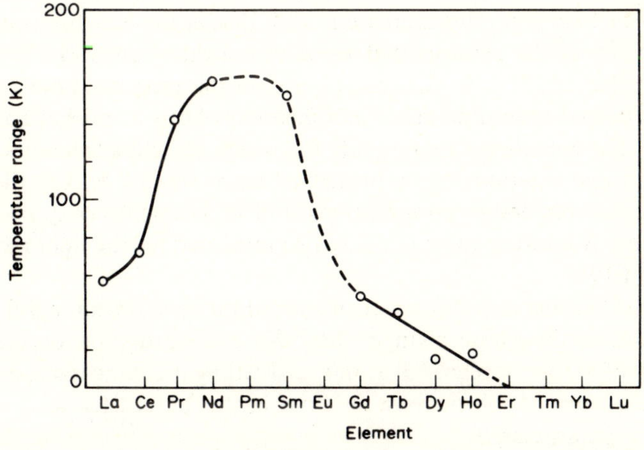

Fig. 2.8 The temperature range of the bcc structure for the rare earth metals.

non-existent at erbium and the following elements.

The thermal expansion coefficients of the rare earth metals have been measured by various research groups with interests both in the high and low temperature regions. Much of the work below room temperature has been carried out near to and below the paramagnetic Curie temperatures and as such the experiments have been directed at an investigation of the effects of magnetic ordering on the crystal structure. The most spectacular observation was that for dysprosium [37], which as a result of the large magnetostatic energies involved in the antiferromagnetic phase change shows a lattice distortion to an orthohombic structure. This change, as well as the other low temperature results, are discussed further in Chapter 4.

Above 300°K, and therefore above the highest magnetic ordering temperature for the pure metals, expansion measurements have shown that the lattice parameter variation with temperature can in general be fitted to the polynomial [9] $R = A+BT+CT^2+DT^3$, where R is a lattice parameter, T the temperature in °C and A, B, C, D are constants.

At the phase transitions in lanthanum appreciable changes occur in the atomic volume. At the hcp–fcc boundary there is a contraction of approximately 0·5% with increasing temperature, while at the fcc–bcc transformation a sudden dilation of the lattice occurs, the magnitude of which is of order 4%. The expansion coefficients of europium and ytterbium are a factor of three larger than most of the other metals, a fact which is attributed to the divalent ionic state in these metals. The value of the coefficient in europium decreases rapidly with temperature increase, possibly due to a gradual excitation of one of the 4*f*-electrons into the conduction band. This would enhance the bonding so causing the lattice to shrink and reducing the magnitude of the expansion coefficient.

These high temperature measurements also indicate that the c/a ratios for all the heavy rare earths increase with increasing temperature due to anisotropic expansion. This increase is such that at the hcp–bcc transformation temperature the extrapolated value of the ratios converge to a nearly constant value of 1·600. In the case of erbium the ratio does not reach this critical value and indeed no cubic modification of this metal has so far been observed. The remaining two metals for which no cubic phase is known, i.e. thulium and lutetium, are reported to reach this value, but the results are based on very long extrapolations and consequently require treating with caution. A limiting value of c/a is also indicated for the light rare earths at $c/a = 1{\cdot}619$.

One must assume that if these critical axial ratios are meaningful in terms of the onset of the phase change, then this transformation occurs by the movement of atoms, or planes of atoms, assisted by the static stress developed as a consequence of the increase in axial lattice parameter relative to the basal plane parameter.

The detailed reasons for the transitions observed in the rare earth series, or indeed for any phase transitions, are not yet fully understood. Nevertheless, empirical correlations of the type described above have been attempted in the past. One of the more promising of these is the general work of Fisher and Dever [38], following the suggestion of Zener [39], to establish a relation between the shear anisotropy ratio $A = c_{44}/c_{66}$ and the occurrence of a thermally induced hcp–bcc transition. Their results, which are given in detail later in this chapter (Section 2.4), show that A has a positive temperature dependence for those metals which undergo the transition and a zero or negative dependence for those which do not. The rare earth metals which they examined were dysprosium and erbium and the related element yttrium, all of which are in the former category and two of which unquestionably show the martensitic transformation. The results for erbium were however somewhat limited. It was not possible to obtain a limiting value of A at which the transition took place. The effects of elastic anisotropy were interpreted in terms of changes in stress within the materials which allow the movement of dislocations necessary for the transition to occur. The mechanism of the structure change originates then in either a series of glissile dislocations that produce the matching of the lattices or stacking faults bounded by partial dislocations.

While this treatment was specifically directed at the hcp–bcc transition it seems probable that the mechanism of the other changes, i.e. hcp–Sm–d–hex–fcc are also associated with elastic stress and, because of the close relation of the various structures, the transitions occur through a slipping process. The energy differences between one structure and another are never large however, and the driving energy for these lower temperature transitions may well be dominated by effects other than those that are elastic in origin. Various proposals have been made concerning these additional terms in the total free energy and these are discussed fully in Section 2.3.

The melting points for all the rare earths have been determined and may be seen to increase in a reasonably smooth way across the series (see Table 2.1). Lanthanum deviates from this variation, a feature which is probably associated with the absence of $4f$ electrons. Cerium, europium and ytterbium have large negative deviations and it is tempting to relate this to the difference of the metallic valence from three. The melting points for Eu and Yb are comparable to those for the alkaline earths, a fact which has been used in the past as part of the evidence for a divalent ionic state in these metals.

2.2.3 *Pressure-induced phase transitions*

Since the early work of Bridgman [40, 41] a great deal of attention has been given to the existence of both solid–solid and solid–liquid phase transitions which can be caused by the application of high pressures to the solids. These have been shown to occur in both the rare earth metals and the intra-rare earth alloys which have been discussed earlier. Much of the work was carried out indirectly by the observation of electrical conductivity [42–44], compressibility [45–47], differential thermal analysis [48], magnetic properties [49. 50], or by the observation of high pressure structures which are retained in a metastable state at atmospheric pressure after the hydrostatic load has been removed [51, 52]. More recently, as improved high pressure techniques have been developed much more quantitative detail has been established concerning these transitions and the state of the material at each side of the phase boundary [53–57].

While it was evident from the observations of Bridgman that structure changes probably occur in the rare earth metals as a result of the change in lattice size caused by compression, it was not until the work of Jayaraman [51, 52] that any direct evidence as to the nature of the high pressure phase was available. In this work it was found that after gadolinium had been subjected to a pressure of 40 kbar at a temperature of 400°C its structure at atmospheric pressure and temperature was the same as that of normal samarium. Similarly a phase change was observed at the same pressure and 300°C in samarium metal, in this case the structure change being from Sm-type to the double hexagonal lattice. Some evidence of this d-hex structure was also observed in the gadolinium specimen but it was not possible to obtain any precise data. More recently however, McWhan and Stevens [58] have observed magnetic changes in gadolinium at about 50 kbar which might be associated with a second transition from the samarium type to the d-hex phase.

On the basis of these experiments, and on the observed transformations from d-hex to fcc in lanthanum [36], praseodymium and neodymium [55], Jarayaman and Sherwood [52] proposed that a sequence of pressure-induced phase transformations occurs in the trivalent rare earth metals. With increasing pressure these polymorphic transitions were in the order hcp–Sm–d-hex–fcc.

As discussed earlier these structures differ only in the stacking of the atomic layers and the overall change from hcp to fcc may be regarded [36] as the result of a series of discrete steps between the extremal structures. If we represent the hcp lattice (stacking sequence AB ...) as h and the cubic structure (stacking sequence ABC ...) as c then the sequence of structures becomes:

$$\text{(hhh...)} - \text{(hhchhc...)} - \text{(hchc...)} - \text{(cccc...)}$$

$$\text{i.e.} \quad \text{hcp} \quad -\tfrac{2}{3}\,\text{hcp}\,\tfrac{1}{3}\,\text{fcc} - \tfrac{1}{2}\,\text{hcp}\,\tfrac{1}{2}\,\text{fcc} \quad -\text{fcc}$$

It will be recognized that this sequence is identical to that discussed in connection with the intra-rare earth binary alloys, for example Gd–Ce and Tb–Pr. In view of this similarity, Jayaraman and his co-workers [59] investigated the behaviour of several such binary systems in an attempt to explore this succession in more detail and to find if the effect of pressure acted in the same way as the addition of a light rare earth to a heavy rare earth. With a few notable exceptions this general pattern was found to be obeyed. These exceptions involve either yttrium or cerium metals, and it would seem that the absence of observed transitions with the former is associated with a much more stable hexagonal structure than occurs in metals such as Ce, Pr, and Nd. Lundin's work with alloys [16, 17] also showed this tendency. In some of the alloys in which both yttrium and cerium are present the observed decomposition into the constituent elements is interpreted as the result of the excitation of a cerium $4f$-electron to a $5d$-state, the resultant change in ion size leading to an unstable alloy system. McWhan and Stevens [60] have observed comparable polymorphic transitions (hcp–Sm type) in measurements of the high pressure magnetic properties of alloys of the heavy rare earths with each other.

The effects of pressure are also comparable to those of alloying in that the c/a ratio of both the pure metals and random alloy systems is increased as a result of increasing pressure, due to anisotropic compressibility. McWhan [58] has given the results of high pressure X-ray measurements for Gd, Tb, Dy, and Ho which show this increase. In this work all the samples transformed to the samarium structure, the limiting value of c/a in the case of Gd and Tb being around 1·62–1·63, while for Dy and Ho it was appreciably smaller, being less than 1·60. The experimental determinations are somewhat uncertain however, and the exact c/a ratios are probably in some doubt. Indeed, McWhan [60] more recently has shown that in some cases the change in c/a may be very much less than that previously reported.

The p–T phase diagrams of many of the elements have been given by Klement and Jayaraman [61] in their review article covering the phase relations and structures of the elements. Those for the rare earth metals are shown in Fig. 2.9. The normally trivalent metals show the structure sequence very clearly, and in addition it is worth noticing the behaviour of the bcc structure with increasing atomic number. In the light elements this phase appears to increase its stability at atmospheric pressure but nevertheless is still unstable with respect to increasing pressure. Samarium shows some evidence for a retention of the phase to the highest pressures and this is carried forward to the heavy rare earths in which the general trend is towards an enhanced stability.

The behaviour of europium and ytterbium is similar to that of the alkaline earths barium and strontium as might have been expected from the earlier discussions. Also in this figure is shown the phase diagram for cerium. The d-hex phase of this metal shows a very small stability region and its place is rapidly replaced with increasing pressure by the α–γ transition boundary. This boundary ends in a critical point at about 550°K, and 20 kbar [62–64]. At temperatures in excess of 600°K, the transformation α–γ appears to be a reasonably rapid but smooth one across the extrapolated transition line. This extrapolated transition intersects the melting curve where it passes through a minimum, and since liquid cerium shows little density variation with pressure in contrast to the fcc density variation, this minimum has been taken to be a consequence of the rapid density variation of the cubic lattice

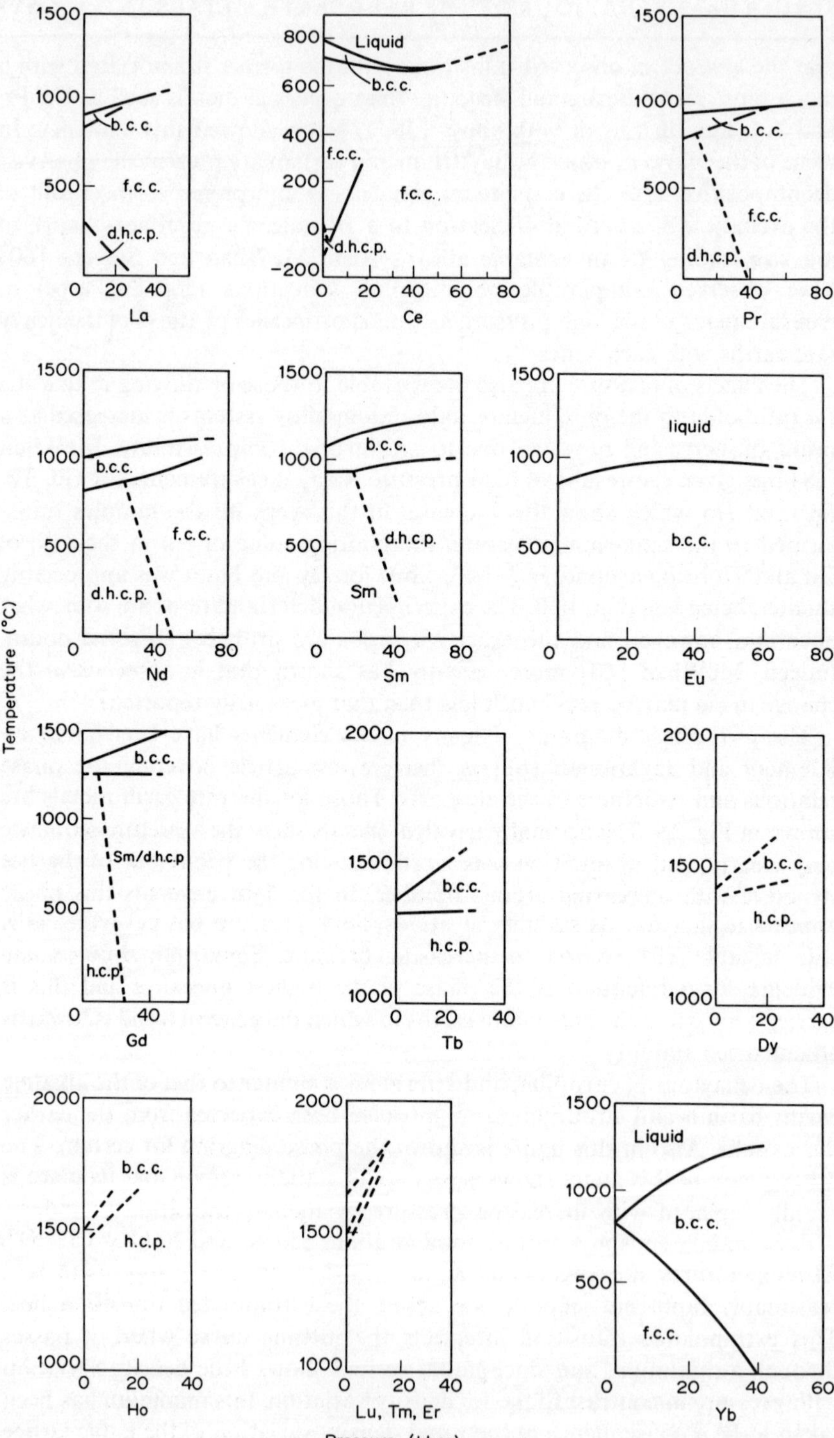

Fig. 2.9 The *p–T* phase diagrams of the rare earth elements; after [61].

in the vicinity of the fusion curve. This conclusion is based on the measurement of magnetic susceptibility through the melting point and is in contrast to the suggestions of a *4f–5d* electronic excitation which have been made by other authors. Franceschi and Olcese [65] have recently detected a new phase (α'–Ce) existing above 50 kbar at room temperature and being produced in an α–α' transition. They report a valence of four for this structure and estimate effective valences of 3·11 and 3·67 for the γ and α phases respectively.

2.3 Discussion of the structural behaviour

In the previous sections we have described the structures of the rare earth metals in elemental form and the effects on these structures of temperature, pressure and alloying. We have seen that a trend exists for a series of crystallographic transitions of the form hcp–Sm-type–d-hex–fcc for all the variables and that this is the same series which exists across the rare earth elements with decreasing atomic number. While this sequence occurs in the direction fcc–hcp for increasing density in the case of the pure metals (low Z to high Z), decreasing temperature and the alloying of a light rare earth with increasing concentrations of a heavy rare earth, the reverse is true for observations of the high pressure behaviour. That is, with increasing pressure and hence again increasing density, the series is crossed in the reverse order hcp–fcc.

These changes represent a remarkable set of experimental data in which the same systematic structure change occurs in a family of metals which are closely related electronically, as a function of three and possibly four separate variables. Whether or not these parameters act in a similar way in producing the phase transitions is not yet clear, although as we have seen, attempts have been made at correlating the appearance of the different structures with a variety of physical properties of the metals.

Attempts to understand the stability of simple metals show that while the energy of any single structure is large, the difference between the energies of the different close packed modifications of a single metal are small and indeed are often comparable to those of a free electron gas. In metals in which other energy terms must be considered, such as magnetic materials, the problem is intensified because of the variety of these additional terms. In the case of the rare earth metals it is probably sufficient to consider the total energy of the unstrained metal as being the sum of five terms, i.e.

$$E_T = E_{\mathrm{ce}} + E_{4f} + E_{\mathrm{d-d}} + E_{Q-Q} + E_{\mathrm{exch}}. \qquad (2.1)$$

Here these terms involve respectively, the energies of the conduction electrons, any contributions to the bonding arising from *4f* wavefunction overlap, magnetic dipole–dipole and quadrupole–quadrupole interactions, and the exchange between magnetic moments arising from the polarization of the conduction electrons.

While E_{ce} is undoubtedly the dominant term, the difference in its value for the closely related structures observed in the rare earth elements is probably extremely small and it is likely that one or more of the other contributions to the total energy may be the decisive factors in determining the final structure of any given metal. Several of these terms have been considered separately for the metals and evidence has been presented to show that in each case they bear some relation to the observed structures.

The major term, E_{ce}, has recently been derived by Hodges [66] using pseudopotentials and the method of Blandin, Friedel and Saada [67] in which contributions to the total energy from pairs of close packed planes is considered. In order to obtain the relative stability of the close packed structures the energy (Φ) necessary to move a pair of planes from equivalent to inequivalent positions was derived under conditions of constant density and axial ratio. The energy difference between two structures a and b then becomes:

$$E_a - E_b = \sum_{m=1}^{\infty} (c_m^a - c_m^b)\Phi(mh), \tag{2.2}$$

where c_m is the fraction of mth nearest planes (of separation mh) in equivalent positions. Table 2.2 shows the c_m values for the four structures of the rare

Table 2.2 The fraction c_m of mth nearest neighbour planes in equivalent positions for the four common rare earth structures [66].

	c_1	c_2	c_3	c_4	c_5	c_6	c_7	c_8	c_9
hcp	0	1	0	1	0	1	0	1	0
Sm	0	$\frac{2}{3}$	0	$\frac{1}{3}$	$\frac{1}{3}$	0	1	0	1
d-hex	0	$\frac{1}{2}$	0	1	0	$\frac{1}{2}$	0	1	0
fcc	0	0	1	0	0	1	0	0	1

earth series obtained by Hodges. The form of $\Phi(mh)[\Phi(x) = x^{-2}\sin\{x \times (4k_f^2 - g_\perp^2)\}^{\frac{1}{2}}$ for $2k_f > |g_\perp|]$ is such that $\Phi(2h) > \Phi(3h) > \Phi(4h)$, etc., and under these conditions, using the values of c_m in Table 2.2, we see that the fcc phase will be energetically preferable to the hcp phase if $\Phi(2h)$ is negative and vice versa for $\Phi(2h)$ positive. If now $\Phi(2h)$ is such that it decreases through zero due to the effect of some experimental variable, then the $c_2\Phi(2h)$ may cease to be the dominant term and intermediate phases may have the lowest energy. Hodges considers the d-hex, hcp, fcc stabilities and within this framework:

$$\begin{aligned} E_{d\text{-}hex} - E_{hcp} &= \tfrac{1}{2}\Phi(2h) + \tfrac{1}{2}\Phi(6h) + \ldots \\ E_{fcc} - E_{d\text{-}hex} &= \tfrac{1}{2}\Phi(2h) - \Phi(3h) + \Phi(4h) + \ldots, \end{aligned} \tag{2.3}$$

With $\Phi(6h)$ small, as may be expected from the form of $\Phi(mh)$, the d-hex and hcp phases will be equally stable for $\Phi(2h) = 0$, and with $\Phi(2h)$ greater or less than zero the stable phase will be the d-hex and hcp respectively. If however, at $\Phi(2h) = 0$, the quantity $-\Phi(3h)+\Phi(4h)$ is less than zero, the fcc lattice will form directly from the hcp lattice. On the other hand for $-\Phi(3h)+\Phi(4h)$ positive the d-hex structure will form, to be followed later by the fcc phase. Similar arguments may be used to derive the conditions necessary for the stability of the intermediate samarium type phase. These both may be summarized as follows:

$$\begin{aligned} &\text{d-hex:} && -\Phi(3h)+\Phi(4h)+\ldots > 0 \\ &\text{Sm-type:} && \tfrac{2}{3}\Phi(4h)-\tfrac{1}{3}\Phi(5h)+\ldots < 0. \end{aligned} \tag{2.4}$$

These critical values involve a variety of assumptions, not least those concerning the behaviour of $\Phi(3h)$, $\Phi(4h)$, etc., but nevertheless this approach is extremely valuable in indicating conditions under which structure changes may occur. Before quantitative results can be evaluated however, it will be necessary to determine the form of the pseudopotentials for the various elements of the rare earth series as these are not currently available.

By examining the form of $\Phi(mh)$ for various electron ratios, Hodges was unable to obtain a clear demarkation for the samarium structure. He suggests that in these structures, since one is involved with interactions between planes at large separations then the fine detail of the Fermi surface will become an important factor, as it also appears to be in determining the stability of the magnetic spin structures observed in these metals (see Chapter 4).

The correlation between the c/a ratio and observed structure is also considered in this work, and it is shown that if the $m = 2$ term dominates the higher m terms then $(c/a)-(c/a)' \propto \Delta c_2$ where $(c/a)'$ is the ideal axial ratio and Δc_2 is the difference between c_2 (see Table 2.2) for the observed phase and the fcc structure. Comparison with the experimental results of various series shows that the prediction is remarkably well obeyed.

Both the 4f and dipole–dipole terms have been considered previously by Gschneidner [12, 68]. In his early work it was concluded that 4f bonding did not provide the direct cause of the structure differences because of the small 4f radius values of the elements compared with their metallic radius. Rather, maximum values of the dipole–dipole interactions were evaluated, neglecting the angle dependent term which arises because of the random orientation of dipoles. As might be expected from the ionic moment values this energy is relatively small for the light rare earths compared to that for the heavy elements, there being almost an order of magnitude difference in the values. On the basis of this it was proposed that the large magnitude of this term for the elements europium to thulium stabilized their respective structures predominantly to the hcp phase in preference to the more complex structures at the beginning of the series. The existence of the simple hexagonal phase in lutetium which carries no dipole moment was not considered. Later [68] the ratio values of the metallic radius to the 4*f* radius of the elements were used to derive a qualitative explanation of the structures in terms of 4*f* bonding. The various structures were shown to exist in certain ranges of values of this ratio namely fcc, ratio $< 3{\cdot}24$, d-hex, ratio between 3·28 and 3·54, Sm, ratio 3·60 to 3·65 and hcp, ratio $> 3{\cdot}68$. It would seem that these results are little more than a further correlation of crystal structure with physical properties and if indeed a 4*f* contribution to the bonding does exist much more extensive and quantitative investigation is necessary to reveal its form.

Some indication that the dipole–dipole interaction may affect the alloy structures for those systems crystallizing in the hcp modification is provided by the c/a ratios of alloys which span, or almost span, the hcp structure of the heavy rare earths. Depending on which of the heavy metals is involved in the alloy, the c/a values lie on a curve which is displaced from that for the pure elements in the direction of the ideal c/a ratio (see Fig. 2.5). On this basis the suggestion would be somewhat similar to that of Gschneidner, in that the dipole interaction leads to a decrease in the axial ratio by an amount which was moment dependent. Within this framework the higher than average value obtained for lutetium is understandable, and it would also suggest that perhaps its value for this element is the normal for the hcp metals. It would be interesting to examine the energy involved in a term of the form:

$$E_{\text{d-d}} = g_J^2 \mu_B^2 J(J+1) \sum_{j \neq 0} \frac{3z_j^2 - r_j^2}{r_j^5} \tag{2.5}$$

which represents the dipole–dipole interaction in terms of the energy at a reference site ($j = 0$) in a dipole field originating from the randomly oriented moments on neighbouring lattice sites.

No quadrupole–quadrupole calculations have been reported in the literature, most probably because of the small magnitude of this term. However, in view of the likely energy differences between the d-hex and Sm structures it is possible that its effect is not entirely negligible, and it would again be informative to evaluate this interaction in a manner similar to that suggested for the dipole term.

The ionic moments have an indirect exchange interaction between them arising through the polarization of the conduction electrons (see Section 4.1), this being the energy term E_{exch}. This is the interaction responsible for the highly complex magnetic structures which are observed in these metals and which are described fully in other sections of this book. In its simplest form (see Chapter 4) the model assumes a free electron system and on this basis Rocher [69] has attempted to account for the crystal structures of the rare earths. In order to evaluate the energy in the paramagnetic temperature region it is most convenient again to derive a molecular field representing the effect of all neighbouring ions. Rocher gives this field as:

$$H_{\mathrm{m}} = -\frac{9\pi}{4}\frac{Z^2 A_0^2}{g_J \mu_{\mathrm{B}} \Omega^2 E_{\mathrm{f}}} \sum_n \phi(2k_{\mathrm{f}} R_{\mathrm{on}}) J_n, \tag{2.6}$$

where Z is the ionic charge, Ω the atomic volume, E_{f} the Fermi energy and J_n the angular momentum of the nth ion. The function $\phi(x)$ is given by

$$\phi(x) = \frac{x \cos x - \sin x}{x^4}, \quad \text{where } x = 2k_{\mathrm{f}} R_{\mathrm{on}},$$

with k_{f} being the Fermi momentum and R_{on} the separation of the reference ion from the nth ion. A_0 is the exchange constant between the spins of the ion and a conduction electron. This exchange mechanism is discussed at length in Section 4.1.

The energy of the system is then given by

$$E_{\mathrm{exch}} = -\tfrac{1}{2}\frac{C}{T} H_{\mathrm{m}}^2$$

$$= -\frac{J(J+1)}{6kT}\left(\frac{9\pi Z^2 A_0^2}{4\Omega^2 E_{\mathrm{f}}}\right)^2 \left\{\sum_n \phi(2k_{\mathrm{f}} R_{\mathrm{on}})\right\}^2, \tag{2.7}$$

and for a random spin orientation this may be reduced to

$$E_{\mathrm{exch}} = -\frac{1}{6kT}\left(\frac{9\pi Z^2 A_0^2}{4\Omega^2 E_{\mathrm{f}} J(J+1)}\right)^2 \sum_n \phi^2(2k_{\mathrm{f}} R_{\mathrm{on}}). \tag{2.8}$$

This energy, evaluated to $n = 3$ for the fcc and hcp structures increases in magnitude with increasing c/a but nevertheless favours the cubic structure. Consequently Rocher assumed that the basic hexagonal crystalline form of the heavy rare earths appears as a result of the detailed form of the *s–d* conduction band. This hexagonal structure was taken to have an axial ratio of 1·57 in the absence of exchange interactions, and the positive deviations of the experimental ratios from this value were attributed to the exchange interaction which is favoured by large c/a values. The exchange energy is of the same order of magnitude as the activation energy of the hexagonal cubic transition in lanthanum but its exact value requires a more complete theory of the *4f–4f* interaction as discussed in Chapter 4. Noskova [70] has also attempted to relate the lattice parameters and their temperature dependence to the exchange interaction.

At present we are not in a position to fully account for the structure sequences observed in the rare earth series, although the evidence suggests that the dominant factor is the behaviour of the conduction electrons in forming a metallic bond. Whether the samarium structure is wholly of this origin is unclear, but it would seem that magnetic moment interactions are at least capable of providing sufficient bonding energy to affect the detailed lattice parameters and perhaps to stabilize this somewhat anomalous structure.

2.4 Elastic properties

In conclusion it is necessary to consider the elastic behaviour of the rare earth elements as it is these properties which are intimately involved with the polymorphism which has been discussed at length in the earlier parts of this chapter and with the magneto-elastic behaviour to be discussed later.

Single crystal measurements are still something of an exception because of the relatively large oriented samples required for the investigation of the elastic moduli using ultrasonic techniques. Consequently much of the following is restricted to a discussion of the classical elastic moduli, i.e. compressibility, shear modulus, etc. Data obtained before 1965, using a variety of techniques, has been well covered in a number of reviews [71, 72] and the room temperature values listed by these authors are shown in Tables 2.3 to 2.5. Much of the early work was carried out above room temperature and consequently the results were unaffected by the magnetic ordering processes. Recently, Rosen [73–75] has published the results of an exhaustive examination of the elastic moduli and ultrasonic attenuations of most of the metals between 4·2 and 300°K.

In this work the constants were determined using a 10 MHz ultrasonic pulse technique to measure the longitudinal (C_l) and transverse (C_t) sound velocities. In terms of these the elastic constants are found using the relations given in Table 2.4.

Table 2.3 The elastic moduli of the rare earth metals.

	Young modulus ($\times 10^6$ kg cm^{-2})	Shear modulus ($\times 10^6$ kg cm^{-2})	Poisson's ratio	Compressibility ($\times 10^6$ cm^2 kg^{-1})
Lanthanum	0·387	0·152	0·288	40·37
Cerium (γ)	0·306	0·122	0·248	40·97
Praseodymium	0·332	0·138	0·305	32·08
Neodymium	0·387	0·148	0·306	30·02
Samarium	0·348	0·129	0·352	33·36
Europium	0·182	0·787		66·63
Gadolinium	0·573	0·277	0·259	25·59
Terbium	0·586	0·233	0·261	24·6
Dysprosium	0·644	0·259	0·243	25·52
Holmium	0·684	0·272	0·255	24·72
Erbium	0·748	0·302	0·238	23·88
Thulium				24·71
Ytterbium	0·182	0·071	0·284	73·93
Lutetium				23·85

The results for all the metals studied show a complex temperature dependence with often quite large anomalies occurring in the vicinity of known magnetic transitions (for the details of these transitions, see Chapter 4). It is rather difficult to select a typical set of results to represent the observed values, but the results for polycrystalline holmium, shown in Fig. 2.10 demonstrate many of the features observed for the other metals.

With decreasing temperature the values of the Young (E) and shear (G) moduli are usually found to increase, passing through sharp minima in the vicinity of magnetic transition temperatures. In the case of dysprosium and terbium the minima at the Curie temperature are large and appear almost as discontinuities in the overall variation and as such have been taken as evidence for first order phase changes. For dysprosium this is almost certainly associated with the known orthorhombic lattice distortion which occurs as the metal passes through the antiferro- to ferromagnetic transition (see Chapter 4).

These observed minima in E and G are associated with lattice softening and correspondingly the adiabatic compressibility is observed to increase at these temperatures. With the exceptions already noted the overall behaviour is in agreement with the Landau–Lifshitz theory of second order transitions [76].

Table 2.4 Relations employed in the evaluation of elastic constants from ultrasonic measurements. C_l and C_t are the longitudinal and transverse velocities of sound. C_m is the average velocity of sound ($C_m = \{\frac{1}{3}[(2/C_t^3)+(1/C_l)]\}^{\frac{1}{3}}$). ρ is the density, M is the atomic weight, and N is Avogadro's number. A_n is the amplitude of the nth pulse and t the transit time between successive pulses.

Shear modulus	G	=	ρC_t^2
Young modulus	E	=	$2G(1+\sigma)$
Compressibility	t_s	=	$3(1-2\sigma)/E$
Poisson's ratio	σ	=	$[2-(C_l \cdot C_t)^2]/2[1-(C_l/C_t)^2]$
Debye temperature	θ_D	=	$h/k(3N\rho/4\pi M)^{\frac{1}{3}} C_m$
Ultrasonic attenuation	α	=	$20 \log_{10}(A_1/A_n)/(n-1)t$

Table 2.5 The Debye temperatures of the rare earth metals derived from ultrasonic measurements [73–75] and specific heat determinations.

	La	Ce	Pr	Nd	Sm	Eu	Gd	Tb	Dy	Ho	Er	Tm	Yb	Lu
Debye temperature θ_D (°K)	154		146·5	163·2	169·5	118	184	177	179	194·5	192		118·1	210

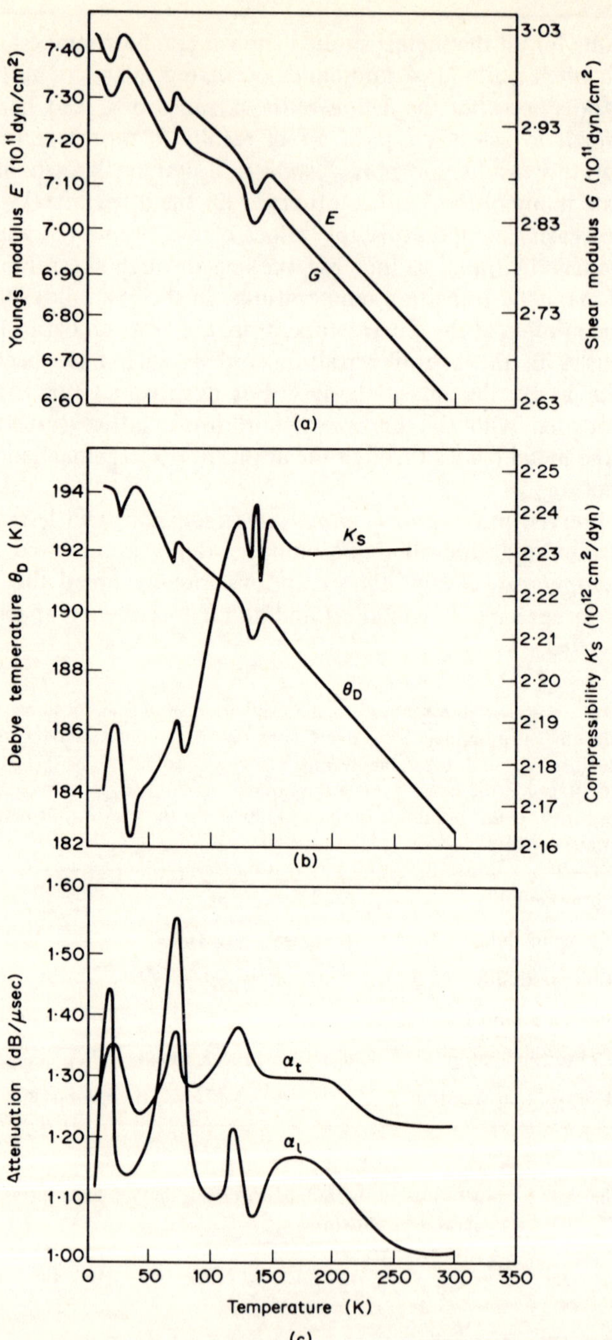

Fig. 2.10 The temperature dependence of some elastic parameters for polycrystalline holmium [74]. (a) Young and shear moduli, (b) adiabatic compressibility and Debye temperature, (c) longitudinal and transverse ultrasonic attenuation coefficients.

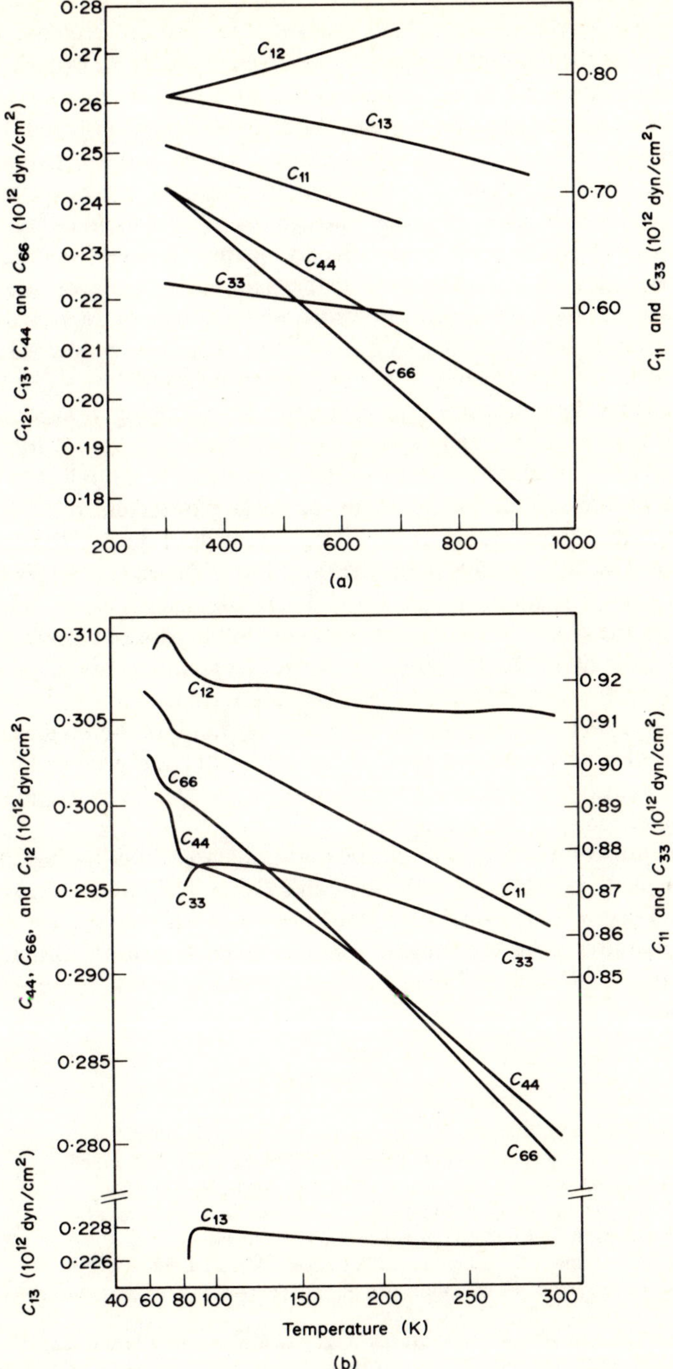

Fig. 2.11 The temperature variation of the principal elastic constants of [38]. (a) dysprosium and (b) erbium.

Landau and Khalatnikov [77] showed that near to magnetic ordering temperatures the interaction of the critical spin fluctuations with the acoustic phonons will give rise to an increase in the ultrasonic attenuation at the low temperature side of these transitions. In all cases this is observed. However, for some metals additional peaks are found, for example in gadolinium at 15°K, which do not coincide with known magnetic characteristics. It is generally possible in these cases though, to identify anomalies in other properties with these extra peaks. Broad maxima in both longitudinal and transverse ultrasonic attenuations in the paramagnetic region are seen for several metals over a temperature range of some tens of degrees above the upper ordering temperature and it is believed that these may arise from the development of spin clusters in the paramagnetic matrix.

The Debye temperatures (θ_D) represent an averaging procedure of the acoustic velocities [78] and as such can be expected to have a temperature variation which is closely related to that of the elastic moduli. The value of θ_D used in the calculation of many physical parameters of solids is the limiting Debye temperature given by the intercept of the $\theta_D v T$ curve on the θ_D axis at $T = 0$°K. At this point the third law of thermodynamics indicates that the curve should have zero slope. These predictions are in good agreement with the observed variations, although in the case of terbium θ_D is still rising rapidly at 4·2°K. The limiting Debye temperatures are given in Table 2.5 along with those obtained from other techniques.

The ultrasonic attentuation in the neighbourhood of the magnetic transitions has also been reported by Luthi *et al* [79–81] and attempts have been made to associate its variation with both volume and linear magnetoelastic coupling of the spin phonon system.

Determination of the tensor elastic constants are rather limited, the early measurements being those of Fisher and Dever [38] on dysprosium and erbium. These workers were concerned primarily with the c_{44}/c_{66} ratio as a means of studying the martensitic hcp–bcc phase transition, although their observations also included the c_{11}, c_{12}, c_{13} and c_{33} terms. These results are shown in Fig. 2.11.

References

[1] W. Klemm and H. Bommer, *Z. anorg. u. allgem. Chem.*, **231**, 138 (1937).
[2] H. E. Nigh, S. Legvold and F. H. Spedding, *Phys. Rev.*, **132**, 1092 (1963).
[3] R. M. Bozorth and T. Wakiyama, *J. Phys. Soc. Japan*, **18**, 97 (1965).
[4] F. H. Ellinger and W. H. Zachariasen, *J. Am. Chem. Soc.*, **75**, 7650 (1953).
[5] A. H. Daane, R. E. Rundle, H. S. Smith and F. H. Spedding, *Acta. Cryst.*, **7**, 532 (1954).
[6] R. T. Weiner and G. V. Raynor, *J. Less Common Metals*, **1**, 309 (1959).
[7] K. A. Gschneidner, R. O. Elliott and R. R. McDonald, *J. Phys. Chem. Solids*, **23**, 255 (1962).
[8] K. A. Gschneidner and J. T. Waber, *J. Less Common Metals*, **6**, 354 (1964).
[9] F. H. Spedding, J. J. Hanak and A. H. Daane, *J. Less Common Metals*, **3**, 110 (1961).
[10] W. Hume-Rothery, *Inst. Metals. Monograph and Report Series*, No. 1, 222 (1936).

[11] K. A. Gschneidner and J. T. Waber, Paper presented at Rare Earths Symposium. Annual Meeting Am. Soc. Metals. Chicago (November 1959).
[12] K. A. Gschneidner, *The Rare Earths*, Eds. F. H. Spedding and A. H. Daane, Wiley (1961).
[13] K. A. Gschneidner, *Rare Earth Alloys*, Van Nostrand (1961).
[14] L. S. Darken and R. W. Gurry, *Physical Chemistry of Metals*, McGraw Hill (1953).
[15] F. H. Spedding, R. M. Valletta and A. H. Daane, *Trans. A.S.M.*, **55**, 483 (1962).
[16] J. F. Nachman, C. E. Lundin and G. B. Rauscher, Tech. Rep. No. 1. Contract No. Nonr. 3661(02). Denver Research Inst. (1963).
[17] C. E. Lundin and D. Klodt, Paper presented at Rare Earths Symposium. Annual Meeting Am. Soc. Metals. Chicago (1959).
[18] I. V. Burov, V. I. Chechernikov, E. M. Savitskii and Pop Iuliu, *Russ. J. Inorg. Chem.*, **9**, 1401 (1964).
[19] I. R. Harris, C. C. Koch and G. V. Raynor, *J. Less Common Metals*, **11**, 436 (1966).
[20] V. A. Finkel and V. V. Vorob'ev, *Soviet Physics JETP*, **26**, 1086 (1968).
[21] R. Vogel and H. Klose, *Z. Metallkunde*, **45**, 633 (1954).
[22] I. V. Burov, V. E. Terekhova and E. M. Savitskii, *Russ. J. Inorg. Chem.*, **8**, 1408 (1963).
[23] W. C. Thoburn, S. Legvold and F. H. Spedding, *Phys. Rev.*, **110**, 1298 (1958).
[24] J. D. Speight, I. R. Harris and G. V. Raynor, *J. Less Common Metals*, **15**, 317 (1968).
[25] B. J. Beaudry, M. Michel, A. H. Daane and F. H. Spedding, *Rare Earth Research III*. Ed. LeRoy Eyring, p. 247, Gordon and Breach (1965).
[26] I. R. Harris and G. V. Raynor, *J. Less Common Metals*, **17**, 336 (1969).
[27] C. J. McHargue and H. L. Yakel, *Acta. Met.*, **8**, 637 (1960).
[28] K. A. Gschneidner and R. Smoluchowski, *J. Less Common Metals*, **5**, 374 (1963).
[29] M. S. Rashid and C. J. Altstetter, *Trans. Met. Soc. AIME*, **236**, 1649 (1966).
[30] W. H. Zachariasen, quoted by A. W. Lawson and T. Tang, *Phys. Rev.*, **76**, 301 (1949).
[31] L. Pauling, quoted by A. F. Schuch and J. H. Sturdivant, *J. Chem. Phys.*, **18**, 145 (1960).
[32] F. Barson, S. Legvold and F. H. Spedding, *Phys. Rev.*, **105**, 418 (1957).
[33] D. T. Eash and O. N. Carlson, *Trans. ASM*, **52**, 1097 (1960).
[34] E. D. Gibson and O. N. Carlson, *Trans. ASM*, **52**, 1084 (1960).
[35] A. E. Miller and A. H. Daane, *Trans. Met. Soc. AIME*, **230**, 568 (1964).
[36] A. Jayarman, *Phys. Rev.*, **139**, A690 (1965).
[37] F. J. Darnell and E. P. Moore, *J. Appl. Phys.*, **34**, 1337 (1963).
[38] E. S. Fisher and D. Dever, *Trans. Met. Soc. AIME*, **239**, 48 (1967).
[39] C. Zener, *Elasticity and Anelasticity of Metals*. University of Chicago Press, Chicago (1948).
[40] P. W. Bridgman, *Proc. Am. Acad. Arts. Sci.*, **76**, 71 (1948).
[41] P. W. Bridgman, *Proc. Am. Acad. Arts. Sci.*, **82**, 83 (1953).
[42] L. F. Vereshchagin, A. A. Semerchan and S. V. Popova, *Soviet Physics, Doklady*, **6**, 609 (1961).
[43] H. D. Stromberg and D. R. Stephens, *J. Phys. Chem. Solids*, **25**, 1015 (1964).
[44] R. A. Stager and H. G. Drickamer, *Science*, **139**, 1284 (1963).
[45] R. I. Beecroft and C. A. Swenson, *J. Phys. Chem. Solids*, **15**, 234 (1960).
[46] F. H. Spedding, J. J. Hanak and A. H. Daane, *Trans. AIME*, **212**, 379 (1958).
[47] D. R. Stephens, *J. Phys. Chem. Solids*, **25**, 423 (1964).
[48] A. Jayaraman, *Phys. Rev.*, **135**, A1056 (1964).
[49] L. B. Robinson, F. Milstein and A. Jayaraman, *Phys. Rev.*, **134**, A187 (1964).
[50] D. B. McWhan, P. C. Souers and G. Jura, *Phys. Rev.*, **143**, 385 (1966).
[51] A. Jayaraman and R. C. Sherwood, *Phys. Rev. Letts.*, **12**, 22 (1964).
[52] A. Jayaraman and R. C. Sherwood, *Phys. Rev.*, **134**, A691 (1964).
[53] A. W. Lawson and T. Y. Tang, *Phys. Rev.*, **76**, 301 (1949).
[54] D. B. McWhan and W. L. Bond, *Rev. Sci. Inst.*, **35**, 626 (1964).
[55] G. J. Piermarini and C. E. Weir, *Science*, **144**, 69 (1964).
[56] J. C. Jamieson, *Science*, **145**, 572 (1964).

[57] E. A. Perez-Albuerne, R. L. Clendenen, R. W. Lynch and H. G. Drickamer, *Phys. Rev.*, **142**, 392 (1966).
[58] D. B. McWhan and A. L. Stevens, *Phys. Rev.*, **139**, A682 (1965).
[59] A. Jayaraman, R. C. Sherwood, H. J. Williams and E. Corenzwit, *Phys. Rev.*, **148**, 502 (1966).
[60] D. B. McWhan and A. L. Stevens, *Phys. Rev.*, **154**, 438 (1967).
[61] W. Klement and A. Jayaraman, *Prog. Solid State Chem.*, **3**, 289 (1967).
[62] Ye. G. Ponyatovskiy, *Soviet Physics, Doklady*, **3**, 498 (1958).
[63] A. Jayaraman, *Phys. Rev.*, **137**, A180 (1965).
[64] B. L. Davis and L. H. Adams, *J. Phys. Chem. Solids*, **25**, 379 (1964).
[65] E. Franceschi and G. L. Olcese, *Phys. Rev. Letts.*, **22**, 1299 (1969).
[66] C. H. Hodges, *Acta. Met.*, **15**, 1787 (1967).
[67] A. Blandin, J. Friedel and G. Saada, Proc. Toulouse Conf. on Dislocations. *J. Phys.*, **C3**, 128 (1967).
[68] K. A. Gschneidner and R. M. Valletta, *Acta. Met.*, **16**, 477 (1968).
[69] Y. A. Rocher, *J. Phys. Chem. Solids*, **23**, 759 (1962).
[70] L. M. Noskova, *Fiz. Met. Met.*, **25**, 397 (1968).
[71] C. R. Simmons, *The Rare Earths*, eds. F. H. Spedding and A. H. Daane, Wiley (1961) p. 428.
[72] K. A. Gschneidner, *Solid State Phys.*, **16**, 279 (1964).
[73] M. Rosen, *Phys. Rev.*, **166**, 561 (1968).
[74] M. Rosen, *Phys. Rev.*, **174**, 504 (1968).
[75] M. Rosen, *Phys. Rev.*, **180**, 540 (1969).
[76] L. D. Landau and E. M. Lifshitz, *Statistical Physics.* Pergamon, New York (1958), p. 430.
[77] L. D. Landau and S. M. Khalatnikov, *Doklady Akad. Nauk. SSSR*, **96**, 469 (1954).
[78] H. B. Huntington, *Solid State Phys.*, **7**, 213 (1958).
[79] R. J. Pollina and B. Luthi, *Bull. Am. Phys. Soc.*, **13**, 616 (1968).
[80] B. Luthi and R. J. Pollina, *Phys. Rev.*, **167**, 488 (1968).
[81] R. J. Pollina and B. Luthi, *Phys. Rev.*, **177**, 841 (1969).

3

Band Structures

Many interesting properties of the rare earth metals, particularly for those with more than half-filled $4f$ shells, arise from the interactions between the magnetic moments of the ions. It is usually assumed that the $4f$ shells are sufficiently compact, and well screened by the $5s5p$ closed shells, for the direct overlap of $4f$ electrons on neighbouring sites to be negligible. The interaction is therefore an *indirect* one in which the $6s6p5d$ electrons, and also the electrons of the closed $5s5p$ shells, participate. The outermost $6s6p5d$ electrons form the conduction band leaving the rare earth ions in the solid tripositive. The properties of this band play a dominant role in determining the interactions between the ions, and hence the magnetic properties of the metals. They also, of course, determine their electric, optical, and transport properties, as for any other metal.

Here we shall be concerned with the properties of the heavy rare earth metals only, for these metals show the most interesting properties, and consequently they have been the most extensively studied. All these metals have the hcp structure. However, it should be noted in passing that a band structure calculation has been reported [17] for the bcc divalent Eu metal, and brief details have been given [40] of calculations for the double hexagonal close packed metals Nd and Pr.

The simplest model of the conduction band, in which the electrons are assumed to be perfectly free, has had remarkable success in the theory of metals. For the rare earth metals its application to the determination of the indirect exchange interaction between ions, known as the Ruderman and Kittel [1], Kasuya [2], Yosida [3] (RKKY) interaction, has led to an understanding of many of their properties. In view of this the free electron approximation will often be adopted in subsequent chapters. The RKKY interaction is described at the beginning of Chapter 4, but here we note that it assumes the ionic spins do not greatly disturb the electrons, so that low order perturbation theory can be used. Consequently, for many purposes it is sufficient to determine the properties of the electrons in the absence of the spins. In the ordered state the periodic spin structures give rise to important new features of the band structure, e.g. *superzone boundaries*, but these are more conveniently described in the relevant sections of the next chapter.

The free electron theory gives a useful guide to the band structures of metals, particularly of the simpler metals. The theory is discussed in Section

3.1 together with some group theoretical ideas which are useful for a qualitative understanding of the results. Group theory also gives rise to the standard notation for describing energy bands. Unfortunately the free electron theory is unable to account quantitatively for many simple electronic properties of rare earth metals, e.g. the electronic contribution to magnetic moment [4] and specific heat [5, 6, 7, 8]. Further, it has been shown [9] that the magnetic ordering to be observed depends sensitively upon the geometry of the Fermi-surface. It is therefore essential to introduce the periodic lattice potential and to obtain the best band structures possible.

Some of the methods currently used in band structure calculations are reviewed in Section 3.2. The earliest calculations for the rare earths [11, 12] were made using the method of augmented plane waves (APW), and results of these are described in Section 3.3. More recently relativistic augmented plane wave calculations (RAPW) have been made for several rare earth metals by Loucks and co-workers [14–20], and their results, particularly for Fermi surfaces, are significantly different from the non-relativistic ones. The method is discussed in Section 3.4, and the results obtained are described in Section 3.5. Finally, the only *direct* measurements, using positron annihilation experiments, which have been made on the Fermi surfaces of the rare earths are described in Section 3.6.

3.1 Symmetry considerations and the free electron theory

The potential in which an itinerant electron moves is periodic, satisfying $V(\mathbf{r}) = V(\mathbf{r}-\mathbf{R})$, where $\mathbf{R}$ is any direct lattice vector. As a result the wave-function for an electron without spin, has the Bloch form

$$\psi_{\mathbf{k}}(\mathbf{r}) = u_{\mathbf{k}}(\mathbf{r})e^{i\mathbf{k}\cdot\mathbf{r}}, \tag{3.1}$$

where $u_{\mathbf{k}}$ is periodic in the lattice, i.e. $u_{\mathbf{k}}(\mathbf{r}) = u_{\mathbf{k}}(\mathbf{r}-\mathbf{R})$. The wavevector $\mathbf{k}$, which specifies the state of the electron, is not unique since the addition of any reciprocal lattice vector $\mathbf{K}$ to it gives a wavenumber which may equally be used to specify the same state, e.g. $\mathbf{k}+\mathbf{K}_n$. In order to label the states uniquely it is necessary to restrict $\mathbf{k}$ to a unit cell of the reciprocal lattice, i.e. to within the Brillouin zone. Within the zone the number of allowed values of $\mathbf{k}$ is equal to the number of unit cells in the crystal lattice. For each value of $\mathbf{k}$ there are many energy levels, or energy bands $E_n(\mathbf{k})$, which can accommodate at most two electrons of opposite spin per unit cell of the crystal lattice. The integer n labels the bands. This representation of $E(\mathbf{k})$ is known as the reduced zone scheme. In reciprocal space the energy bands are continuous functions of $\mathbf{k}$, are periodic such that $E_n(\mathbf{k}) = E_n(\mathbf{k}+\mathbf{K}_n)$, and have the full symmetry of the crystal lattice point group. This latter fact greatly simplifies energy band calculations, since it is only necessary to calculate $\mathbf{E}(\mathbf{k})$ for $\mathbf{k}$ values within a small portion of the Brillouin zone, the energy in other parts being obtained by symmetry.

The heavy rare earth metals have a hexagonal close packed structure. The primitive translation vectors are

$$\mathbf{a}_1 = (a, 0, 0), \qquad \mathbf{a}_2 = \left(-\frac{a}{2}, \frac{\sqrt{3}a}{2}, 0\right), \qquad \mathbf{a}_3 = (0, 0, c), \tag{3.2}$$

when referred to the orthogonal axes of Fig. 3.1, and there are two atoms per

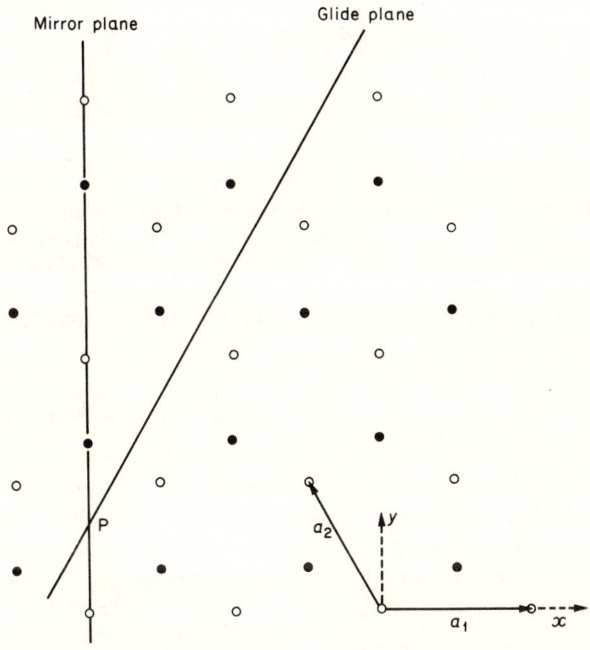

Fig. 3.1 An illustration of the hexagonal close packed lattice showing a glide plane and two of the primitive translation vectors. The open circles lie in a plane $\frac{1}{2}c$ above the dots. A screw axis passes through point P along a line perpendicular to the plane of the paper.

unit cell. The symmetry operations of the lattice, which leave it invariant, cannot be factored into those of translations and those of a point group because there are screw and glide operations which combine both types. There are 24 operations, forming the space group D_{6h}^4, of which 12 are identical to those of the point group D_{3h} when the origin is chosen to be point P (Fig. 3.1), which is not a lattice point. The screw axis and the three glide planes also pass through P. The screw operation consists of a rotation of $\pi/3$ and a displacement of $c/2$, while the glides consist of displacements of $c/2$ followed by reflection. The reader is referred to Jones [21] for details.

The reciprocal lattice is hexagonal and the Brillouin zone is the prism, Fig. 3.2, formed by the planes which bisect the lines joining the origin $\mathbf{K} = 0$ to the nearest reciprocal lattice points. The primitive translation vectors of

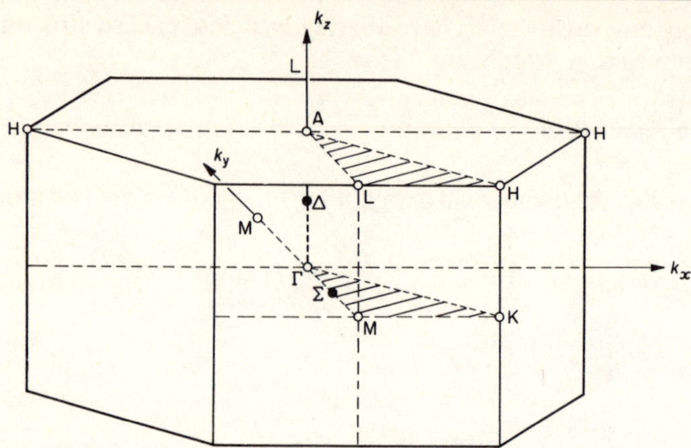

Fig. 3.2 The Brillouin zone for the hexagonal close packed lattice showing symmetry points and the cartesian coordinates of reciprocal space with respect to which the primitive vectors are defined. The shaded areas indicate a portion of the zone which has $1/24^{\text{th}}$ of the total volume.

the reciprocal lattice, referred to the orthogonal axes k_x, k_y and k_z of Fig. 3.2, are

$$\mathbf{b}_1 = \frac{2\pi}{a}\; 1, \frac{1}{\sqrt{3}}, 0\;. \qquad \mathbf{b}_2 = \frac{2\pi}{a}\left(0, \frac{2}{\sqrt{3}}, 0\right), \qquad \mathbf{b}_3 = \frac{2\pi}{c}(0, 0, 1). \tag{3.3}$$

The labelled points lie on one, or more, axes of symmetry and are known as symmetry points. Different points which are marked with the same letter are said to be equivalent, in that they are separated by a reciprocal lattice vector, and therefore represent the same $\mathbf{k}$ state. The symmetry points, with positive coordinates, are

$$\begin{aligned} &\Gamma = (0, 0, 0), &&\mathrm{A} = \left(0, 0, \frac{\pi}{c}\right), &&\mathrm{M} = \left(0, \frac{2\pi}{\sqrt{3}a}, 0\right), \\ &\mathrm{L} = \left(0, \frac{2\pi}{\sqrt{3}a}, \frac{\pi}{c}\right), &&\mathrm{K} = \left(\frac{2\pi}{3a}, \frac{2\pi}{\sqrt{3}a}, 0\right), &&\mathrm{H} = \left(\frac{2\pi}{3a}, \frac{2\pi}{\sqrt{3}a}, \frac{\pi}{c}\right). \end{aligned} \tag{3.4}$$

The simplest wavefunctions which have the Bloch form (3.1) are those with $u_{\mathbf{k}}(r) =$ constant, i.e. plane waves, and these correspond to the case $V(\mathbf{r}) = 0$. The energy of an electron with wavenumber $\mathbf{k}$ is just $E(\mathbf{k}) = \hbar^2 k^2/2m$. If $|\mathbf{k}|$ is unrestricted in value the energy band in any given direction in reciprocal space is parabolic in form. This is well known for free electrons. One effect of a weak crystalline potential $V(\mathbf{r})$ is to produce energy gaps in the parabola at the Brillouin zone surface and, in general, also at the planes which perpendicularly bisect the vectors from Γ to other reciprocal lattice points.

In the reduced zone scheme the free electron parabolic energy band becomes folded so that it is contained in the Brillouin zone, and the structure looks quite complicated. For a given vector **k** there are many energy bands given by

$$E_n(\mathbf{k}) = \frac{\hbar^2}{2m}|\mathbf{k}+\mathbf{K}_n|^2. \tag{3.5}$$

These are often a useful guide to the band structure expected in solids. If **k** is written as $(2\pi/a)(k_x, k_y, ak_z/c)$ and $\mathbf{K}_n$ as $(l_n\mathbf{b}_1 + m_n\mathbf{b}_2 + n_n\mathbf{b}_3)$, then for the hexagonal close packed lattice

$$E_n\frac{2ma^2}{h^2} = (k_x + l_n)^2 + \left\{k_y + \frac{1}{\sqrt{3}}(l_n + 2m_n)\right\}^2 + \left(\frac{a}{c}\right)^2(k_z + n_n)^2. \tag{3.6}$$

The lowest energy bands are shown in Fig. 3.3 for certain symmetry directions, and have been calculated assuming $c/a = \sqrt{8/3}$. It should be noted that the bands may have high degeneracy at the symmetry points.

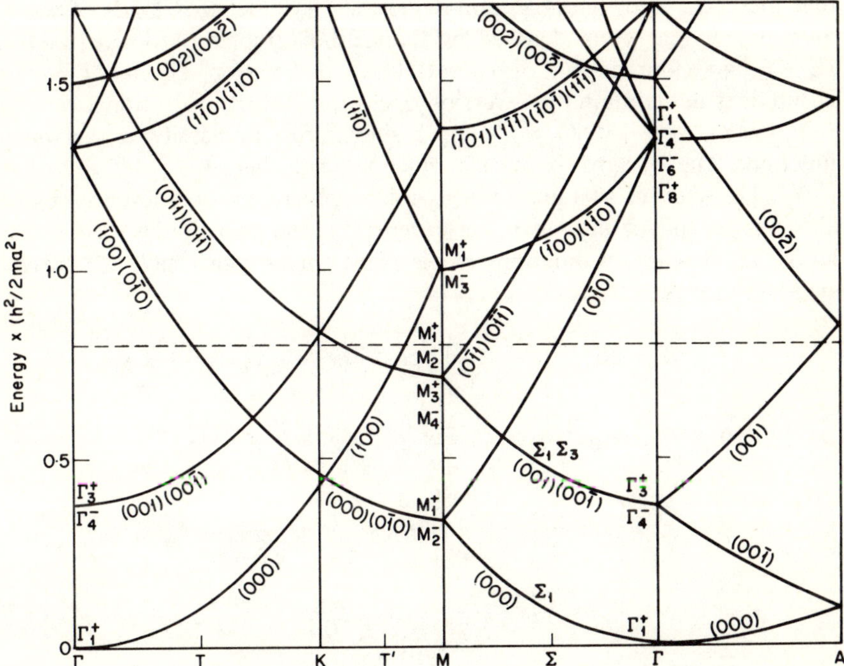

Fig. 3.3 The lowest free electron energy bands of a hexagonal close packed lattice with ideal c/a ratio. The bands are labelled by the integers l_n, m_n, n_n which specify the reciprocal lattice vectors, as in equation (3.6). The broken line indicates the position of the Fermi energy for Gd.

3.1.1 *Finite crystal potential*

The most important effect of a non-zero crystal potential is to mix plane waves which were originally degenerate and to remove that degeneracy, at least partially. The degeneracy remaining is determined by the symmetry of the point in the zone which is being considered. For a general **k** value there will be no symmetry so that the energies are non-degenerate in the absence of spin. However, for a special point there may be many symmetry operations which leave **k** invariant, or transform it into an equivalent point. These operations are said to form the *group of the wavevector*, and this determines the allowed possible degeneracies, and symmetries of the wavefunctions, at that particular point. As an example consider the band at Γ with energy $\frac{4}{3}(h^2/2ma^2)$. For $V(\mathbf{r}) = 0$ the energy is six-fold degenerate and the wavefunctions are

$$\exp\left\{\pm\frac{2\pi}{a}\left(x\pm\frac{y}{\sqrt{3}}\right)\right\}, \quad \exp\left\{\pm\frac{4\pi}{\sqrt{3}a}y\right\}.$$

The point Γ has the highest symmetry in the zone and the wavevector group consists of 48 operations [21, 22]. For a finite $V(r)$, group theory indicates that there are eight possible, non-degenerate, symmetrized kinds of wavefunctions at that point denoted by the symbols (representations) $\Gamma_1^\pm$, $\Gamma_2^\pm$, $\Gamma_3^\pm$, $\Gamma_4^\pm$, and four doubly degenerate kinds of function, $\Gamma_5^\pm$ and $\Gamma_6^\pm$. The group does not contain the inversion operation, but the operation $x \to -x$, $y \to -y$, $z \to -z - c/2$, Q say, plays a similar role in classifying the wavefunctions. The $\pm$ signs on the Γ's above indicate that $Q\psi^+ = \psi^+$, $Q\psi^- = -\psi^-$. The effect of $V(r)$ upon the six-fold degenerate energy level discussed above is to split it into two non-degenerate, and two doubly degenerate levels. The linear combinations of the plane waves which have the correct symmetry for these states are:

$$\begin{aligned}
\Gamma_1^+ : \psi &= \cos\frac{2\pi}{a}\left(x-\frac{1}{\sqrt{3}}y\right)+\cos\frac{2\pi}{a}\left(x+\frac{1}{\sqrt{3}}y\right)+\cos\frac{4\pi}{\sqrt{3}a}y,\\
\Gamma_4^- : \quad &= \sin\frac{2\pi}{a}\left(x-\frac{1}{\sqrt{3}}y\right)-\sin\frac{2\pi}{a}\left(x+\frac{1}{\sqrt{3}}y\right)+\sin\frac{4\pi}{\sqrt{3}a}y,\\
\Gamma_5^+ : \quad &= \begin{cases}\cos\frac{2\pi}{a}\left(x-\frac{1}{\sqrt{3}}y\right)-\cos\frac{2\pi}{a}\left(x+\frac{1}{\sqrt{3}}y\right),\\ \cos\frac{2\pi}{a}\left(x+\frac{1}{\sqrt{3}}y\right)-\cos\frac{4\pi}{\sqrt{3}a}y,\end{cases}\\
\Gamma_6^- : \quad &= \begin{cases}\sin\frac{2\pi}{a}\left(x-\frac{1}{\sqrt{3}}y\right)+\sin\frac{2\pi}{a}\left(x+\frac{1}{\sqrt{3}}y\right),\\ -\left(\sin\frac{2\pi}{a}\left(x-\frac{1}{\sqrt{3}}y\right)+\sin\frac{4\pi}{\sqrt{3}a}y\right.\end{cases}
\end{aligned} \tag{3.7}$$

The determination of the order and the separations of these levels is a matter for detailed calculation. If each of the functions (3.7) is expanded in spherical harmonics it is found that the first terms are, in spectral notation, $\psi(\Gamma_1^-) \rightarrow$ s, $\psi(\Gamma_4^-) \rightarrow$ p, $\psi(\Gamma_5^+) \rightarrow$ d and $\psi(\Gamma_6^-) \rightarrow$ p. These are leading terms so that the overall symmetries are lower. Even so, if the potential is attractive near the origin it would be expected that the *s*-like state $\psi(\Gamma_1^+)$ has the lowest energy.

For symmetry points A and L, on the hexagonal face, group theory shows that the energy levels must be at least doubly degenerate [22]. The effect can be seen for point A in Fig. 3.3. Further, if time reversal symmetry is considered it may be shown that the levels are doubly degenerate at any **k** value on that face. The bands are said to *stick together*. A consequence is that no energy gap can exist across these faces, and it is sometimes convenient to consider a zone in **k** space larger than the Brillouin zone. Use may be made of the Jones' zone, which is defined as the smallest region bounded on all sides by planes of energy discontinuity, but often the *double zone* is preferred. In this scheme the zone containing the first energy band, sometimes referred to as the first zone, is sandwiched between the two halves of the (second) zone containing the second band, so that the faces AHL are in contact. Similarly the third zone is situated between the halves of the fourth zone, and so on. The double zone is illustrated in Figs. 3.5 and 3.10.

In considering **k** values along a line of symmetry, the symmetry of any one point must be compatible with any other. That is, the wavevector group of one point must be the same as, or contain as a subgroup, or be a subgroup of, the group of the other points. The symmetry types of special points which are compatible are given in tables [21, 23]. As an example, in the direction ΓM it is possible for bands with symmetries Σ_1 and Σ_3 to cross, but not two bands which both have symmetry Σ_1.

3.1.2 *Spin–orbit coupling*

Electron spin is usually ignored in energy band calculations, it being assumed that each band is doubly degenerate, containing two electrons of opposite spin. However, for the rare earths the spin–orbit interaction can be appreciable ($\sim$ 0·1 eV) and it must be included. Its effect is to remove some of the remaining degeneracy. As discussed in Section 1.1.7 the interaction is essentially a relativistic effect, and should properly be treated as such, but it is sometimes sufficient to include the approximate term (1.8) in the single particle Schrödinger equation. The inclusion of spin doubles the number of operations in each wavevector group, which is then known as a *double group*, but in general the number of extra allowed types of symmetrical function is not doubled. For example, at Γ the double group has six extra representations, $\Gamma_7^{\pm}$, $\Gamma_8^{\pm}$ and $\Gamma_9^{\pm}$. At energy $\frac{4}{3}(h^2/2ma^2)$, group theory predicts [23] that the states Γ_1^+ and Γ_4^- remain doubly degenerate, spin included of course, but that both four-fold degenerate states, Γ_5^+ and Γ_6^-, are split into two doubly degenerate states. In particular it should be noted that the degeneracy over the face AHL is lifted except along the line AL. Hence there will be energy gaps on this face and consequently the double zone scheme is less convenient.

3.1.3 *The Fermi surface*

The conduction electrons fill the lowest energy bands up to some maximum energy, the Fermi energy E_f, and for the heavy rare earth metals E_f intersects the third and fourth bands. The surface bounding the occupied region of **k**-space is known as the Fermi surface. For free electrons, assuming $V(\mathbf{r}) = 0$, the occupied region in the extended zone scheme is a sphere of radius k_f, where $\hbar^2 k_f^2/2m = E_f$. A small crystalline potential distorts the spherical surface, particularly near the planes of energy discontinuity, and the effective area of the Fermi surface is reduced; see Fig. 3.4. In Fig. 3.5 the free electron Fermi surface is shown in the reduced double zone scheme, where it appears more complicated. As remarked above, if spin–orbit coupling is present there will be energy gaps on the faces AHL, and the

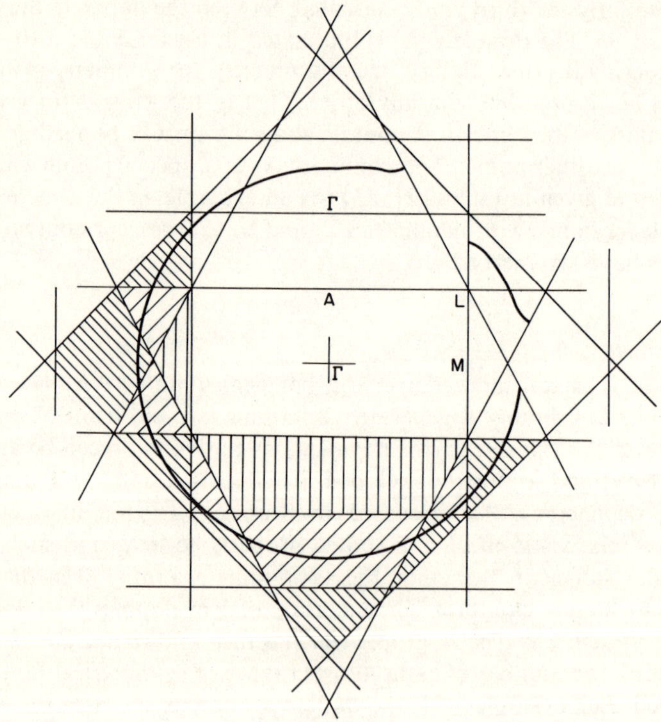

Fig. 3.4 The circle represents the free electron Fermi surface in the extended zone scheme. The lines indicate possible planes of energy discontinuity; note there is no discontinuity on the plane AL(H). The broken curves indicate the effect of a finite crystal potential. The shaded regions indicate portions of higher zones up to the fourth. In the reduced zone scheme these fold into the first zone.

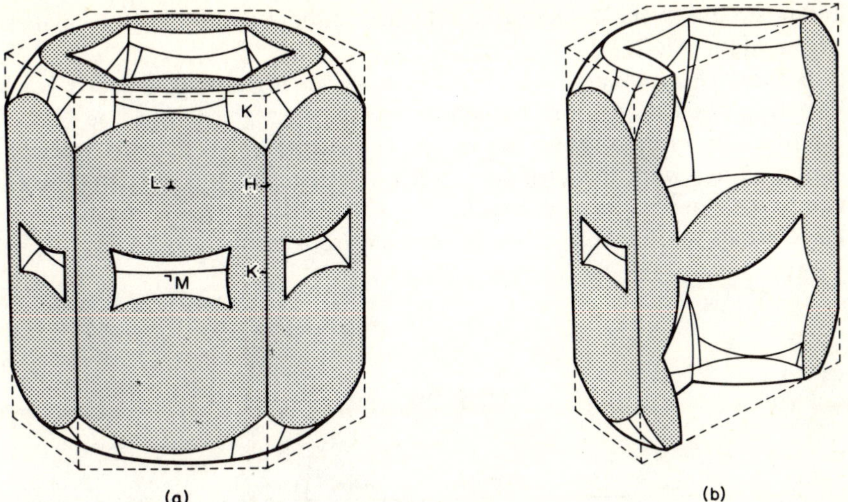

Fig. 3.5 (a) The free electron Fermi surface in the double zone scheme of a trivalent hexagonal close packed metal with ideal c/a ratio; (b) is a section through the surface. After ref. [42].

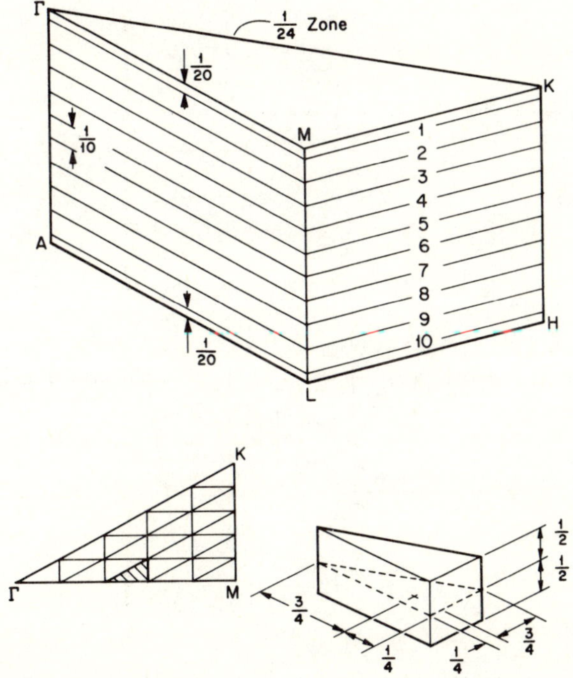

Fig. 3.6 A $1/24^{\text{th}}$ portion of the Brillouin zone, showing a typical division into 250 subunits used in determining the energy at general points in the zone. After ref. [20].

Fermi surface in the double zone will show discontinuities on these planes, except along AL.

Because of symmetry, it is only necessary in performing detailed band structure calculations of rare earths to consider a segment which is 1/24 of the Brillouin zone, as shown in Fig. 3.6. The intersections of the free electron Fermi surface with the surfaces of such a segment look quite complex, as can be seen, in plan form, in Fig. 3.7. It is convenient to present the results of calculations more realistic than for free electrons in this way.

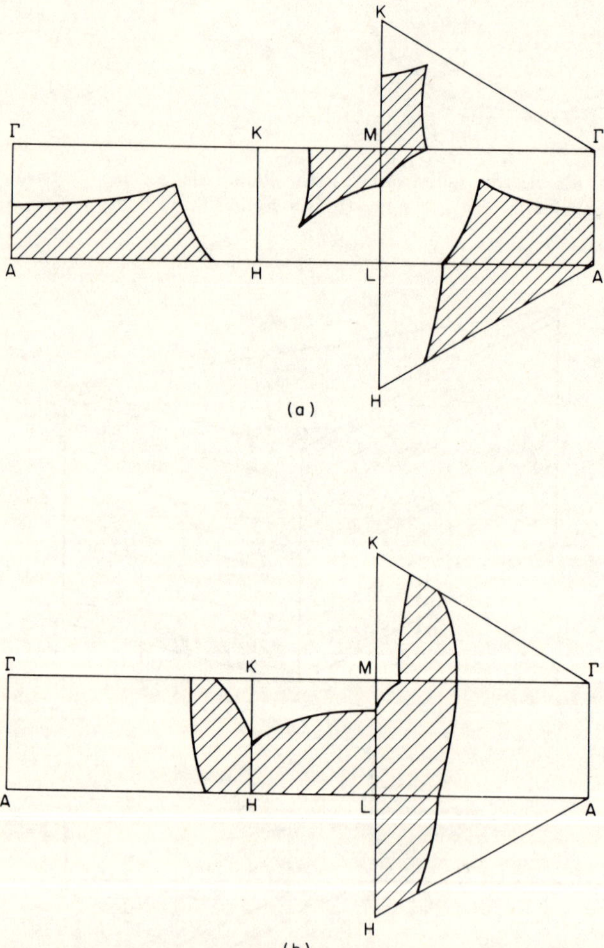

Fig. 3.7 An artistic impression showing the intersections of the free electron Fermi surface of Fig. 3.5 with the planes of symmetry bounding the $1/24^{th}$ zone. (a) Shaded area represents holes in the third zone; (b) shaded area represents electrons in the fourth zone.

3.2 Non-relativistic methods of band structure calculation

The methods available for energy band calculations at the present time have been comprehensively reviewed by Callaway [24], and reference should be made to that work for details. In principle there are two ways of constructing single particle crystal wavefunctions $\psi_{\mathbf{k}}(\mathbf{r})$. The first is to employ localized, atomic-like orbitals centred about each lattice site, and the second is to expand $\psi_{\mathbf{k}}(\mathbf{r})$ in plane waves. Since in metals the electrons of interest are itinerant, forming a conduction band, the second method is most appropriate, and only plane wave methods will be considered here. However, it should be noted that the cellular method, which is here classified as a method of the first kind, is often useful for determining the position of the bottom of the lowest energy band of interest, ($\mathbf{k} = 0$), and hence cohesive energies.

The crystal wavefunctions for a given $\mathbf{k}$ may quite generally be expanded in basis functions $\phi_{\mathbf{k}}(\mathbf{r})$ which have the Bloch form, i.e.

$$\psi_{\mathbf{k}}(\mathbf{r}) = \sum_{\mathbf{K}_n} \alpha(\mathbf{k}, \mathbf{K}_n)\, \phi_{\mathbf{k}+\mathbf{K}_n}(\mathbf{r}), \tag{3.8}$$

where $\mathbf{k}$ is restricted to the Brillouin zone, and the coefficients $\alpha(\mathbf{k}, \mathbf{K}_n)$ are determined from the requirement that $\psi_{\mathbf{k}}$ satisfy the Schrödinger equation

$$\left(-\frac{\hbar^2}{2m}\nabla^2 + V(\mathbf{r})\right)\psi_{\mathbf{k}}(\mathbf{r}) = E(\mathbf{k})\psi_{\mathbf{k}}(\mathbf{r}). \tag{3.9}$$

Since $V(\mathbf{r})$ is periodic it is also convenient to make the expansion

$$V(\mathbf{r}) = \sum_{\mathbf{K}_n} V(\mathbf{K}_n) e^{i\mathbf{K}_n\cdot r} \tag{3.10}$$

If there are m atoms per unit cell at positions $\boldsymbol{\rho}_j (1 \leq j \leq m)$, and such that $\boldsymbol{\rho}_1 = 0$, the Fourier coefficients may be written

$$V(\mathbf{K}_n) = \sum_{j=1}^{m} V_j(\mathbf{K}_n) e^{i\mathbf{K}_n\cdot\boldsymbol{\rho}_j}.$$

Even for atoms of the same type it may be expected that all the $V_j(\mathbf{K}_n)$ are not the same, because each atom in the cell is situated differently with respect to its neighbours. However, it is usually assumed that the $V_j(\mathbf{K}_n)$ are all identical, in which case

$$V(\mathbf{K}_n) = V_1(\mathbf{K}_n) \sum_{j=1}^{m} e^{i\mathbf{K}_n\cdot\boldsymbol{\rho}_j} = V_1(\mathbf{K}_n) F(\mathbf{K}_n), \tag{3.11}$$

where this defines the *structure factor* $F(\mathbf{K}_n)$. For the hcp lattice it is often convenient to choose the origin of the unit cell midway between the two atoms, so that $F(\mathbf{K}_n)$ is real. Explicitly $F(\mathbf{K}_n) = 2\cos(\mathbf{K}_n\cdot\boldsymbol{\rho}/2)$, where $\boldsymbol{\rho} = \frac{1}{3}\mathbf{a}_2 + \frac{2}{3}\mathbf{a}_2 + \frac{1}{2}\mathbf{a}_3$.

The simplest basis functions $\phi_{\mathbf{k}}$ are plane waves. In this case, on substituting equation (3.8) into (3.9), multiplying by $e^{-i(\mathbf{k}+\mathbf{K}_m)\cdot\mathbf{r}}$, and integrating over variable $\mathbf{r}$, equation (3.9) becomes:

$$[E^0(\mathbf{k}+\mathbf{K}_n)-E(\mathbf{k})]\alpha(\mathbf{k},\ \mathbf{K}_n) = \sum_{\mathbf{K}_m} V(\mathbf{K}_m+\mathbf{K}_n)\ \alpha(\mathbf{k},\mathbf{K}_m) = 0 \qquad (3.12)$$

where $E^0(\mathbf{k}) = \hbar^2 k^2/2m$. This equation is one of a set of linear equations for the α's which must be solved simultaneously. Solutions only exist for certain values of $E(\mathbf{k})$ which make the secular determinant vanish, and these are the estimates of the band energies at the selected $\mathbf{k}$ value. The dimension of the secular equation to be solved is equal to the number of coefficients α in expansion (3.8) However, as discussed in the previous section, at the $\mathbf{k}$ values of the special points in the Brillouin zone $\psi_{\mathbf{k}}(\mathbf{r})$ must have the symmetry of the wavevector group. For these points it is therefore possible to save much labour by first using group theory to combine the degenerate plane waves of a given energy into functions $\phi_{\mathbf{k}}(\mathbf{r})$, e.g. equations (3.7), which have the required symmetry properties. In this way, for a given number of plane waves, the dimension of the secular equation is reduced. The number of terms in expansion (3.8) should be increased until the energy converges to a constant value. The main disadvantage of using simple plane waves is that this convergence is slow, so that very many terms may be required. The physical reason is that while the electrons are free electron-like between atoms, near the nuclei the potential has the form of a deep well, and $\psi_{\mathbf{k}}(\mathbf{r})$ varies rapidly. Many plane waves are needed in order to reproduce these oscillations.

3.2.1 *The orthogonalized plane wave method*

This (OPW) method [25] attempts to remove the above difficulty by expanding $\psi_{\mathbf{k}}(\mathbf{r})$ in the functions

$$\phi_{\mathbf{k}}(\mathbf{r}) = e^{i\mathbf{k}\cdot\mathbf{r}} - \sum_i \beta_i \chi_{i\mathbf{k}}, \qquad (3.13)$$

where the $\chi_{i\mathbf{k}}$ are Bloch functions constructed out of localized core states χ_i:

$$\chi_{i\mathbf{k}} = \sum_{\mathbf{R}_n} e^{i\mathbf{k}\cdot\mathbf{R}_n}\chi_i(\mathbf{r}-\mathbf{R}_n). \qquad (3.14)$$

The functions $\chi_{i\mathbf{k}}$ are solutions of the Schrödinger equation for the crystal corresponding to electrons localized on each atom and which form a very narrow energy band. Since an itinerant state $\psi_{\mathbf{k}}(\mathbf{r})$ is a solution of the same Schrödinger equation, but corresponding to a higher energy, it must be orthogonal to all the $\chi_{i\mathbf{k}}$, i.e.

$$\langle \psi_{\mathbf{k}}, \chi_{i\mathbf{k}} \rangle = 0.$$

For $\psi_{\mathbf{k}}$ constructed from (3.8) and (3.13) this condition can be satisfied automatically provided $\beta_i = \langle \chi_{i\mathbf{k}}, e^{i\mathbf{k}\cdot\mathbf{r}} \rangle$. In this case (3.13) is said to be an *orthogonalized plane wave*:

$$\phi_{\mathbf{k}}(\mathbf{r}) = e^{i\mathbf{k}\cdot\mathbf{r}} - \sum \langle \chi_{i\mathbf{k}}, e^{i\mathbf{k}\cdot\mathbf{r}} \rangle \chi_{i\mathbf{k}}. \tag{3.15}$$

Because $\chi_{i\mathbf{k}}$ is small between atoms, $\phi_{\mathbf{k}}$ is free electron-like in these regions. On the other hand near the nuclei the $\chi_{i\mathbf{k}}$ ensure that $\phi_{\mathbf{k}}$ is correctly orthogonal to the core states and has the required oscillations.

The crystal wavefunctions $\psi_{\mathbf{k}}$ are expanded in the $\phi_{\mathbf{k}}$ as in equation (3.8). The secular equation for the coefficients $\alpha(\mathbf{k}, \mathbf{K}_n)$ may again be set up, but there is an added complication now in that the $\phi_{\mathbf{k}}$ are not mutually orthogonal. For special $\mathbf{k}$ vectors, linear combinations of the $\phi_{\mathbf{k}}$ which have the correct point symmetry can be formed, and these contain exactly the same coefficients as used for simple plane waves.

In applications the OPW method is most appropriate for metals in which the electrons can unambiguously be separated into core and itinerant groups. This condition is well fulfilled for simpler metals, such as the alkalis, which have almost spherical Fermi surfaces, so that the conduction electrons may be expected to have energy bands similar to those of *nearly free electrons* (NFE). However, this separation is difficult for partially filled d-bands, so that the method is much less suitable for the rare earths ($5d$), or for the transition elements ($3d$). In practice the most serious difficulty of the OPW method is that correct core functions χ_i are not known. In order to be mutually orthogonal, the basis functions must be solutions of the same Hamiltonian. This means that the core states should be eigenfunctions of the crystal Hamiltonian, whereas usually they are only known for the atomic, Hartree–Fock Hamiltonian. In calculating correct core states for the heavier elements there is the disadvantage that there are a lot of them.

The OPW method has been formulated in a different way by Phillips and Kleinman [26]. They have shown that the ϕ_k can be replaced by simple plane waves provided a non-local repulsive potential, called the *pseudo-potential*, is added to $V(\mathbf{r})$. The advantage of the method is that the basis functions are plane waves, and it is particularly useful for NFE metals. The disadvantages are that the pseudo-potential depends upon energy and momentum, and hence is different for each state, and also that it is not uniquely defined. The pseudo-potential is often treated empirically.

3.2.2 *Augmented plane waves*

An alternative method to that of OPW's is to use *augmented plane waves* [27, 14] (APW'S) which are compounded of spherical waves near the nuclei, and plane waves between nuclei. Explicitly an APW may be written as

$$\phi_{\mathbf{k}}(\mathbf{r}) = \theta(\mathbf{r}-\mathbf{r}_i)\, e^{i\mathbf{k}\cdot\mathbf{r}} + \theta(\mathbf{r}_i - \mathbf{r}) \sum_{l,m} C_{lm} Y_{lm}(\hat{\mathbf{k}}) R_l(\xi, r) \tag{3.16}$$

where

$$\left.\begin{aligned} \theta(r) &= 1, \quad r > 0 \\ &= 0, \quad r < 0 \end{aligned}\right\}.$$

The form of (3.16) implies that the potential about a nucleus, $U(r)$, is spherically symmetric out to a distance r_i and zero, or constant, beyond that radius. The crystal potential constructed from such potentials,

$$V(\mathbf{r}) = \sum_{\mathbf{R}} U(\mathbf{r}-\mathbf{R})$$

is known as a *muffin-tin* potential. The functions $R(\xi, r)$ are taken to be solutions of the radial Schrödinger equation, in atomic units,

$$\frac{\mathrm{d}^2}{\mathrm{d}r^2}(rR_l)+\left\{\xi - U(r) - \frac{l(l+1)}{r^2}\right\}rR_l = 0. \tag{3.17}$$

The coefficients C_{lm} are chosen to make $\phi_{\mathbf{k}}$ continuous on the sphere of radius r_i. This is achieved by expanding the plane waves in spherical waves,

$$e^{i\mathbf{k}\cdot\mathbf{r}} = 4\pi \sum_l i^l j_l(kr) Y_{lm}(\hat{\mathbf{r}}) Y^*_{lm}(\hat{\mathbf{k}}),$$

so that matching at $r = r_i$ gives

$$C_{lm} = 4\pi i^l Y^*_{lm}(\hat{\mathbf{k}})\frac{j_l(kr)}{R_l(\xi, r)}. \tag{3.18}$$

The slope $(\partial\phi_{\mathbf{k}}/\partial r)_{r_i}$ is not continuous and it contributes kinetic energy to the APW. In consequence, on expanding the crystal wavefunction $\psi_{\mathbf{k}}$ in APW'S, as in (3.8), it is found that matrix elements entering the secular equation contain a contribution from $\partial\phi/\partial r$ at the surface $r = r_i$. Explicitly the matrix elements of the effective single particle Hamiltonian $\mathscr{H}$ between two APW'S $\phi_{\mathbf{k}_n}$ and $\phi_{\mathbf{k}_N}$ are:

$$\langle \mathbf{k}_n|\mathscr{H}|\mathbf{k}_N\rangle = (\mathbf{k}_n\cdot\mathbf{k}_N - E)\left[\Omega\delta_{n'N} - 4\pi r_i^2 \frac{j_l(|\mathbf{k}_n+\mathbf{k}_N|r_i)}{|\mathbf{k}_n-\mathbf{k}_N|}\right]$$
$$+4\pi\sum_{l=0}^{\infty}(2l+1)P_l(\widehat{\mathbf{k}_N\cdot\mathbf{k}_n})j_l(k_n r_i)j_l(k_N r_i)\frac{R'(E, r_i)}{R(E, r_i)} \tag{3.19}$$

where Ω is the volume of the unit cell and $R' = \partial R/\partial r$. The matrix elements have been reduced to this relatively simple form by assuming that ξ, in radial equation (3.17), is the same as the crystal energy E.

Since the matrix elements are themselves functions of the energies which are to be determined it is necessary to evaluate the secular determinant as a function of E, and then find the values of E which make it zero. These are the estimates of the energy bands. The convergence of the reciprocal lattice expansion is expected to be good, essentially because natural wavefunctions are used in each region of the potential. As previously, at special points in the zone it is possible to form symmetrized APW's by taking the appropriate linear combinations of spherical waves inside r_i, and of plane waves outside. In practice, with the use of modern computers, this is often not done.

The APW method is the only one, in both its non-relativistic and relativistic forms, which has been applied to the band structure calculations of the rare earths. A critique of the method has been given by Loucks [14]. Its advantages compared with the OPW method are as follows: (i) there is no problem of classifying the electrons as localized and itinerant, (ii) the muffin-tin potential is easier to construct than the potential used in OPW calculations. The disadvantages are: (i) that the matrix elements are functions of E, so that solution of the secular equation is not straightforward, (ii) for APW's the potential about each nucleus must be spherically symmetric, but there is no such restriction for OPW's.

Another method, which is similar to that of APW's, has been developed by Korringa [28] and Kohn and Rostoker [29], and is known as the Green function method. A muffin-tin potential is assumed, but emphasis is placed upon the inner wavefunctions, and modified (phase shifted) spherical waves are fitted in the outer regions. The basic problem is then to satisfy the periodic boundary condition. Reference should be made to Callaway [24] for details. The matrix elements are more complex than those for APW's, but the secular equation is smaller. The APW method is more convenient for numerical work if several different crystal structures are to be considered [14].

3.3 Results of non-relativistic APW calculations

For most rare earths the spin–orbit interaction is important and in order to obtain reliable band structures it is necessary to perform relativistic calculations. Such calculations will be described in the following sections. However, for metals with S-state ions, such as Gd, and for Sc and Y, which are similar to the heavier rare earths, non-relativistic calculations may be sufficiently good. APW calculations have been made for Gd [11], Sc [19], Y [20] and also for Tm [12, 13] and Ho [30], but in the latter two cases relativistic effects may be expected to be appreciable [18]. Here attention will be focused on Gd and Y.

A basic problem in APW calculations is to determine the spherically symmetric potential within the sphere of radius r_i about a nucleus, i.e. $U(r)$ of equation (3.17). Usually r_i is chosen to be half the separation of the two ions in the unit cell of the hcp lattice. Typically, for Y, which has lattice constants $a = 6{\cdot}871$ and $c = 10{\cdot}89$ (atomic units), this choice gives $r_i = 2{\cdot}858$. It is convenient to choose the origin of the unit cell half-way between the two atoms so that the structure factor $F(K_n)$ is real. Within each sphere, $U(r)$ is constructed from wavefunctions of the neutral atom at its centre together with spherically averaged contributions from the neighbouring atoms. Exchange is included through the Slater $\rho^{1/3}$ term, equation (1.3). Details are given in Loucks' book [14].

In a band structure calculation for Gd, Dimmock and Freeman [11] considered two different central potentials. The first was constructed for the configuration $4f^7 6s^2 5d$, with exchange included through the Slater $\rho^{1/3}$ term, and the second from Hartree–Fock wavefunctions [31] for the configuration $4f^7 6s^2$, plus an atomic $5d$ wavefunction. Their calculations indicate that the energies of the conduction bands are insensitive to the potential, differing by $\sim$ 0·02 Ry.

Similar energy bands have also been obtained by the same authors [13] for Tm, and these are shown in Fig. 3.8. Comparison with Fig. 3.3 indicates

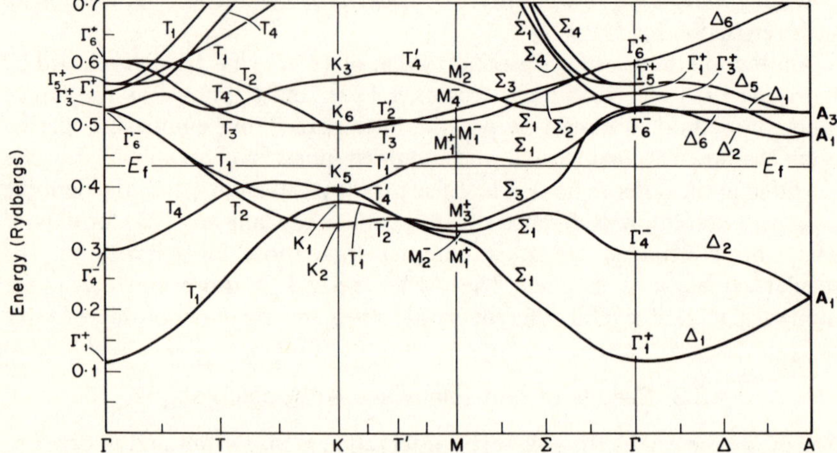

Fig. 3.8 Non relativistic augmented plane wave energy bands for Tm, after ref. [13]. E_f is the Fermi energy.

that the bands are considerably modified from the free electron ones, most importantly in that there are many more flatter bands at higher energies which are mainly of d-character.

In an APW calculation [20] for Y the 1/24 zone was divided into 10 layers, each of which was subdivided into 250 triangular regions; see Fig. 3.6. The eigenvalues were determined at each of the central points of these small regions using an expansion for $\psi_{\mathbf{k}}$ containing 22 APW's. The reciprocal lattice vectors used are shown in Table 3.1. With six valence electrons per unit cell the electrons occupy the 750 lowest, doubly degenerate energy states, and the highest occupied state, i.e. the Fermi energy, was found to be 0·337 Ry above the lowest, 0·137 Ry at Γ. The intersections of the Fermi surface with the faces of the 1/24 zone are shown in Fig. 3.9.

Non-relativistic APW calculations tend to produce very similar Fermi surfaces for all the heavy rare earth metals [18], and in Fig. 3.10 is shown a typical surface in the double zone scheme. The surface is actually that determined for Tm [13], but relativistic effects are important in that case; on the other hand, it is probably a reasonable representation of the surface for Gd. The Fermi energy of Gd is calculated [11] to be 0·25 Ry from the

Table 3.1 Reciprocal vectors used [20] in an APW expansion for Y. The notation is that of Loucks and corresponds to $(ijk) = (2\pi/a)\{i\,\mathbf{b}_3 + j\mathbf{b}_1 + k\mathbf{b}_2\}$ when the primitive vectors are defined by equation (3.3).

$(\bar{2}\ 0\ 0)$	$(\bar{2}\ \bar{1}\ 1)$
$(\bar{1}\ 0\ 0)$	$(\bar{1}\ \bar{1}\ 1)$ $(\bar{1}\ \bar{1}\ 0)$ $(\bar{1}\ 0\ \bar{1})$ $(\bar{1}\ 1\ \bar{1})$ $(\bar{1}\ 1\ 0)$ $(\bar{1}\ 0\ 1)$
$(0\ 0\ 0)$	$(0\ \bar{1}\ 1)$ $(0\ \bar{1}\ 0)$ $(0\ 0\ \bar{1})$ $(0\ 1\ \bar{1})$ $(0\ 1\ 0)$ $(0\ 0\ 1)$
$(1\ 0\ 0)$	$(1\ \bar{1}\ 1)$ $(1\ \bar{1}\ 0)$ $(1\ 0\ \bar{1})$
$(2\ 0\ 0)$	

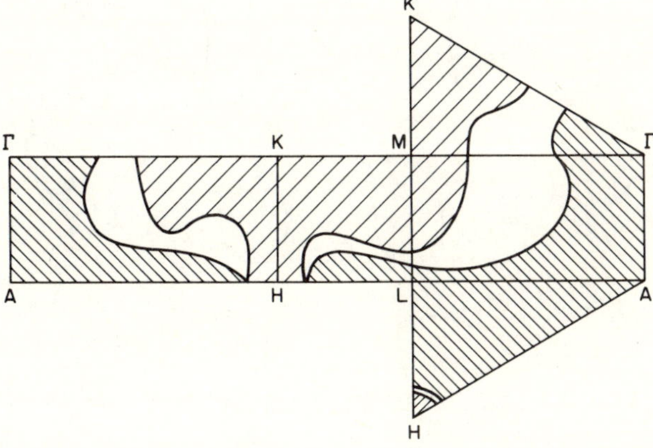

Fig. 3.9 Intersections of the Fermi surface of Y with the faces of the $1/24^{\text{th}}$ zone. The broad cross hashing indicates holes in the third zone. The other shaded area refers to electrons in the fourth zone. After ref. [20].

bottom of the conduction band. Compare the free electron value of 0·54 Ry.

An important property of a conduction band is the number of available states per atom, i.e. different $\mathbf{k}$ vectors, $N(E)\mathrm{d}E$ say, which have energy between E and $E+\mathrm{d}E$. $N(E)$ is called the *density of states*. The computed [11] density of states curve for Gd is shown in Fig. 3.11. The bands just below the Fermi energy are *s–d* like in character and have a narrow band width. They contribute a large number of states and increase $N(E)$ well above the free electron value. At the Fermi surface, for example, it is found that $N(E_\mathrm{f})$ is 1·8 electrons per atom per eV, compared with the free electron value of 0·6 $(\mathrm{eV})^{-1}$.

The density of states at the Fermi surface is important in determining such properties as the moment of the conduction electrons induced by an ionic spin μ_c, and the electronic contribution to the specific heat, γ. Both quantities are directly proportional to $N(E_\mathrm{f})$. The induced moment depends also upon the exchange integral between 4*f* and conduction electrons, and for Gd there is some uncertainty in the value this has. A direct comparison of the induced

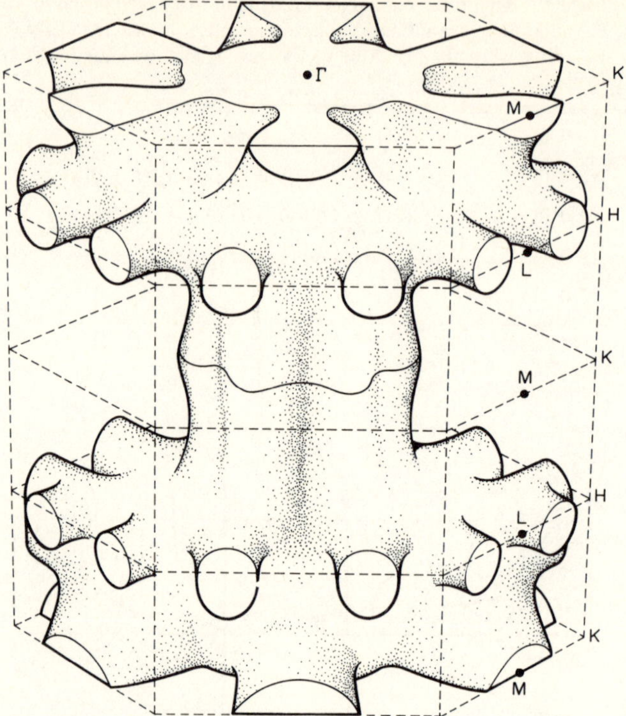

Fig. 3.10 The Fermi hole surface for Tm in the double zone scheme.

moment with the observed extra saturation moment of 0·5 Bohr magnetons is thus difficult, but see Section 4.6. For the same metal the calculated electronic contribution to the specific heat is 4·2 mJ mole^{-1} deg^{-2}. While this is three times larger than the free electron result, it is still a factor of two smaller than the experimental result. Similar results for γ were found for Y, and it has been suggested that the discrepancies could arise due to neglect of electron–phonon contributions [36].

3.4 The relativistic augmented plane wave method

The earliest attempts to include relativistic effects in band structure calculations introduced one or both of the spin–orbit coupling and mass correction terms discussed in Section 1.1.7; see references cited in [14]. More recently relativistic OPW [34] and APW [15] calculations have been made using solutions of the Dirac equation, and have given results which agree favourably with existing experimental data. These plane wave calculations closely follow the non-relativistic ones in principle. For rare earths the relativistic augmented plane wave method (RAPW) has exclusively been employed [15, 20], and only these calculations are described here. A detailed exposition of the

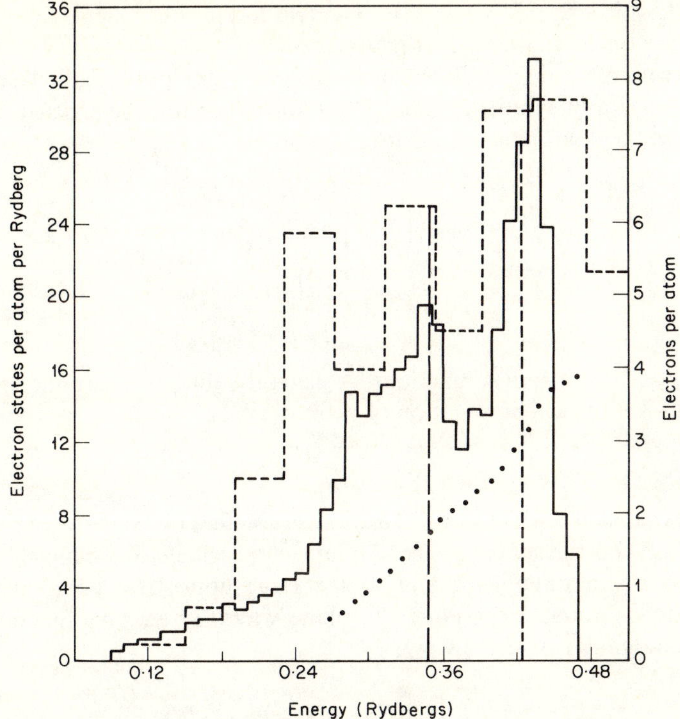

Fig. 3.11 The density of states for Gd calculated by the augmented plane wave method. The broken curve is the non-relativistic result [11] and the solid curve is the result of a relativistic calculation [18]. The dots give the integrated density of states, in units on the right, for the relativistic calculation. Vertical lines indicate Fermi energy.

basic relativistic electron theory required is given by Rose [33].

Relativistic calculations are complicated by the fact that the solutions of the Dirac equation [33],

$$(\boldsymbol{\alpha}\cdot\mathbf{p}+\beta+V)\psi = W\psi \tag{3.20}$$

have four components. In the RAPW method $V(r)$ is a muffin-tin potential. In regions where $V = 0$ the solutions of equation (3.20) are the plane waves

$$\psi_{n,\pm} = \left(\frac{k_n^*+1}{2k_n^*}\right)^{\frac{1}{2}} \begin{bmatrix} \chi_{+\frac{1}{2}} \\ \chi_{-\frac{1}{2}} \\ \dfrac{\boldsymbol{\sigma}\cdot\mathbf{k}_n}{k_n^*+1}\chi_{+\frac{1}{2}} \\ \dfrac{\boldsymbol{\sigma}\cdot\mathbf{k}_n}{k_n^*+1}\chi_{-\frac{1}{2}} \end{bmatrix} e^{i\mathbf{k}_n\cdot\mathbf{r}} \tag{3.21}$$

where $\mathbf{k}_n = \mathbf{k}+\mathbf{K}_n$, $k_n^* = (k_n^2+1)^{\frac{1}{2}}$. The two lower components are approximately v/c times smaller than the upper ones.

For regions inside a unit cell such that $r < r_i$ the potential V is assumed to be spherically symmetric, and (3.20) may therefore be written in polar form, and the solutions are, c.f. equation (1.30)

$$\psi_\kappa^\mu = \frac{1}{(2l+1)^{\frac{1}{2}}}\left\{\begin{array}{l} g_\kappa(r)\begin{bmatrix}(l+\mu+\frac{1}{2})^{\frac{1}{2}}\chi_{+\frac{1}{2}}\,Y_l^{\mu-\frac{1}{2}}\\(l-\mu+\frac{1}{2})^{\frac{1}{2}}\chi_{-\frac{1}{2}}\,Y_l^{\mu+\frac{1}{2}}\end{bmatrix},\\ if_\kappa(r)\begin{bmatrix}-(l-\mu+\frac{1}{2})^{\frac{1}{2}}\chi_{+\frac{1}{2}}\,Y_l^{\mu-\frac{1}{2}}\\(l+\mu+\frac{1}{2})^{\frac{1}{2}}\chi_{-\frac{1}{2}}\,Y_l^{\mu+\frac{1}{2}}\end{bmatrix},\end{array}\right\} \tag{3.22}$$

where g and f are radial functions. A general solution inside the spherical potential is a linear combination of these orbitals.

$$\psi_{n,m} = \sum_{\kappa,\mu} A_{\kappa\mu}^{nm}\psi_\kappa^\mu. \tag{3.23}$$

The expansion coefficients $A_{\kappa\mu}^{nm}$ are chosen so that (3.23) matches the plane waves (3.21) on the sphere $r = r_i$. All four components of ψ cannot be made continuous simultaneously, and so the large upper two components are chosen to be equal. Expanding the plane waves in angular functions one finds, analogously to (3.19)

$$A_{\kappa\mu}^{nm} = 4\pi i^l\left(\frac{k_n^*+1}{2k_n^*}\right)^{\frac{1}{2}} C(l\tfrac{1}{2}, \mu-m, m)\, Y_l^{\mu-m}(\hat{\mathbf{k}}_n)\frac{j_l(k_n r_i)}{g_\kappa(r_i)} \tag{3.24}$$

where the C's are Clebsch–Gordon coefficients.

The crystal wavefunction may be expressed in terms of the RAPW's by expanding in the reciprocal lattice vectors, analogous to equation (3.8), and a secular equation may be set up for $W = E+mc^2$, where here m is the rest mass in the region $V = 0$. The matrix elements of the Dirac Hamiltonian between RAPW'S $\psi_{n,m}$ and $\psi_{N,M}$ are given by [15]

$$\begin{aligned}\langle n, m|\mathscr{H}|N, M\rangle = {}& \{\tfrac{1}{2}(k_N^2+k_n^2)-E\}\delta_{mM}\Big\{\Omega\delta_{nN}-4\pi r_i^2\\ & \frac{j_l[|\mathbf{k}_N-\mathbf{k}_n|r_i]}{|\mathbf{k}_N-\mathbf{k}_n|}\Big\}+4\pi r_i^2\sum_\kappa D_\kappa(nm:NM)\{j_l(k_n r_i)\\ & j_l(k_N r_i)\frac{cf_\kappa(r_i, E)}{g_\kappa(r_i, E)}-\tfrac{1}{2}S_\kappa[k_N j_l(k_n r_i)j_{l'}(k_N r_i)\\ & +k_n j_l(k_N r_i)j_{l'}(k_n r_i)]\},\quad \text{(a.u.)}\end{aligned} \tag{3.25}$$

where

$$\begin{aligned}D_\kappa(nm:NM) = {}& 4\pi\sum_\mu C(l\tfrac{1}{2}; \mu-m, m)C(l\tfrac{1}{2}; \mu-M, M)\,Y_l^{\mu-m\,*}(\hat{\mathbf{k}}_n)\times\\ & Y_l^{\mu-m}(\hat{\mathbf{k}}_N).\end{aligned} \tag{3.26}$$

S_κ means 'sign of κ', and $l' = \kappa - 1(\kappa > 0)$, $l' = -\kappa(\kappa < 0)$. Terms have been neglected which are of order $e^4 = (1/137)^2$ compared with unity. The first two terms of (3.25) are similar to the non-relativistic results (3.19), except that here there is a sum over both possible spin orientations, and the logarithmic derivative of R is replaced by the ratio of the two radial functions f_κ and g_κ. The major complication compared with the non-relativistic case is that the D_κ are complex quantities, and it is therefore necessary to solve a complex secular equation for the band energies E.

3.5 Relativistic calculations for the heavy rare earths

Relativistic augmented plane wave calculations for the heavy rare earth metals Gd, Dy and Er, and also Lu, have been made by Keeton and Loucks [16, 18]. A calculation for Tb has been made by Jackson [39] using identical methods. The details in the following discussion are taken from Keeton and Loucks. They constructed the spherical symmetric potential $U(r)$ about a lattice site from atomic charge densities which were themselves obtained from relativistic self-consistent field calculations for the atoms [35]. Exchange was included using the Slater $\rho^{1/3}$ approximation, equation (1.3), both for the atomic and muffin-tin potentials. As in all band structure calculations, there is some uncertainty in which atomic configurations are most appropriate in constructing the potential, and Keeton and Loucks employed two potentials for both Dy and Er to investigate how sensitive their results were to the potential. In these two cases the atomic configurations used were $4f^9 5d^1 6s^2$ and $4f^{10} 6s^2$, denoted DyI and DyII respectively, and $4f^{11} 5d^1 6s^2$ and $4f^{12} 6s^2$ denoted ErI and ErII respectively. For Gd and Lu the atomic configurations $4f^7 5d^1 6s^2$ and $4f^{14} 5d^1 6s^2$ respectively were employed.

Energy bands were calculated at 60 points in the 1/24 zone using 28 vectors in the reciprocal lattice expansion. The results of the relativistic calculations differ significantly from the non-relativistic ones at energies just below the Fermi energy, as is illustrated for Lu in Fig. 3.12, and hence may give important differences in the predicted Fermi surfaces. The main relativistic effect is the lifting of degeneracies by the spin–orbit interaction. The splittings so produced may be understood qualitatively from group theory, but the exact positions of the levels is determined by all the relativistic effects. These are all included in RAPW calculations since the Dirac equation is employed from the outset.

In computing the density of states and Fermi surface the number of reciprocal lattice vectors used was increased to 32 in order to improve convergence of the eigenvalues to within 0·005 Ry. Results for the Fermi surface of Gd is illustrated in Fig. 3.13 where its intersections with the surfaces of the 1/24 zone are shown. For this series of elements the major differences in Fermi surfaces occur around the symmetry points M and L, and Loucks has broadly referred to the surfaces as being of either of the extreme types

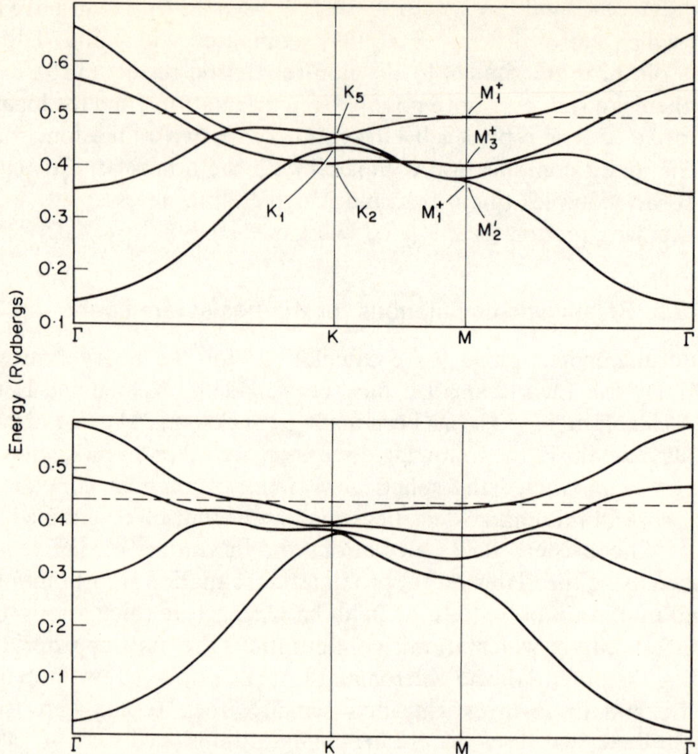

Fig. 3.12 A comparison of non-relativistic energy bands (upper) and relativistic energy bands (lower) for Lu metal. The broken lines indicate the approximate positions of the Fermi energy.

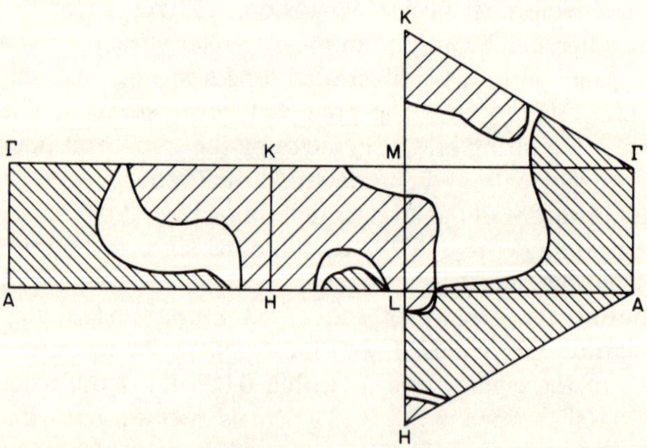

Fig. 3.13 Intersections of the Fermi surface with the faces of the 1/24th zone for Gd. The broad cross-hashing refers to holes in the third zone, and the other shaded region refers to electrons in the fourth zone.

Gd or Y. In the double zone scheme, the Gd-type surface is characterized by there being one arm at M, and two arms near L; cf. Fig. 3.10. If the arm at M is removed, and the two arms at L joined by a 'thin' sheet, so that it looks like a webbing, then the surface becomes Y-type; see Fig. 3.9. The transition from Gd to Y like Fermi surfaces is able to account for many observed properties of the heavy rare earths, as will be discussed in the next chapter. The variation of the webbing feature for the different metals is shown in Fig. 3.14.

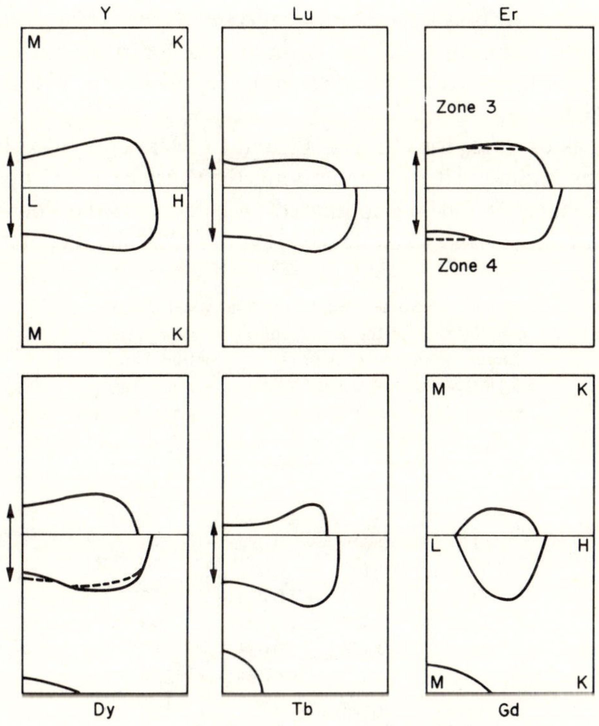

Fig. 3.14 Cross-sections of the Fermi surfaces of different metals illustrating the webbing feature. The arrows are the wavevectors associated with the periodic spin structures in the ordered state, and are discussed in Section 4.3.2. (After refs. [18] and [39].)

The energy bands near the Fermi energy are very flat, so that the Fermi surface depends sensitively upon them, and it might be expected that the exact choice of potential would be important. In the case of Er the results for ErI and ErII indicate that this is not so, but for Dy the results show a marked dependence upon the potential used, particularly near M, where for DyI there is an arm, while for DyII there is none. It appears therefore that Dy is near the transition point between Gd and Y types, and that the choice of potential may not be so critical for the other elements.

In passing it should be noted that the fact that the bands are so flat at energies near E_f can have important consequences for the Fermi surfaces in the *ordered state.* In the simplest case of ferromagnetic order, as in Gd, the effect of the exchange interaction between the ionic moments and the conduction electrons is to split the spin-up and spin-down bands [41]. The Fermi energy readjusts accordingly, thereby giving rise to the spin polarization. For Gd, at low temperatures, the splitting is $\sim 0{\cdot}004$ Ry. It is clear from Fig. 3.8 for Tm that changes in E_f of this magnitude for the two bands will result in very dissimilar Fermi surfaces for the spin-up and spin-down electrons, and the splitting is therefore important in determining the magnetic properties. The situation is more complicated for a periodic magnetic structure because then extra gaps occur in the energy bands. This is discussed in Section 4.3.3.

The densities of states for Gd, Dy, Er and Lu were computed by Keeton and Loucks by finding, with interpolations, the energies at 567 points in the 1/24 zone. Results for Gd are compared with those of the non-relativistic calculation in Fig. 3.11 and it can be seen that there are significant differences.

Table 3.2 Some results of relativistic APW calculations for the Fermi energy in Ry, E_f, the energy at the bottom of the conduction band E_0 in Ry, and the density of states at the Fermi energy $N(E_f)$ in states/Ry/atom. After refs. [18] and [39].

	E_f	E_0	$N(E_f)$
Gd	0·425	0·082	28·5
Tb	0·452	0·08	32·4
DyI	0·450	0·075	27·7
DyII	0·505	0·104	24·3
ErI	0·452	0·053	24·3
ErII	0·519	0·091	23·6
Lu	0·453	0·018	25·5

The Fermi energies for the four metals are shown in Table 3.2. In particular for Gd the width of the conduction band, 0·34 Ry, is larger than the value of 0·25 Ry obtained with the APW method, and also the relativistic value of $N(E_f) = 2{\cdot}1$ electrons per atom eV^{-1} is larger than the non-relativistic value of 1·8 eV^{-1}, although still too small to account for the observed electronic specific heat.

3.6 Positron annihilation experiments

The calculated band structures and Fermi surfaces considered in this chapter have been those of the heavy rare earth metals in the paramagnetic state, for it is in this state that the relationship between Fermi surface and magnetic structures may be understood. In order to investigate these Fermi surfaces experimentally it is necessary to work at relatively high temperatures, above the magnetic ordering temperatures, and this severely restricts the number of techniques which can be used. A further problem is that rare earth metal crystals are relatively impure, so that many other otherwise possible conventional methods can not be employed. Because of these difficulties, the only *direct* experimental evidence to date which can provide a check of the validity of the calculated Fermi surfaces, is that obtained from positron annihilation studies by Williams and Mackintosh [37], and these are briefly described here. The general subject of positron annihilation in solids has been reviewed, for example, by Wallace [38].

On entering a solid it is generally assumed that a positron rapidly loses energy and becomes thermalized. On annihilating with an electron, energy $\simeq 2mc^2$ is released, and in order to conserve momentum two γ-ray photons are usually emitted in almost opposite directions. If the annihilating particles have relative momentum $\mathbf{p}$, the angle between the γ-rays, projected onto a plane parallel to the z-axis, is $\theta = p_z/mc$. If the whole momentum $\mathbf{p}$ is ascribed to the electron, then it is possible to investigate the annihilation of electrons with a particular p_z value by counting the pairs of γ-rays detected, simultaneously, which have angular correlation θ.

The probability that an electron with wavevector $\mathbf{k}$ will annihilate with a positron of wavevector $\mathbf{k} = 0$, to produce two photons of total momentum $\mathbf{p}$, is proportional to [38]

$$F(\mathbf{p}, \mathbf{k}) = \left|\int \psi(\mathbf{k}, \mathbf{r})\psi_+(\mathbf{r})\, e^{-i\mathbf{p}\cdot\mathbf{r}} d\mathbf{r}\right|^2$$

where ψ and ψ_+ are the electron and positron wavefunctions respectively. The quantity usually observed experimentally is the rate of counting at slits parallel to the z-axis, with given angular correlation θ, and this is proportional to:

$$N(\theta) = \iint dp_x dp_y \sum_{|\mathbf{k}|<k_f} F(\mathbf{p}, \mathbf{k}) \tag{3.27}$$

It is clear from the outset that the information to be gained from this function is extremely limited. Even assuming that ψ_+ is accurately known, it still only contains information about the wavefunctions ψ, which will be distorted by the positrons, and the states occupied, through averages over the unit cell and the Fermi sea. In the simplest approximation, in which ψ and ψ_+ are respectively plane waves and a constant everywhere, and the Fermi surface is a sphere, it is readily shown that $N(\theta)$ takes the inverted parabolic form

$$N(\theta) \propto (p_f^2 - p_z^2) \tag{3.28}$$

where $p_z = mc\theta$. In general one expects a tail at high θ due to the effect of the atomic core electrons. It is from this kind of expression that some information, namely p_f in the z-direction, may be obtained. Experimental results are shown for $N(\theta)$ for Gd and Y, along the c-axis, in Fig. 3.15, and it is clear that the free electron expression (3.28) is inadequate for both.

In the case of Y, a detailed calculation of $N(\theta)$ has been made by Loucks [20], based upon his APW calculation for that metal, and the results (Fig. 3.15b) are in qualitative agreement with the observed angular distribution of photons. The theory is based upon the independent particle model, in which the mutual Coulombic attraction of electron and positron is neglected. A (repulsive) muffin-tin potential was constructed for the positron using the same APW sphere as for the electrons, but neglecting exchange. Outside the sphere ψ_+ was taken to be unity, and inside a ground state solution was obtained numerically. The latter, of course, tends to zero as the distance from the nucleus decreases, since the positron is repelled by the positive ionic charge. The electron wavefunction ψ was approximated by a linear combination of APW'S containing up to 32 terms, and $N(\theta)$ finally computed. The distinctive feature of the distribution is the hump at about $\theta = 3$ mrad. This arises from the contribution of the d-band, which provides a large number of occupied states for values of k_z near the Fermi momentum. In

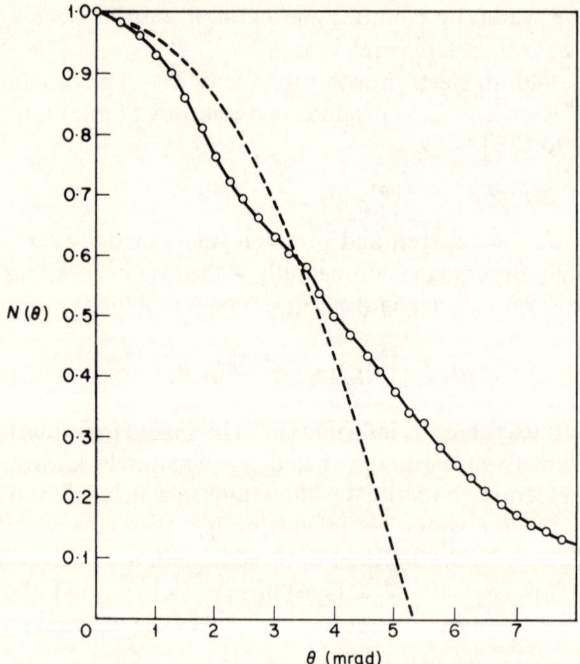

Fig. 3.15(a) The angular distribution of photon coincidences $N(\theta)$ along the c-axis for (a) Gd, experimental results and (b) Y, where the dots are experimental points, and the solid line is from theory. In both cases $T = 300°K$ and the dashed curve is the free electron parabola. (After ref. [37].)

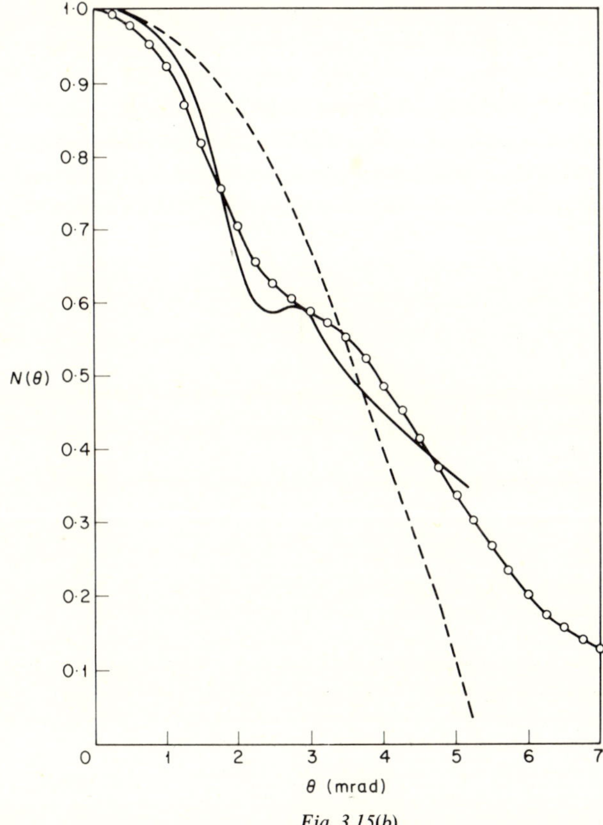

Fig. 3.15(b)

practice, because of the finite resolution of any experiment, the observed hump will be less pronounced, and this must be corrected for. This has been done in Fig. 3.15b. There also exists the possibility that the hump is partly smeared out by the effect of imperfections.

The experimental $N(\theta)$ distribution for Gd differs from that of Y in that it decreases relatively much more slowly with angle at small θ, and the hump is much less marked. The distributions for Er, Ho, Dy and Tb are intermediate between those of Y and Gd, lending support to the results of Louck's calculations, which indicate a gradual transition in Fermi surface type between those two elements.

The agreement here between experiment and theory encourages the belief that the calculated Fermi surfaces are realistic ones. It must, however, be borne in mind that the positron annihilation experiments are not sensitive to the detailed structure of the Fermi surface. Further, the assumption of an independent particle model may give rise to appreciable errors in the theory, since the electronic structure of the metal is certainly distorted in the locality of the positron, so as to shield its positive charge.

It would be of considerable interest to produce rare earths of sufficient purity for conventional experiments, such as the de Haas van Alphen effect, to be made. However, these measurements must be made at low temperatures, and there will still exist a problem of investigating the Fermi surfaces in the paramagnetic regions. In the meanwhile there are other experimental results, e.g. the magnetic properties, which can be explained in terms of the calculated Fermi surfaces, and which lend further indirect support for the validity of these surfaces; see Section 4.3.

References

[1] A. Ruderman and C. Kittel, *Phys. Rev.*, **96**, 99 (1954).
[2] K. Yosida, *Phys. Rev.*, **106**, 893 (1957).
[3] T. Kasuya, *Progr. Theor. Phys.* (Kyoto) **14**, 45 (1956).
[4] H. Nigh, S. Legvold and F. H. Spedding, *Phys. Rev.*, **132**, 1092 (1963).
[5] L. D. Jennings, R. E. Miller and F. H. Spedding, *J. Chem. Phys.*, **33**, 1849 (1960).
[6] A. Berman, M. W. Zemansky and H. A. Borse, *Phys. Rev.*, **109**, 70 (1958).
[7] O. V. Lounasmaa, *Phys. Rev.*, **126**, 1352, 1357 (1962).
[8] O. V. Lounasmaa, *Phys. Rev.*, **133**, A219 (1964).
[9] W. E. Evenson and S. H. Liu, *Phys. Rev. Letts.*, **21**, 432 (1968).
[10] J. M. Ziman, *Principles of the Theory of Solids*, C.U.P. London (1964).
[11] J. O. Dimmock and A. J. Freeman, *Phys. Rev. Letts.*, **13**, 750 (1944).
[12] A. J. Freeman, J. O. Dimmock and R. E. Watson, *Phys. Rev. Letts.*, **16**, 94 (1966).
[13] R. E. Watson, A. J. Freeman and J. O. Dimmock, *Phys. Rev.*, **167**, 497 (1968).
[14] T. L. Loucks, *Augmented Plane Wave Method*, W. A. Benjamin Inc., New York (1967).
[15] T. L. Loucks, *Phys. Rev.*, **139**, A1333 (1965).
[16] S. C. Keeton and T. L. Loucks, *Phys. Rev.*, **146**, 429 (1966).
[17] O. K. Andersen and T. L. Loucks, *Phys. Rev.*, **167**, 551 (1968).
[18] S. C. Keeton and T. L. Loucks, *Phys. Rev.*, **168**, 672 (1968).
[19] G. S. Fleming and T. L. Loucks, *Phys. Rev.*, **173**, 685 (1968).
[20] T. L. Loucks, *Phys. Rev.*, **144**, 504 (1966).
[21] H. Jones, *The Theory of Brillouin Zones and Electronic States in Crystals*, North Holland Publishing Co. §33, 2nd ed. (1962).
[22] C. Herring, *J. Franklin Institute*, **233**, 525 (1942).
[23] R. J. Elliott, *Phys. Rev.*, **92**, 280 (1954).
[24] J. Callaway, *Energy Band Theory*, Academic Press (1964).
[25] C. Herring, *Phys. Rev.*, **57**, 1169 (1940).
[26] J. C. Philips and L. Kleinman, *Phys. Rev.*, **116**, 287 (1959).
[27] J. C. Slater, *Phys. Rev.*, **51**, 846 (1937).
[28] J. Korringa, *Physica*, **13**, 392 (1947).
[29] W. Kohn and N. Rostoker, *Phys. Rev.*, **94**, 1411 (1954).
[30] R. W. Williams, T. L. Loucks and A. R. Mackintosh, *Phys. Rev. Letts.*, **16**, 168 (1966).
[31] A. J. Freeman and R. E. Watson, *Phys. Rev.*, **127**, 2058 (1962).
[32] F. Herman and S. Skillman, *Atomic Structure Calculations*, Prentice Hall, Inc., New Jersey (1963).
[33] M. E. Rose, *Relativistic Electron Theory*, John Wiley and Sons, Inc., New York (1961).
[34] P. Soven, *Phys. Rev.*, **137**, A1706 (1965).
[35] D. Liberman, J. T. Waber and D. T. Cromer, *Phys. Rev.*, **137**, A27 (1965).
[36] K. Krebs, *Phys. Letts.*, **6**, 31 (1963).
[37] R. W. Williams and A. R. Mackintosh, *Phys. Rev.*, **168**, 679 (1968).

[38] P. R. Wallace, *Solid State Phys.*, **10**, 1 (1960).
[39] C. Jackson, *Phys. Rev.*, **178**, 949 (1969).
[40] G. S. Fleming, S. H. Liu and T. L. Loucks, *J. Appl. Phys.*, **40**, 1285 (1969).
[41] R. E. Watson, A. J. Freeman and J. P. Dimmock, *Phys. Rev.*, **167**, 497 (1968).
[42] W. A. Harrison, *Phys. Rev.*, **118**, 1190 (1960).

4

Magnetic Properties

A vast literature now exists dealing with the magnetic properties of the rare earth metals and intra-rare earth alloys. This includes the details of measurements of almost all magnetic phenomena and of other properties which are influenced by the behaviour of the spin systems in both the ordered and disordered states. There exist theoretical interpretations of these results, several catalogues of observations pertaining to the magnetically ordered state, and one or two reviews attempting to draw together in a single analysis the dominant features relating to the experimental and theoretical work on these metals [1–3, 7, 11–13, 122, 154, 199].

Of the seventeen metals being considered within this work *eleven* exhibit some form of ordered spin structure in the temperature range from 4·2°K to 300°K. It is these ordered states that have been the focus of the majority of work carried out on the pure metals during the past decade, since specimens of reasonable purity became available to us due largely to the extensive extraction and purification programme conducted by Spedding and his colleagues at Ames, Iowa. Relatively little work has been carried out in the paramagnetic state since the early coverage of many of the metals by Lock [4, 5], Leipfinger [6], and Arajs and Colvin [7] and a considerable gap exists in the literature now that single crystals are available of much higher purity than those used in the earlier work. The present status of our knowledge and understanding of all the magnetic and related properties will be discussed in the following.

4.1 The RKKY interaction

It is generally supposed that a rare earth metal is composed of the trivalent ions embedded in a sea of conduction electrons. That the 4*f* electrons are tightly bound is supported by results of measurements of $\langle r^{-3} \rangle$ from hyperfine fields, and of high temperature magnetic susceptibilities, both of which are similar to those of the free ions. Typically, for the heavy series, the mean radius of the 4*f* shell is one tenth of the interionic spacing, so that significant direct overlap between 4*f* orbitals on neighbouring ions is negligible, and they can play little in chemical bonding. It is usually assumed that the 4*f* electrons form a narrow energy band which is below the conduction band and is unable to mix with it, but see the following subsection.

Although the lattice spacings between ions in the metals are not so different from those in the salts, the magnetic couplings between them are two orders of magnitude stronger in the metals, and magnetic order is possible up to 300°K. The exchange interaction responsible for this coupling between $4f$ orbitals is an indirect one which involves the polarization of the conduction electrons. Such an interaction is often oscillatory and has long range, and is therefore capable of giving rise to a variety of periodic spin structures, such as are indeed observed in the heavy rare earth metals. The theory is associated with the names of Ruderman, Kittel, Kasuya and Yosida [8–18], (RKKY), and is described in some detail in the following sections. Reviews have been given by, for example, Rocher [11], de Gennes [12], Kasuya [13] and Kittel [14].

The other interactions between ions are small. Of them the most important is the Coulomb electrostatic interaction, and the dominant dipole term may be incorporated in a crystalline field. Its effect is to align the anisotropic charge clouds of the $4f$ electrons, and hence their magnetic moments, in some preferred direction relative to the crystal axes. Together with a much smaller interaction between quadrupole moments of the $4f$ electrons it gives rise to magnetocrystalline anisotropy, which in turn is important in determining the precise forms of the observed spin structures in the ordered state [15, 16]. All other interactions, e.g. direct exchange, or coupling between moments and the anisotropic part of the polarized conduction electron cloud are usually assumed small.

4.1.1 *The s–f interaction*

The exchange integral between the $4f$ electrons of an ion and a conduction electron, known as the s–f interaction, may be expected to take the form

$$A(\mathbf{k},\mathbf{k}') = \sum_{i,\|\,\text{spins}} \iint \psi^*(\mathbf{r}',\mathbf{k})\Psi_{4f}(\mathbf{r}_1,\mathbf{r}_2,\ldots\mathbf{r}_i\ldots\mathbf{r}_n)\frac{e^2}{|\mathbf{r}_i-\mathbf{r}'|}\times$$
$$\psi(\mathbf{r}_i,\mathbf{k}')\Psi_{4f}(\mathbf{r}_1,\mathbf{r}_2,\ldots\mathbf{r}'\ldots\mathbf{r}_n)\mathrm{d}\mathbf{r}'\mathrm{d}\mathbf{r}_1,\mathrm{d}\mathbf{r}_2,\ldots\mathrm{d}\mathbf{r}_n, \qquad (4.1)$$

where $\psi(\mathbf{r},\mathbf{k})$ is the Bloch wavefunction of the conduction electron. The $4f$ electron wavefunction Ψ_{4f} consists of a determinant of the central field single particle functions $R_{4f}(r_i)Y_3^m(\hat{\mathbf{r}}_i)$. By expanding both the Bloch waves (cf. Section 3.2.2) and the factor $|\mathbf{r}_i-\mathbf{r}'|^{-1}$ in spherical harmonics, a multipole expansion of $A(\mathbf{k},\mathbf{k}')$ may be obtained in which the odd order terms are usually zero because of inversion symmetry. The zero order term was first evaluated by Liu [17] using plane waves for $\psi(\mathbf{r},\mathbf{k})$, and he showed it could be written in the form of an isotropic interaction between the conduction electron spin $\mathbf{s}$ and the local ionic spin $\mathbf{S}$, i.e. in the form $A(\mathbf{k},\mathbf{k}')\mathbf{s}\cdot\mathbf{S}$. This leading term is a good approximation provided the wavelength of the conduction electron ($\sim k_f$) is large compared with the size of the $4f$ shell ($\langle r_{4f}\rangle$), but for the rare earths this condition is not well satisfied ($k_f\langle r_{4f}\rangle$

$\sim 0{\cdot}5$). Consequently Kaplan and Lyons [18] considered the second, anisotropic term in the multipole expansion and, assuming a wavevector dependence $A(\mathbf{k}, \mathbf{k}') = A(\mathbf{k}-\mathbf{k}')$, they find it is only $\sim 10\%$ of the isotropic term. However, Specht [19] has shown that the removal of the restriction upon the wavevector dependence increases the magnitude of this second term by a factor of four.

The conventional exchange energy (4.1) always favours the parallel alignment of the $4f$ spins and the conduction electrons spins, so that the conduction electron moment adds to the ionic moment. For the rare earth metals this is observed to be the case. However, experiments on alloys, such as rare earth–Al_2 alloys [20], indicate that in some circumstances the s–f interaction produces antiparallel alignment of $\mathbf{S}$ and $\mathbf{s}$. Watson *et al* [21, 22] have explained this in terms of interband mixing between the $4f$ states and the conduction electron states. The case of Gd has been considered in detail since this has a half-filled $4f$ shell and its local moment is spherical (^8S). The occupied $4f$ states which are situated in a narrow band well below $(-U/2)$ the conduction band and all have up spin, say, may mix with an unoccupied spin-up (plane wave) conduction electron state above the Fermi surface. That is to say, in general there will be non-zero matrix elements $\langle \Psi_{4f}|\mathscr{H}|\psi_{\mathbf{k}}\rangle$ where, in principle, $\mathscr{H}$ is the total Hamiltonian for the crystal. The effect of this mixing is to raise the conduction electron energy at the Fermi surface. Similarly an occupied spin-down conduction electron state may mix with the unoccupied spin-down $4f$ electron states which are situated well above $(+U/2)$ the conduction band, thereby lowering the energy of the conduction electron. The net effect is to lower the energies of the spin-down conduction electrons with respect to the spin-up electrons, which is opposite to the effect of conventional exchange. The interband exchange interaction, $A'(\mathbf{k}, \mathbf{k}')$ say, is the sum of two terms of the form

$$\sum_{f\,\text{electrons}} \langle \mathbf{k}|\mathscr{H}|f\rangle\langle f|\mathscr{H}|\mathbf{k}'\rangle/\{U/2 \pm E(\mathbf{k})\},$$

where constants are omitted. If $\mathscr{H}$ is spherically symmetric then the matrix elements are only non-zero for the $l = 3$ term in the plane wave expansion, and the interaction has the anisotropic form $A'(\mathbf{k}, \mathbf{k}') = Y_3^0(\mathbf{k}\cdot\mathbf{k}'/kk')G(k, k')$, with no isotropic contribution. A major difficulty in comparing quantitative predictions of the theory with experiment is that the interaction is much weaker than conventional exchange. This is because $U/2 \sim 10$ eV so that the energy denominators are large.

The use of plane waves in estimating $A(\mathbf{k}, \mathbf{k}')$ and $A'(\mathbf{k}, \mathbf{k}')$ is such a severe approximation (cf. Chapter 3) that the results of the theories discussed above may be in very serious error. There is also another source of error [23] in neglecting the effect of the closed $5s^2$ and $5p^6$ shells which lie outside the $4f$ electrons. These shells are expected to be perturbed both electrostatically and by interband mixing so as to screen the $4f$ orbitals from the conduction electrons. In view of the uncertainties arising from the various approximations the exchange integral is usually assumed to be isotropic and a function of $|\mathbf{k}'-\mathbf{k}| = q$ only. In the RKKY theory, which has been most widely employed, $A(q)$ is taken to be a constant, A_0 say. Here we shall be primarily concerned with the RKKY interaction.

For some applications it is more convenient to use the exchange integral in direct **r**-space, i.e. to use

$$A(\mathbf{r}) = \sum_{\mathbf{q}} A(\mathbf{q})\,\mathrm{e}^{i\mathbf{q}\cdot\mathbf{r}}, \tag{4.2}$$

where the wavevector $\mathbf{q} = \mathbf{k}'-\mathbf{k}$ is in the extended zone scheme, anticipating that the conduction electrons will be represented by plane waves. The exchange interaction between a conduction electron at **r**, spin **s**, and an ionic spin **S** at position **R** may then be written

$$\mathscr{H}_{\mathrm{sf}} = -A(\mathbf{r}-\mathbf{R})\mathbf{s}\cdot\mathbf{S}. \tag{4.3}$$

In the RKKY approximation $A(\mathbf{r}-\mathbf{R}) = A_0\delta(\mathbf{r}-\mathbf{R})$.

4.1.2 *Polarization of the conduction electrons*

The effect of a single ionic spin **S** upon the conduction electrons may be calculated using perturbation theory. Because $\mathscr{H}_{\mathrm{sf}}$ is spin dependent, conduction electrons of different spin respond to the perturbation differently. For example if $A(\mathbf{r}) < 0$ electrons of up spin (+) prefer to be near **S**, while those of down spin (−) do not. As a result the electron gas is polarized.

In first order perturbation theory the wavefunctions are

$$\psi_{\pm}^{(1)}(\mathbf{r},\mathbf{k}) = \psi_{\pm}(\mathbf{r},\mathbf{k}) + \sum_{\mathbf{k}',s} \psi_s(\mathbf{r},\mathbf{k}')\frac{\langle\psi_s(\mathbf{r},\mathbf{k}')|\mathscr{H}_{\mathrm{sf}}|\psi_{\pm}(\mathbf{r},\mathbf{k})\rangle}{E(\mathbf{k})-E(\mathbf{k}')}, \tag{4.4}$$

where $s = \pm$ refers to the electronic spin, and $E(\mathbf{k})$ is the unperturbed energy of an electron in state **k**. The densities of electrons with + and − spins are the sums of $|\psi_{\pm}^{(1)}(\mathbf{r},\mathbf{k})|^2$ over all filled states, and the spin polarization $P(\mathbf{r})$ is the difference between them, i.e.

$$P(\mathbf{r}) = \sum_{\mathbf{k}} f(\mathbf{k})\{|\psi_{+}^{(1)}(\mathbf{r},\mathbf{k})|^2 - |\psi_{-}^{(1)}(\mathbf{r},\mathbf{k})|^2\}, \tag{4.5}$$

where $f(\mathbf{k})$ is the occupation number for states **k**. Assuming that the ψ's are plane waves, $P(\mathbf{r})$ may readily be evaluated to first order in the exchange constant.

Working in **r**-space, and at $T = 0°\mathrm{K}$, so that only states $|\mathbf{k}| \leq k_{\mathrm{f}}$ are occupied, (4.4) may be evaluated in terms of the free electron Green's function [24] $\exp(ik|\mathbf{r}-\mathbf{r}'|)/4\pi|\mathbf{r}-\mathbf{r}'|$ to give

$$P(\mathbf{r}) = \frac{9\pi S_z Z^2}{2\Omega^2 E_{\mathrm{f}}}\int \phi(2k_{\mathrm{f}}|\mathbf{r}-\mathbf{r}'|)A(\mathbf{r}')\mathrm{d}\mathbf{r}', \tag{4.6}$$

where Z is the number of conduction electrons per atom, Ω is the atomic volume, and

$$\phi(x) = \frac{\sin x - x \cos x}{x^4}. \tag{4.7}$$

It can be seen that the RKKY result, which follows trivially with $A(\mathbf{r}) = A_0\delta(\mathbf{r})$ has a long range oscillatory behaviour falling off as $\cos(2k_f r)/r^3$ at large r. The oscillations arise when the Fermi surface is sharply defined by the Fermi–Dirac function $f(\mathbf{k})$. For a non-degenerate electron gas, where the Fermi surface is not sharp, there are no oscillations in $P(\mathbf{r})$. In the limit $r \to 0$, ϕ diverges, indicating that the contact potential $A_0\delta(r)$ is not physically realistic, i.e. $A(\mathbf{r})$ should have a finite range which is the same as saying $A(\mathbf{q})$ should not be a constant. This divergence does not affect the interaction energy between two localized spins and the oscillations at larger r which are important in determining magnetic structures are reasonably well predicted. Watson and Freeman [23] have shown that the introduction of a more realistic exchange interaction $A(q)$ not only removes the divergence at $r \to 0$ but near that point it alters the sign of $P(\mathbf{r})$ (to negative) from that given by (4.6).

In generalizing equation (4.6) to include $A(\mathbf{q})$, and in order to work with Bloch states in the reduced zone scheme, it is convenient to work in $\mathbf{q}$-space. For the particular case of free electrons (4.4) and (4.5) give

$$P(\mathbf{r}) = \sum_{\mathbf{q}} A(\mathbf{q}) \sum_{\mathbf{k}} \left\{ \frac{f(\mathbf{k}) - f(\mathbf{k}+\mathbf{q})}{E(\mathbf{k}) - E(\mathbf{k}+\mathbf{q})} \right\} e^{i\mathbf{q}\cdot\mathbf{r}} \tag{4.8}$$

$$= \sum_{\mathbf{q}} A(\mathbf{q}) \chi(\mathbf{q}) e^{i\mathbf{q}\cdot\mathbf{r}}, \tag{4.9}$$

where this defines $\chi(\mathbf{q})$. In evaluating $\chi(\mathbf{q})$ explicitly, care is needed when $q \to 0$, as then the energy denominator vanishes. However, it can be shown [10, 25] that if the summation is replaced by an integration, the correct result is obtained by taking its principal value. Then:

$$\chi(\mathbf{q}) = \frac{mk_f}{4\pi^2\hbar^2} \left(1 + \frac{4k_f^2 - q^2}{4k_f q} \ln \left| \frac{q + 2k_f}{q - 2k_f} \right| \right), \tag{4.10}$$

and is just the Fourier transform of $\phi(2k_f r)$:

$$\sum_{\mathbf{q}} \chi(\mathbf{q}) e^{i\mathbf{q}\cdot\mathbf{r}} = \frac{9\pi Z^2}{2\Omega^2 E_f} \phi(2k_f r). \tag{4.11}$$

More will be said of the physical significance of $\chi(\mathbf{q})$ in Section 4.1.4.

4.1.3 *The exchange interaction*

The polarization produced by one ionic spin at $\mathbf{R}_i$ will interact with another spin at $\mathbf{R}_j$ through $\mathscr{H}_{sf}$. Using second order perturbation theory, the effective exchange interaction between the two spins is

$$\mathscr{H}_{ij} = -\sum_{\mathbf{k},\mathbf{k}'} |\langle \mathbf{k}'|\mathscr{H}_{\mathrm{sf}}|\mathbf{k}\rangle|^2 \left\{\frac{f(\mathbf{k})-f(\mathbf{k}+\mathbf{q})}{E(\mathbf{k})-E(\mathbf{k}+\mathbf{q})}\right\}, \tag{4.12}$$

where the occupation numbers are again such as to restrict intermediate states to $|\mathbf{k}'| > k_{\mathrm{f}}$, and occupied states to $|\mathbf{k}| \leq k_{\mathrm{f}}$. Substituting for $\mathscr{H}_{\mathrm{sf}}$ from (4.2) and (4.3) and using plane waves, $\mathscr{H}_{ij}$ may be written as

$$\mathscr{H}_{ij} = -\mathscr{J}(\mathbf{R}_i - \mathbf{R}_j)\mathbf{S}_i \cdot \mathbf{S}_j \tag{4.13}$$

where

$$\mathscr{J}(\mathbf{R}_i - \mathbf{R}_j) = \sum_{\mathbf{q}} A^2(\mathbf{q})\chi(\mathbf{q})\,\mathrm{e}^{i\mathbf{q}\cdot(\mathbf{R}_i - \mathbf{R}_j)}. \tag{4.14}$$

In the RKKY approximation $A(\mathbf{q}) = A_0$, so that by (4.11)

$$\mathscr{J}(\mathbf{R}_i - \mathbf{R}_j) = \frac{9\pi Z^2 A_0^2}{2\Omega^2 E_{\mathrm{f}}}\,\phi(2k_{\mathrm{f}}|\mathbf{R}_i - \mathbf{R}_j|). \tag{4.15}$$

This is a function which behaves like the polarization, being in fact just $A_0 P(\mathbf{R}_i - \mathbf{R}_j)$.

The exchange interaction is always between the *spins* **S**. On the other hand the state of a rare earth ion is specified by its total angular momentum **J** and it is then necessary to project **S** onto **J**, as was discussed in Chapter 1 for the rare earth salts. Explicitly it can be seen that on eliminating **L** between $\mathbf{L}+2\mathbf{S} = g_J\mathbf{J}$ and $\mathbf{L}+\mathbf{S} = \mathbf{J}$ the spin may be replaced by

$$\mathbf{S} = (g_J - 1)\mathbf{J}$$

The factor $(g_J - 1)$ is negative for the light series and positive for the heavy series, and in consequence it should be noted that the interaction between a light and a heavy ion has opposite sign to the $\mathscr{J}$ between them.

The interaction $\mathscr{H}_{ij}$ has the Heisenberg form and may be employed within the Weiss molecular field theory to give for the paramagnetic Curie temperature, for one kind of rare earth ion:

$$k\theta_{\mathrm{p}} = \frac{3\pi Z^2 A_0^2 (g_J - 1)^2 J(J+1)}{\Omega^2 E_{\mathrm{f}}} \sum_{\mathbf{R}_i \neq \mathbf{R}_j} \phi(2k_{\mathrm{f}}|\mathbf{R}_i - \mathbf{R}_j|). \tag{4.16}$$

This expression has commonly been employed to explain the observed Curie temperatures of the rare earth metals, and of some alloys and compounds. The lattice sums involved have been computed by Mattis [26] for the simple lattices as functions of the electron density, and they are oscillatory functions of this parameter. As a result θ_{p} depends sensitively upon the density and in some alloys it may even change sign as the electron concentration is varied over a small range [27].

The interaction between two ions is much more complex if the non-isotropic terms are included in $\mathscr{H}_{sf}$. For example Specht [19] has obtained the result

$$\mathscr{H}_{ij} = [\mathscr{J}'(\mathbf{R}_i-\mathbf{R}_j)+F(\mathbf{R}_i-\mathbf{R}_j)\{(\mathbf{L}_i\cdot\hat{\mathbf{R}}_{ij})^2+(\mathbf{L}_j\cdot\mathbf{R}_{ij})^2-\tfrac{2}{3}L(L+1)\}]\mathbf{S}_i\cdot\mathbf{S}_j, \quad (4.17)$$

where a single ion anisotropy term has been neglected. The functions $\mathscr{J}'$ and F are similar in form to $\mathscr{J}$ given by equation (4.15) and $\hat{\mathbf{R}}_{ij}$ is the unit vector in the direction joining the two ions. He finds that for large separations $|\mathbf{R}_i-\mathbf{R}_j|$, the anisotropic contribution can be of the same magnitude as the isotropic term, although its precise value is sensitive to k_f. However, on summing the interaction over all lattice sites to obtain the total exchange energy, and as required for example in equation (4.16), the anisotropic contribution becomes neligible.

4.1.4 *The generalized susceptibility* $\chi(\mathbf{q})$

The function $\chi(\mathbf{q})$ completely describes the response of a free electron gas to a small perturbation. That this is so for localized spins is clear from (4.9) and (4.12). As a further example consider the application of a non-uniform magnetic field which may be Fourier analysed as

$$H(\mathbf{r}) = \sum_{\mathbf{q}} h_{\mathbf{q}} \cos(\mathbf{q}\cdot\mathbf{r}). \quad (4.18)$$

It interacts with the electrons through the Zeeman term $\mu_B\sum_i \boldsymbol{\sigma}_i\cdot\mathbf{H}$, and the calculation of the second order perturbation energy follows through as in the last section. The induced, paramagnetic moment $m_{\mathbf{q}}$ due to the component $h_{\mathbf{q}}$ is given by $-\partial E/\partial h_{\mathbf{q}}$ and it is found that [14]

$$m_{\mathbf{q}} = \mu_B^2 h_{\mathbf{q}} \chi(\mathbf{q}). \quad (4.19)$$

$\chi(\mathbf{q})$ is therefore interpreted as a generalized susceptibility. For a static field, $q = 0$, it reduces to the Pauli paramagnetic susceptibility. In atomic units $\chi(0) = N(E_f)/2$.

In the case of an electrostatic potential $V(r)$ the electrons respond by being attracted or repelled, and thereby screen V so that it is modified. The resulting self-consistent potential may be shown [28] to have Fourier components $V(\mathbf{q})/\varepsilon(\mathbf{q})$ where the dielectric constant $\varepsilon(\mathbf{q})$ is defined by

$$\varepsilon(\mathbf{q}) = 1+\frac{4\pi e^2}{q^2}\chi(\mathbf{q}). \quad (4.20)$$

Of course, for the case of localized spins there is no electrostatic screening of this kind. Within the free electron approximation the total electron density is everywhere constant.

The explicit expression (4.10) for $\chi(\mathbf{q})$ is that for a Hartree model of the free electron gas, in which exchange and Coulombic interactions between the electrons are neglected. A plot of this function is shown in Fig. 4.1. The effects of electron interactions have been considered by Wolf [29] using

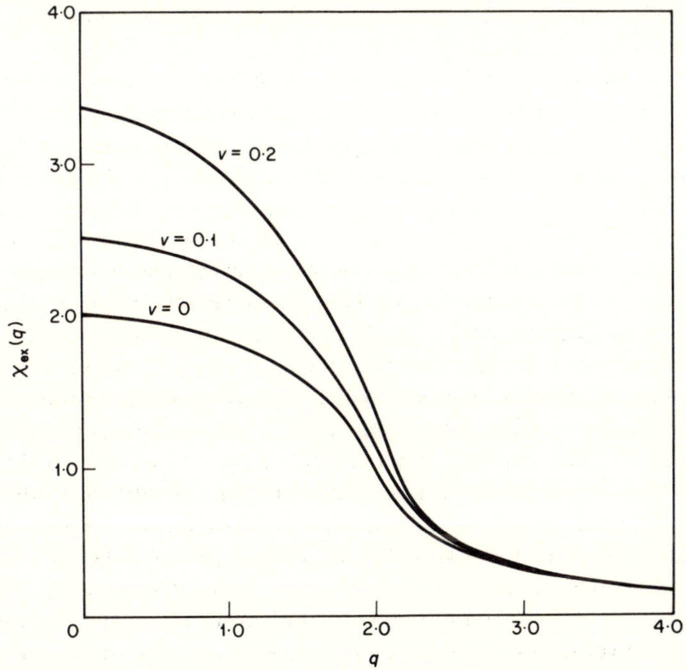

Fig. 4.1 The effect of exchange upon the susceptibility $\chi(q)$, as given by equation (4.21).

approximations equivalent to those of the linearized Hartree–Fock theory [30]. A simple argument which gives the basic result is the following one after Watson and Freeman [23, 37]. The exchange produces an effective magnetic field which is proportional to the induced magnetization, i.e. $m_{\mathbf{q}} = \chi(\mathbf{q})\mu_B^2(h_{\mathbf{q}} + vm_{\mathbf{q}})$, where v is an exchange constant, which in general is dependent on $\mathbf{q}$ and is expected to be positive. The susceptibility with exchange included is therefore

$$\chi_{ex}(\mathbf{q}) = \chi(\mathbf{q})/[1 - \mu_B^2 v\chi(\mathbf{q})]. \tag{4.21}$$

$\chi(\mathbf{q})$ is enhanced by a factor $\sim$ 1–2 at small q, see Fig. 4.1, and the effect upon the RKKY interaction is to increase its range.

4.1.5 *Corrections for scattering and non-spherical Fermi surfaces*

The electrons in a rare earth metal are scattered by impurities and also, through $\mathscr{H}_{sf}$, by any disorder in the spins. These mechanisms limit the mean free path l of the electrons and thereby tend to damp the oscillations of the RKKY interaction. Intuitively it may be expected that the polarization is modified to be

$$P(\mathbf{R}) \sim \phi(2k_f R)e^{-R/l}. \tag{4.22}$$

A detailed theory has been given by de Gennes [32] and experimental evidence for the validity of (4.22) has been obtained by Heeger *et al* [33] from nmr studies of Mn ions in Cu.

The use of spherical Fermi surfaces to account for the properties of the rare earth metals is a gross approximation, and several authors [34, 35, 36] have considered how the RKKY interaction is affected when the Fermi surface assumes other, simple forms. The theoretical work of Roth *et al* [34] shows the following general features. For spheroidal or ellipsoidal surfaces the interaction falls off with distance as R^{-3}, and the period of oscillation is anisotropic, depending upon the length of the wavevector which calipers the Fermi surface in the direction of $\mathbf{R}$. There may be several such periods, corresponding to different parts of the Fermi surface. For cylindrical regions, and more importantly for flat regions of the surface, the range of the interaction may be considerably increased, falling off as R^{-2} and R^{-1} respectively. In dealing with complicated Fermi surfaces it is more convenient to discuss the susceptibiiity function $\chi(\mathbf{q})$. For realistic surfaces χ must be evaluated numerically (cf. Section 4.3.2) but the simple analytical results of Roth *et al* give insight into the behaviour to be expected. In particular it is found that discontinuities occur at the wavevectors $\mathbf{q}_0$ which determine the periodicity of the real space interactions. These $\mathbf{q}_0$ vectors caliper points on the Fermi surface whose normals are in opposite directions. The latter condition implies that the velocities of the electrons in these two states on the surface are antiparallel. In the case of the Fermi sphere $|\mathbf{q}_0| = 2k_f$ and $\chi(\mathbf{q})$ has an infinite slope at that value, a fact referred to as the Kohn anomaly. For a cylindrical Fermi surface the slope in χ at $\mathbf{q}_0$ is discontinuous. For parallel flat surfaces, or any pair of electron and hole surfaces which have the same curvature and are separated by a vector $\mathbf{q}_0$, and which in consequence are said to nest together, the susceptibility has a logarithmic singularity. The importance of these discontinuities in χ are discussed in Section 4.3.2.

4.1.6 *Observable effects of the electron polarization*

Besides giving rise to exchange coupling between ionic moments, the polarization of the conduction electrons manifests itself in other ways. For example it contributes to the total magnetic moment of the metal, and it produces a shift in the g-value of the magnetic ions which can be determined by esr measurements. The polarization also gives rise to a hyperfine field which may be measured by nmr or Mössbauer techniques.

The polarization produced by a single ion adds a moment

$$\mu_c = \mu_B \int P(\mathbf{r})\mathrm{d}\mathbf{r} = \mu_B \frac{3ZA_0}{4\Omega E_f}(g_J - 1)J = \mu_B(g_J - 1)JA_0 N(E_f) \qquad (4.23)$$

to that of the ion. As discussed in Chapter 3, realistic values for the density of states $N(E_f)$ may be several times greater than the free electron value, and with their use the observed extra moments of the rare earths can be accounted for with reasonable values of A_0. Precise quantitative comparison may often be difficult because the experimental value of μ_c is sensitive to the presence of impurities.

If a magnetic field **H** is applied, the term in the total energy per ion of the metal which is linear in H is $(g\mu_B+\mu_c)H$. The g-value of the ion is therefore shifted by an amount

$$\Delta g = (g_J-1)JA_0N(E_f) \tag{4.24}$$

due to the polarization of the electrons. The free electron form of this result has been compared with experimental data for the rare earths by Liu [17] and he finds better than order of magnitude agreement.

The effective magnetic field seen at the site of a nucleus at $\mathbf{R}_i$ is the sum of the external field H, and of terms $K_0H = \mu_B N(E_f)H$ and K_1H, due respectively to the polarization of the electrons by H and by the neighbouring spins. The contribution of the latter is $\langle J_z\rangle \sum_{j\neq i} P(\mathbf{R}_j-\mathbf{R}_i)$, where $\langle J_z\rangle$ is the thermal average of J_z and may be written in terms of the ionic susceptibility χ_{ion} through

$$g_J\mu_B\langle J_z\rangle = \chi_{\text{ion}}H.$$

The total correction to **H** is given by the *Knight shift* $K = K_0+K_1$ which, by (4.6), is

$$K = K_0\left[1+\frac{6\pi Z(g_J-1)A_0\chi_{\text{ion}}}{g_J\mu_B^2}\sum_{j\neq i}\phi(2k_f|\mathbf{R}_j-\mathbf{R}_i|)\right]. \tag{4.25}$$

In principle a determination of K by nmr measurements [20] should give both the sign and magnitude of A_0. However, in practice there is usually a difficulty in assigning a value to k_f, and since the lattice sums are sensitive to this parameter, there may be large uncertainties in A_0 obtained in this way.

4.2 Magnetic bulk properties

4.2.1 *Paramagnetism of the pure metals*

As shown in Chapter 1, the paramagnetic susceptibility of an assembly of ions is given by the Van Vleck theory [37]. If the interactions between ions are neglected the paramagnetic susceptibility of the rare earth metals is the same as that of the free ions. As discussed in Section 1.2.1, Hund's formula (1.37) should be adequate except for samarium and europium where it is necessary to include the Van Vleck term $N\alpha$.

When there are magnetic interactions between the ions, the Curie–Weiss modification of this law describes the behaviour of the susceptibility at high temperatures in the paramagnetic region. This has the form

$$\chi = C/(T-\theta)$$

where θ is the paramagnetic Curie point (see equation (1.89)). In the rare earth metals, which are magnetically highly anisotropic, the paramagnetic Curie point of polycrystalline specimens is best given by $\theta = (\theta_c/3 + 2\theta_b/3)$ with θ_c and θ_b relating to the ordering temperatures parallel and perpendicular to the hexagonal axis.

Since the early investigations of Klemm and Bommer [38, 40] and Trombe [41–43] which formed the basis for subsequent studies of the magnetic properties of the rare earths, most of the experimental work has been carried out below 300°K. While a few authors had reported on individual elements [44–46] the first extensive examination of the paramagnetic behaviour to elevated temperatures was carried out by Arajs and Colvin who in a sequence of publications ([7] and refs. therein) reported on the temperature dependence of the susceptibility in all but three of the metals (La, Yb, Lu) over a temperature range from 300 to 1500°K. These authors found that with the exception of europium, the paramagnetic behaviour of all the elements can be understood in terms of the Van Vleck theory and assuming a hydrogenic energy level structure determined by an effective nuclear charge $(Z-\sigma)e$, where σ is a screening constant. In the case of the light rare earths, only cerium was found to require an appreciable allowance for interactions between the ions, although these could not be ruled out in this early work because of the somewhat insensitive nature of the fitting procedure for small interactions. The effect of interionic coupling was much more marked in the second half of the series and it was necessary to fit to a Curie–Weiss behaviour with θ values varying from 310°K for gadolinium to 17·4°K for thulium. The agreement between the experimental results and the theoretical curves was generally good using screening constant values of $\sigma = 35$ and $\sigma = 34$ for the light and heavy metals respectively. This may be seen from Fig. 4.2 where the results for praseodymium and dysprosium are shown. It is found that these theoretical curves are not highly sensitive to changes in σ. By making an allowance for the conduction electron contribution to the susceptibility, Arajs and Colvin were able to fit all the experimental data to $\sigma = 34$ curves with a conduction electron susceptibility of order $1{\cdot}0 \times 10^{-6}$ g^{-1} cm^3. This value agrees well with the results of earlier workers obtained for lanthanum which has no $4f$ electrons. The upper temperature limit of these measurements was above the melting point of all the light elements and consequently the results covered the normal room temperature crystal structure, the high temperature bcc phase and the liquid state for the elements cerium to europium. Slight anomalies occurred in the variation of susceptibility with temperature at these phase transitions, being most noticeable in the case of praseodymium. These deviations were rarely much larger than the scatter of experimental points however and suggest that there are no ionic interactions in these materials. Unfortunately the phase changes in the heavy rare earth metals occur above 1500°K so that the effects of the martensitic hcp–bcc phase change and of the melting phenomena on the magnetic properties could not be studied in the systems in which interionic coupling was known to be appreciable.

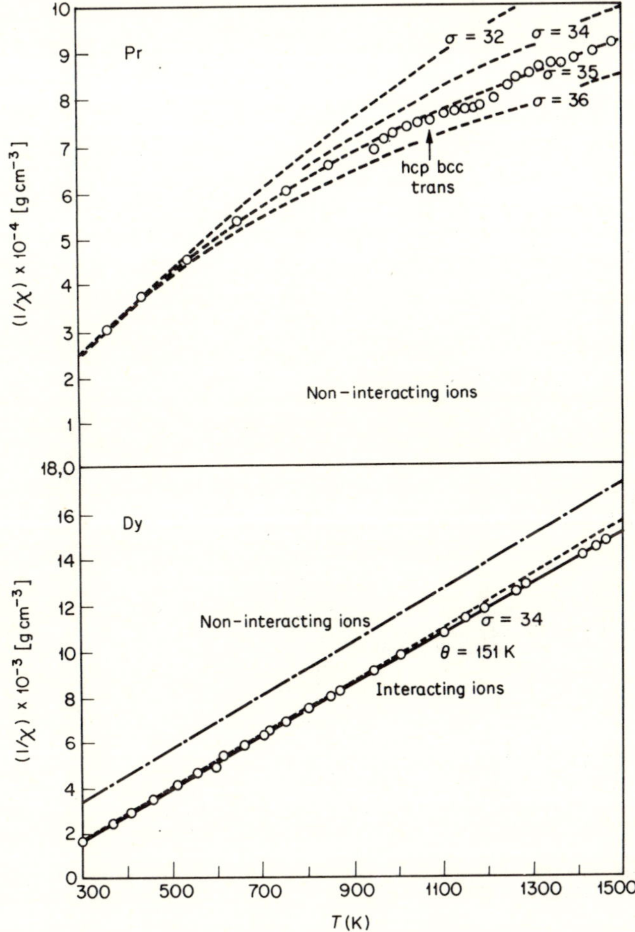

Fig. 4.2 Inverse paramagnetic susceptibility of praseodimium and dysprosium as a function of temperature [7]. Broken curves are theoretical predictions of the Van Vleck theory employing hydrogenic energy levels with a screening constant σ.

Burr and Ehara [46] examined both lanthanum and cerium from room temperature to above the melting points. The behaviour of lanthanum is Pauli paramagnetic but the susceptibility passes through a cusp at the d-hex to fcc transition and increases at both the fcc–bcc and melting transitions. These anomalies are undoubtedly due to the effect of crystalline transformations on the susceptibility, but the magnitudes of the changes differ by a factor of ten from predicted values based on the volume changes observed by Jayaraman at the transitions. The susceptibility of cerium shows an appreciable deviation from Curie–Weiss behaviour only at the melting point. The magnitude of this deviation and the absence of a deviation at the solid–solid

	$g_J(J(J+1))^{\frac{1}{2}}$	Van Vleck	$g_J J$	θ_{basal}	θ_c	θ_{poly}
	μ_B	μ_B	μ_B	°K	°K	°K
La	0	0	0			
Ce	2·54	2·56	2·14			-42 ± 4
Pr	3·58	3·62	3·20			
Nd	3·62	3·68	3·27			4·3
Pm	3·68	2·83	2·40			
Sm	0·85	1·55	0·72			
Eu	0·00	3·40	0·0			
Gd	7·94	7·94	7·0	317	317	310
Tb	9·72	9·70	9·0	239	195	232
Dy	10·64	10·6	10·0	169	121	151
Ho	10·60	10·6	10·0	88	73	85
Er	9·58	9·6	9·0	32·5	61·7	
Tm	7·56	7·6	7·0	17	41	20
Yb	4·53	4·5	4·0			
Lu	0	0	0			

	μ_{eff}	T_N	Spin structure	T^*	Spin structure	T_c
	μ_B	°K		°K		°K
Ce	2·51	12·5				
Pr	3·56					
Nd	3·3	20	hexagonal sites ordered	7·5	cubic sites also ordered	
Pm						
Sm		14				
	μ_B	°K		°K		°K
Eu	8·3					
Gd	7·98					293·2
Tb	9·77	229	helical in basal plane; $w_i = 20{\cdot}5$, $w_f = 18{\cdot}0$			221
Dy	10·64 ($\parallel c$) 10·65 ($\perp c$)	178·5	helical in basa l plane; $w_i = 43$, $w_f = 26{\cdot}5$			85

Table 4.1 A summary of the magnetic properties of the rare earth elements. Magnetic moments in Bohr magnetons and temperatures are in °K. $g_J\{J(J+1)\}^{\frac{1}{2}}$ is the theoretical free ion moment and the 'Van Vleck' column gives the predicted moments using the Van Vleck theory. θ's are the paramagnetic Curie temperatures. μ_{eff} is the moment deduced from inverse susceptibility measurements in the paramagnetic region. T^* is an ordering temperature between

transformation have been accounted for entirely in terms of the effects of strucrutal changes on the *f*-electron paramagnetism through crystal field effects and on the Pauli paramagnetism as obtained in the lanthanum observations. This is in contrast to the previously proposed models of the melting process

Ho	11·2	132	helical in basal plane; $w_i = 50$, $w_f = 30$	20	distorted helic in basal plane; ferromagnetic component along *c*-axis	
Er	9·9	85	modulated moment along *c*-axis; $w_i = 51$	53·5	helix in basal plane; gradual squaring of *c*-axis modulation	19·6
Tm	7·61 ($\parallel c$) 7·62 ($\perp c$)	58	modulated moment along *c*-axis; $w_i = 51$, $w_f = 51$	25	ferrimagnetic antiphase domain (↑↑↑↑ ↓↓↓)	
Yb						

	Spin structure	T_γ	μ (magnetization)	μ (neutron; 4·2°K)	Easy direction at 4·2°K
		°K	μ_B	μ_B	
La					
Ce		12·5		0·62	*c*-axis
Pr					
Nd		{7·2 19		2·3 1·8	b_1-axis
Pm					
Sm		13·6			*c*-axis
Eu			3·0 (140 kOe)	5·9	Moments lie in {100} planes < to *c*-axis
Gd	ferromagnetic; easy direction variable	291·8	7·55	7·0	
Tb	ferromagnetic	{221 227·7	9·34	9·0	
Dy	ferromagnetic	{174 83·5	10·65	9·5	*a*-axis
Ho		{19·4 131·6	10·34	9·9	*b*-axis
Er	helix in basal plane: ferromagnetic component along *c*-axis	{19·9 53·5 84	8·3	9·0	*c*-axis
Tm		55	7·14	7·0	*c*-axis
Yb					
Lu					

T_N and T_c. The T_γ are the λ-point temperatures in the specific heat–temperature curves. μ (magnetization) and μ (neutron) are saturation moments deduced from magnetization and neutron diffraction measurements respectively. Data is incorporated from [51, 52, 57, 60, 69–71, 73–76, 84, 75, 89–91, 93–98, 124, 127–129, 131, 132].

in cerium which postulate the necessity of a 4*f* to 5*d* electron promotion.

The effective ionic moments in the paramagnetic state obtained from the measurements at elevated temperatures and the results of many workers below room temperature are given in Table 4.1. The theoretical values obtained using Hund's rule are also given in this table, and as may be seen agreement is excellent except for the cases of samarium and europium. These may be understood in terms of the Van Vleck theory by considering the small energy separation between the ground state and the first excited state. Thermal population of this higher level causes an admixture of the higher J value and results in an enhanced effective moment. The Van Vleck predictions for the effective moment of samarium and europium are also given in Table 4.1 and a comparison of both theoretical and experimental results is shown in Fig. 4.3.

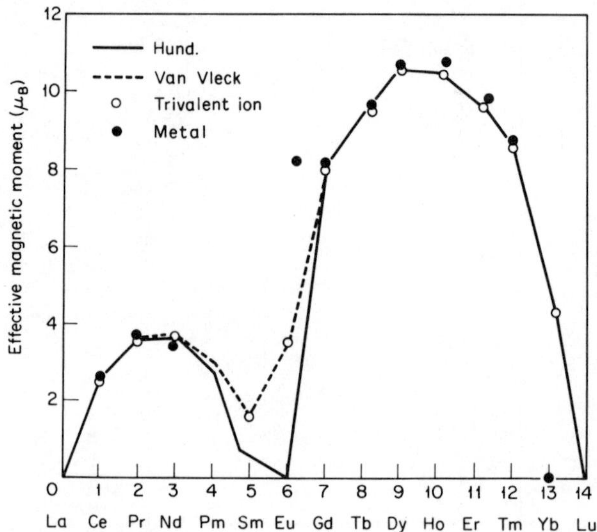

Fig. 4.3 The dependence of the ionic moment on the number of electrons in the $4f$ shell.

4.2.2 *Magnetization in the ordered state*

At low temperatures (< 300°K) the susceptibilities of most of the elements (with the exception of La, Pr, Yb and Lu) deviate from the Curie–Weiss behaviour (or modified Curie–Weiss, $\chi = C/(T-\theta)+K$ in the case of Nd) and exhibit magnetic ordering below some characteristic temperature. Cerium and samarium have maximum susceptibilities at 12·5 and 14·8°K respectively which are Néel points corresponding to the onset of antiferromagnetic ordering. The existence of these transitions, and of transitions in neodymium, has been supported by both specific heat [47–49], and neutron diffraction observations [50–53].

The low temperature susceptibility of praseodymium has been found to become constant below about 5°K [3, 54]. A broad specific heat anomaly [48] reaching a maximum at 30°K is associated with the thermal population of the $4f$ levels which have been split by crystal field effects. The nuclear hyperfine coupling contribution to the heat capacity is found to be only 3% of that calculated on the basis of an exchange field sufficient to give the full tri-positive ionic moment and was originally taken as evidence for the absence of magnetic ordering in this metal. Although the measurements of Cable *et al* [51] suggested that this metal is antiferromagnetic below 25°K, more recently, single crystal neutron diffraction observations [53] show that praseodymium does not form a magnetically ordered state. However, the application of a magnetic field in the basal plane induces a large magnetic moment predominantly on the hexagonal sites.

In the case of neodymium, neutron diffraction measurements [53] show that there are two magnetic transitions. These are due to the independent ordering of the hexagonal and cubic sites in the double hexagonal structure of this metal [53]. At 19°K the atoms of the B and C layers, which have a hexagonal environment, order antiferromagnetically while the A layers with cubic nearest-neighbour arrangement remain disordered until 7·5°K. These transitions are also observable in the susceptibility measurements on poly-crystal [4] and single crystal specimens [55].

Europium, the last of the light rare earths to be discussed in this section, undergoes a first order transition to an antiferromagnetic phase at 94°K [55, 56]. The magnetization shows no anisotropic behaviour and increases linearly with applied field to 140 kG where the moment value is $3\mu_B$ and still far from saturation [57].

The elements from Gd to Yb have been studied much more extensively than the lighter elements, and the magnetic behaviour of both polycrystalline [43, 58–68] and single crystal [57, 69–76] material is now well established. On cooling below room temperature the elements Gd, Tb, Dy, Ho, Er and Tm become ferromagnetic at some temperature characteristic of the individual metal and all but gadolinium transform into this magnetic state from a higher temperature antiferromagnetic phase which forms directly from the paramagnetic region. (Strictly at low temperatures thulium is ferrimagnetic as will be discussed later.) The magnetization curves typically show a maximum at the para-antiferromagnetic transition (T_N) which is followed at lower temperatures by a very sharp increase in the magnetization as the ferromagnetic phase forms. The temperature at which this occurs is in general highly sensitive to the magnetic field applied during the observations and with increasing field approaches T_N. The value of T_N itself is dependent on the applied field (see Fig. 4.4a). The Curie point T_C is taken as the highest temperature at which ferromagnetism is observed in zero applied field.

Corresponding to this field dependence, the magnetization-field isotherm plots show that ferromagnetism may be induced in the specimen in the temperature range between T_N and T_C provided that the field strength is increased to exceed a critical value H_C, where H_C for a particular metal is a function of the temperature.

The behaviour described above is shown in Fig. 4.4 (a and b) where the

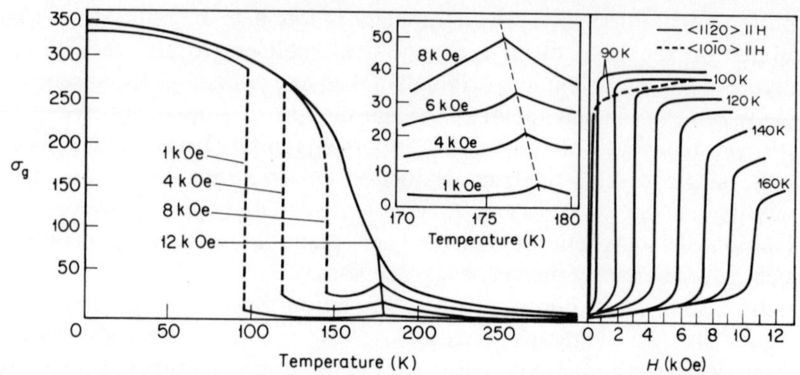

Fig. 4.4 Magnetization of single crystal dysprosium with the magnetic field along an *a*-axis as a function of (a) temperature and (b) field [71]. The inset is an enlargement of the data near the Néel point.

magnetization of single crystal dysprosium is displayed as a function of temperature and field. The form of these curves is characteristic of the four metals, terbium to erbium. The temperature dependence of the critical field strengths has been investigated both directly [77] from high field magnetization measurements and indirectly from a variety of other parameters [78, 79]. The results for dysprosium, holmium, and erbium are shown in Fig. 4.5 from which it may be seen that in general H_C increases linearly with temperature, to a maximum value in the vicinity of the Néel point and then falls rapidly as the transition to paramagnetism occurs.

In measuring fields of less than 27 kOe, thulium [75] shows a distinct Néel point at about 55°K, but at 4·2°K, the magnetization is still less than 15% of that to be expected from the ionic moment of $7\mu_B$. At 27·9 kOe the magnetization increases dramatically and corresponds to 7·07 μ_B per atom at 32·7 kOe. Saturation essentially occurs by 42 kOe, and at 85·4 kOe the moment has reached 7·14 μ_B per atom. The critical field varies little with temperature (see Fig. 4.5) to just greater than 40°K, when it decreases rapidly and does not exist above 55°K.

Gadolinium has no antiferromagnetic phase and transforms directly to the ferromagnetic state of 289°K. Some evidence has been presented in the past [80, 81] for extremely low ($\simeq$ 0·50 Oe) critical fields in this material with a Néel point in the vicinity of 240°K but this has found little support and would appear to arise from inhomogeneous or strained samples containing perhaps the high pressure antiferromagnetic gadolinium phase [82] (samarium structure, see Chapter 2, and Section 4.2.8). It is also possible that these observations may be related to the anomalous behaviour of the easy magnetization direction in this metal [83–87] below 240°K.

The inability to obtain magnetic saturation in magnetic fields of up to 20 kOe in early measurements on polycrystalline samples of the heavy rare earths suggested the existence of large magnetocrystalline anisotropies

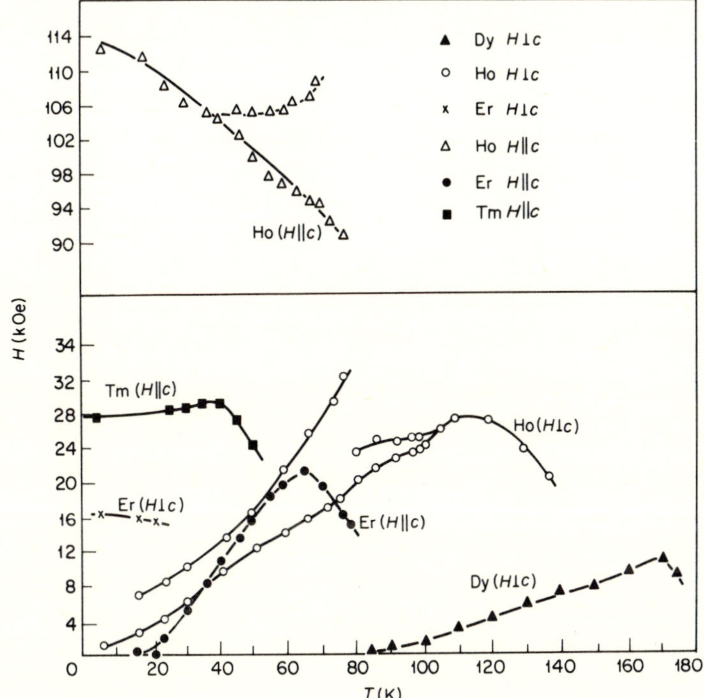

Fig. 4.5 Critical field versus temperature for single crystal dysprosium, holmium, erbium and thulium [75, 77].

which has been confirmed in the later single crystal determinations. This is clear from magnetization measurements along both the *c*- and *a*-axes. The easy direction varies for the different metals and indeed for some metals is a function of temperature (see Section 4.2.6 below). An indication of these easy directions is given in Table 4.1 together with the magnetic transition temperatures and saturation moment values obtained for the applied field parallel to the easy direction. These saturation moments approach and frequently exceed the theoretical values obtained by assuming a complete parallel alignment of the atomic moments of the trivalent ions. A comparison of the experimental and theoretical moments are given in Fig. 4.6 along with values obtained from neutron diffraction studies.

The origin of the magnetically ordered state in the heavy rare earth metals arises from the indirect exchange mechanism described earlier. The predictions of this theory (see Section 4.1.3) that the paramagnetic Curie temperatures will be proportional to a factor $G = (g_J - 1)^2 J(J+1)$ (often referred to as the de Gennes factor) are reasonably well obeyed in practice (see Fig. 4.7), a fact which may be viewed with some surprise in view of the approximate basis of the theory. The conduction electron polarization which

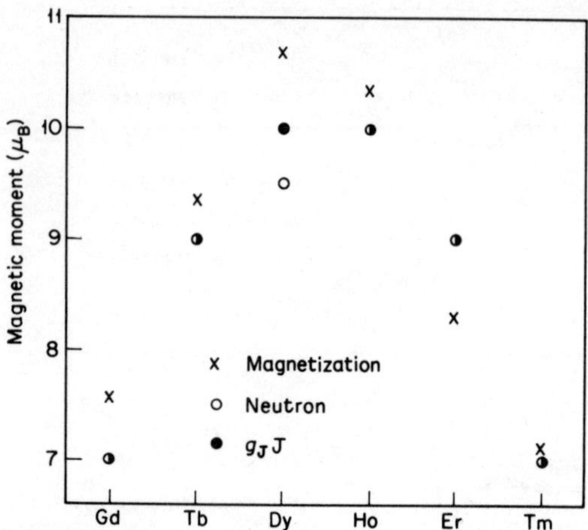

Fig. 4.6 Comparison of the saturation moments of the heavy rare earths obtained from magnetization and neutron diffraction data.

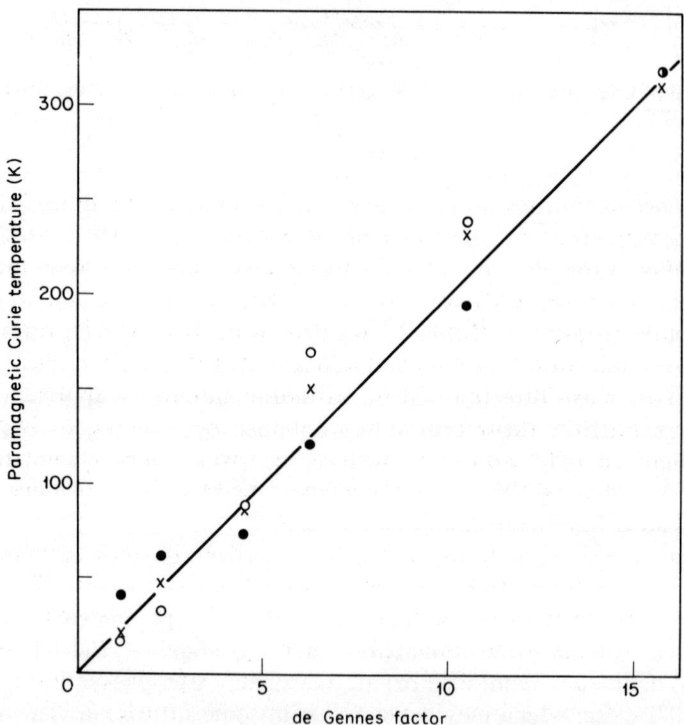

Fig. 4.7 Variation of the paramagnetic Curie temperature with the de Gennes factor (G).

occurs within this exchange mechanism results in an additional contribution (μ_c) to the observed magnetization (see equation (4.23)), and is believed to account for the excessive moment values obtained for the heavy metals. This contribution is not measured during neutron diffraction investigations [85] and results in the discrepancy between the two types of experimental results shown in Fig. 4.6.

The Néel points of the heavy metals are also a function of the de Gennes factor, and are given by $T_N \propto G^{\frac{2}{3}}$. The detailed form of this variation was first given by Weinstein *et al* [88], although other authors had drawn attention to the existence of a general functional relation between the transition temperature and G as the result of alloy studies. These are discussed more fully in Section 4.2.7.

4.2.3 *Specific heat capacity anomalies*

Specific heat measurements have been made on all the elements at temperatures below 300°K. The results for most of the metals [89–90] show the presence of λ-anomalies on an otherwise typically metallic specific heat variation. The low temperature (< 3°K) specific heat values for which the nuclear contribution is of increased importance are discussed in Section 4.5.1 in connection with the hyperfine interactions.

The temperature variation of the specific heat of dysprosium is shown in Fig. 4.8 from which it may be seen that sharp maxima occur at the magnetic transition temperatures 85°K and 179°K respectively. Similar results are observed for the other metals and the location of the maxima in the anomalies may be expected to provide accurate values of the transition temperatures. The values obtained in this way have been given in Table 4.1 along with those obtained by other means.

The entropy change associated with the total magnetic anomaly, i.e. integrated for all peaks, compares favourably for many of the elements with that to be expected by assuming the metal to consist of a collection of a tri-positive free ions. The experimental values are given in Table 4.2 along with the theoretical values $R \ln(2J+1)$. The variation of the magnetic entropy with temperature for dysprosium is shown in Fig. 4.9 on which the location of the two transition temperatures are readily visible. Above the Néel point at 179°K, the slope of this curve remains finite to approximately 300°K and is probably associated with the persistence of some degree of short range magnetic order in the form of clusters. It is interesting that other properties also show evidence of incomplete disordering for appreciable temperature ranges above the upper transition temperature.

4.2.4 *Magnetocrystalline anisotropy*

As has already been mentioned, the early magnetization measurements of the heavy rare earth metals were found to show evidence of large magneto-crystalline anisotropy. Torque measurements of the axial anisotropy [83, 84]

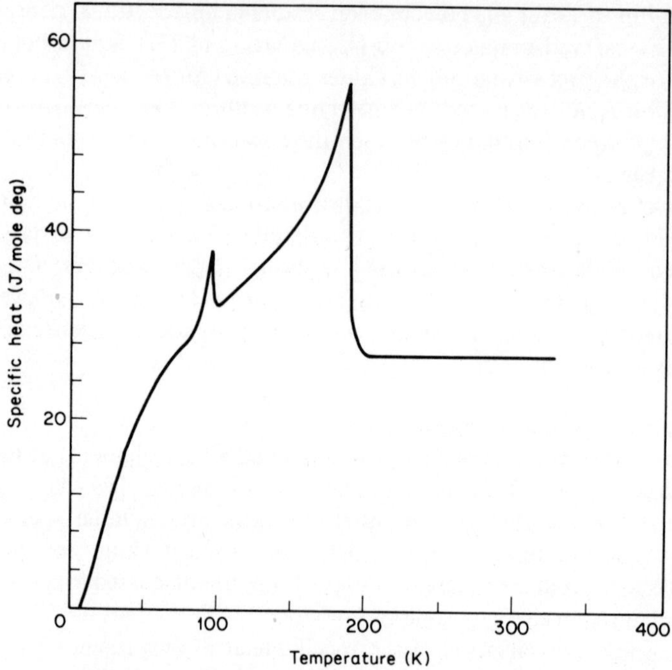

Fig. 4.8 Temperature dependence of the specific heat of dysprosium [95].

Table 4.2 Entropy changes associated with the transition from complete magnetic order at 0°K to the disordered state for the heavy rare earths [201].

	Gd	Tb	Dy	Ho	Er	Tm
$2\ln(2J+1)$	4·13	5·10	5·51	5·63	5·51	5·10
Magnetic entropy $S(T)$ (from $0 \rightarrow T$°K)	3·81 (360°K)	5·03 (360°K)	5·37 (300°K)	5·59 (300°K)	4·93 (320°K)	5·09 (360°K)
Magnetic entropy S_{∞} ($0 \rightarrow T = \infty$)	4·17	5·19	5·52	5·63	4·93	5·09

of gadolinium showed that its magnitude was several times that of the hexagonal transition metal cobalt. The variation of the anisotropy constants K_n ($n = 1, 2, 3, 4$) with temperature for this metal are shown in Fig. 4.10(a). Below T_C the *c*-axis is the easy direction. The observed change in sign of the principal axial term K_1 at about 240°K corresponds to a sudden change in the moment orientation away from the *c*-axis towards the basal plane. The magnitudes of K_2 and K_3 however, are such as to prevent the easy direction ever reaching the basal plane. The variation with temperature of the moment orientation relative to the axial direction is shown in Fig. 4.10 (b). The original results of Graham indicated that for a small interval of temperature the basal

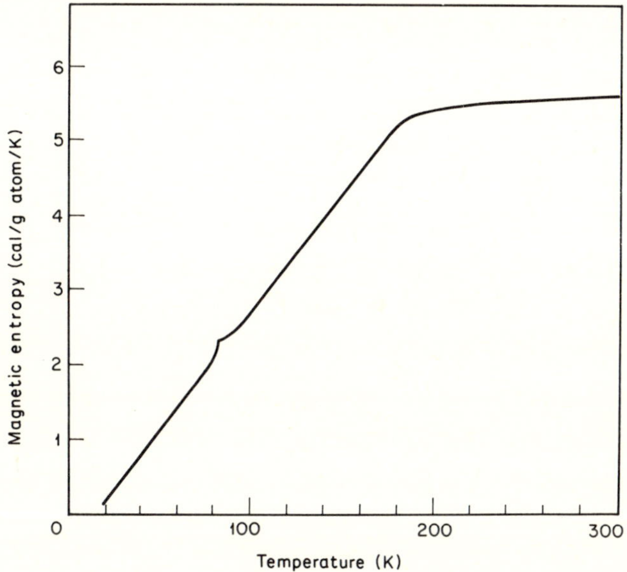

Fig. 4.9 Magnetic entropy of dysprosium as a function of temperature [95].

plane was in fact the easy direction. Subsequent neutron diffraction studies [95–97] have established that this is not so. The basal plane anisotropy measurements made by ourselves [78, 99] and Graham [100] show that the K_4 term is relatively small, becoming appreciable only below about 100°K and indicating a preferred *a*-axis. Combination of these constants with the magnetization results leads to the polar energy diagram shown in Fig. 4.10(c), from which it may be seen that just below 240°K, the moments are prevented from aligning parallel to the basal plane by a small energy maximum in that direction. The overall picture appears to be one of a decreasing axial anisotropy accompanied by a rapidly increasing total energy in the basal plane direction.

In the other heavy metals, saturation in the hard direction has not yet been achieved in fields of up to 150 kOe [57, 76] (except for erbium in which this appears to be the saturating field) and the corresponding, extremely large, anisotropy constants have been difficult to determine. Rhyne and Clark [101], and Levitin and Ponomaryov [102] have succeeded in studying the constants K_1 and K_4 to high applied fields although at no time was magnetic saturation achieved. The results of Rhyne and Clark at low temperatures indicated a value of 5×10^8 erg cm^{-3} (11°K) for the axial constant of both terbium and dysprosium and values of $2{\cdot}4 \times 10^6$ and $7{\cdot}5 \times 10^6$ erg cm^{-3} (4·2°K) for the basal plane constants of these two metals respectively. The measurements of Levitin *et al* [102] were carried out to 100°K in pulsed fields of 150 kOe and the magnitude of the observed axial

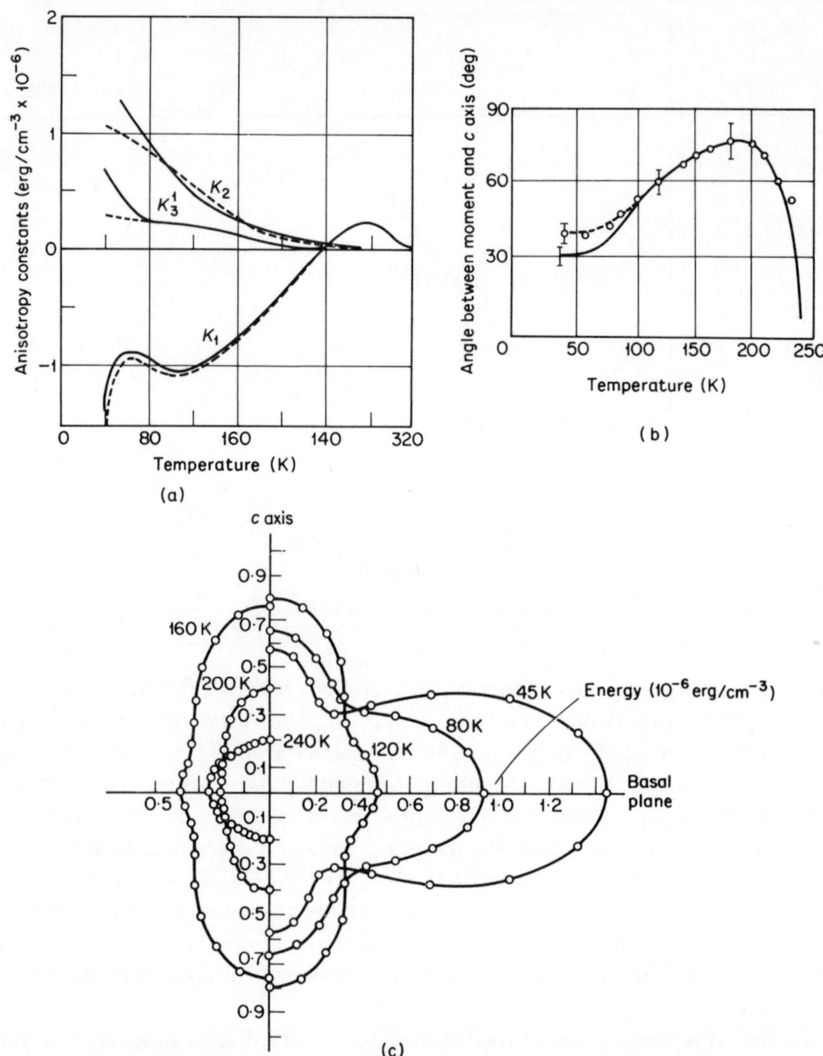

Fig. 4.10 The magnetocrystalline anisotropy behaviour of gadolinium. (a) Variation of the anisotropy constants with temperature; (b) temperature variation of the moment orientation with respect to the *c*-axis [84]; (c) polar energy diagrams in a plane containing the *c*-axis at various temperatures [99].

constant (of order 3×10^8 erg cm^{-3}) is in good agreement with the earlier measurements. Attempts have also been made by Tajima and Chikazumi [103] to overcome the experimental problem of enormous torques and lack of saturation by using a dilute solution of the element in gadolinium. Perhaps somewhat surprisingly, in view of the anomalous behaviour of the magnetic 'easy' directions in pure gadolinium and of the c/a ratio changes on alloying which will affect the crystal field magnitude, their results were also in reasonable agreement with those of Rhyne and Clark.

The temperature dependence of both the axial and basal plane constants of terbium and dysprosium and the basal plane constant of gadolinium are very well represented by the single ion theory originally proposed by Zener [104] and extended by Callen and Callen [105]. The basic assumptions in the theory are that temperature acts on the spin system in such a way as to cause a local deviation of the moments from the direction of the bulk magnetization but leaving the magnitude of the magnetization vector J unchanged. Each spin then samples a temperature-independent anisotropy energy, the sum of which, over all spins, gives the bulk anisotropy energy. The determination of the average value of the local anisotropy requires a knowledge of the distribution of the atomic moments about the mean direction, and was assumed by Zener to be given by a random walk function. Callen and Callen, adopting a Boltzmann distribution, obtained results in agreement with those of Zener at low temperatures and predict a temperature variation of the various constants as:

$$K_l \propto \hat{I}_{(2l+1)/2}\{\mathscr{L}^{-1}(\sigma)\} \rightarrow \sigma^{l(l+1)/2} \quad \text{at low } T, \tag{4.26}$$

where $\hat{I}_\mu(x) = I_\mu(x)/I_{\frac{1}{2}}(x)$, I being the hyperbolic Bessel function, and $\mathscr{L}^{-1}(\sigma)$ the inverse of the Langevin function of the reduced magnetization σ. Fig. 4.11 shows the excellent agreement which has been obtained between the experimental results and these theoretical predictions for dysprosium.

At low field strengths (< 10 kOe), Bly *et al* [106] examined the variation of K_1 in the paramagnetic region and showed that the results could be described by $K_1 \propto (MT)^2 \propto (\chi H)^2 \propto H^2/(T-\theta)^2$ in agreement with the

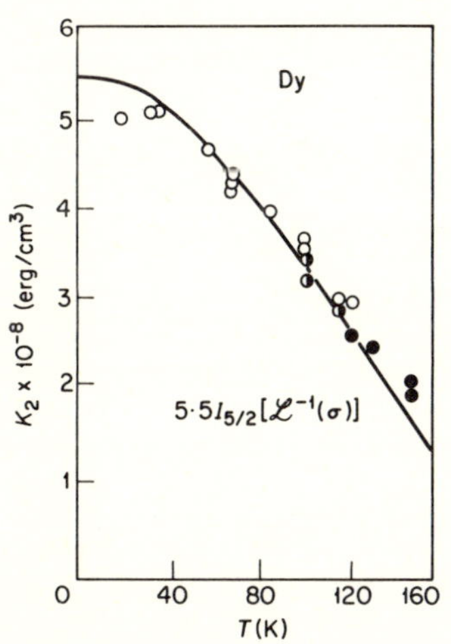

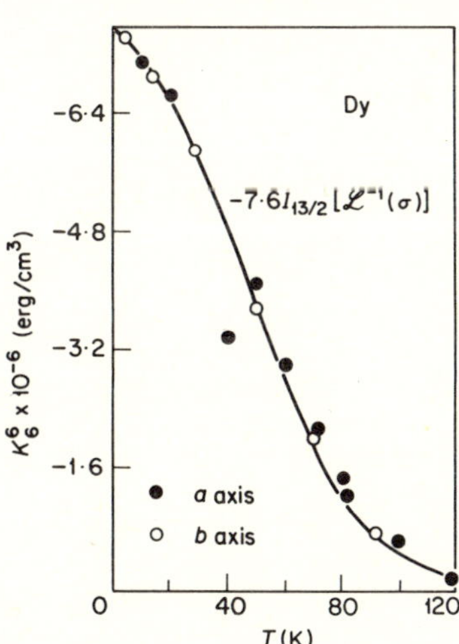

Fig. 4.11 Comparison of experimental asnisotropy constants of dysprosium [101] with the theory of Callen and Callen [105].

work by Callen and Callen. The values of θ employed were the average paramagnetic Curie temperatures given by $\theta = (\theta_{\parallel} + \theta_{\perp})/2$. Also determined in this work, by the observation of highly distorted torque curves, were the critical field values in the helical spin state. These were in good agreement with those of earlier workers using conventional techniques. In this region the change in sign of the basal plane constant of dysprosium with both temperature and applied field strength is taken to indicate changes in the easy direction which correlate with the detailed magnetization measurements of Jew and Legvold [107].

Bly also determines the basal plane anisotropy in holmium at 55°K and found that a twelve-fold term was necessary to describe the observed torque variation in addition to the fundamental six-fold component.

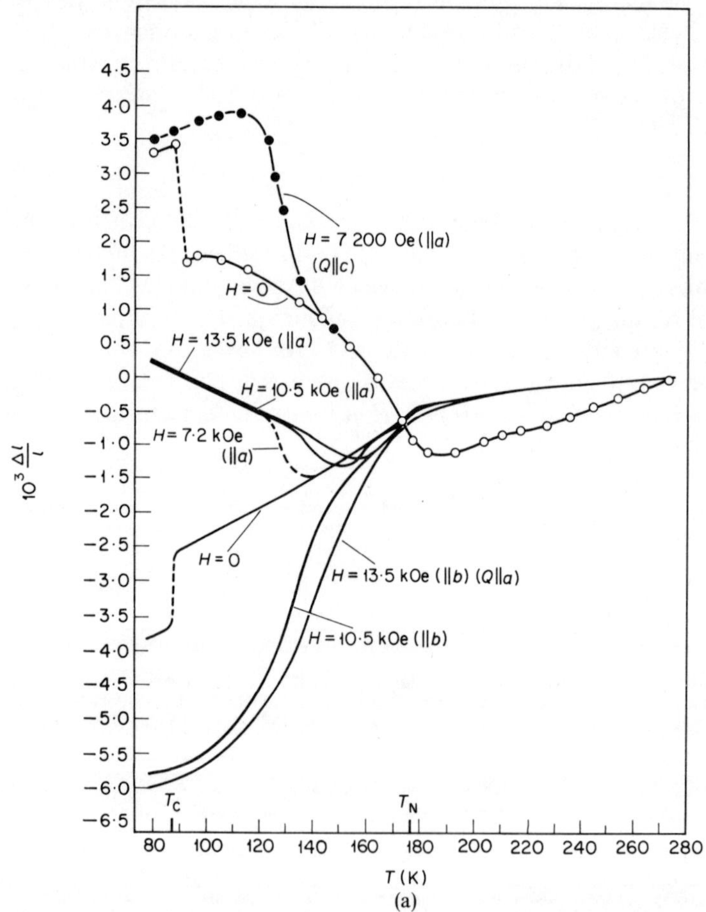

(a)

Fig. 4.12 Linear strains along the *a*-, *b*- and *c*-axes as a function of temperature for the applied fields shown; (a) dysprosium [116], (b) holmium [115]. *Q* indicates the direction of the strain gauge.

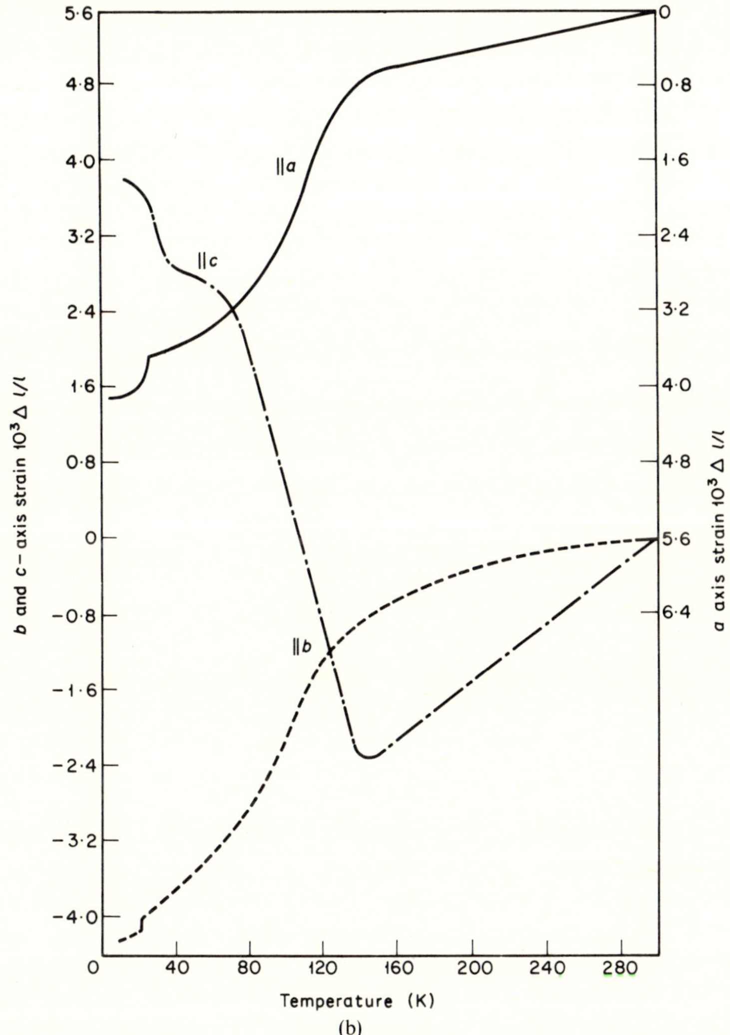

(b)

4.2.5 *Magnetostriction*

As might be expected the magnetostrictive strains which are related to the magnetocrystalline anisotropy through the elastic energy, have also been found to be large in these metals [108–119]. The experimental values of the linear strains in all three major axial directions in dysprosium and holmium are shown in Fig. 4.12(a, b) as a function of temperature. The curves are given for observations both with and without an applied magnetic field. The derived zero field expansion coefficient curve for holmium is also given in Fig. 4.13. In dysprosium the sudden changes in specimen dimensions at 85°K have been shown to arise because of an orthorhombic distortion of the hexagonal lattice [120]. The results of Fig. 4.12(a) indicate that when an

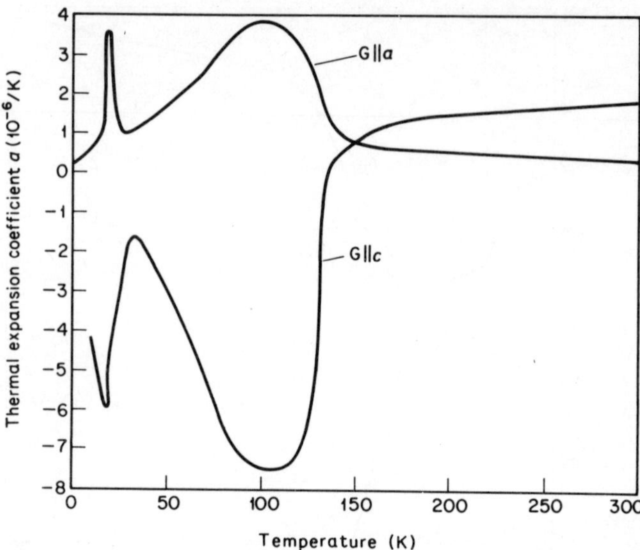

Fig. 4.13 Anomalous thermal expansion coefficients for the *a*- and *c*-axis directions obtained from the data of Fig. 4.12(b) for holmium [115].

applied field in excess of the critical field is applied to the specimen in the antiferromagnetic region these changes occur at higher temperatures and consequently that the structure transition is driven by the enormous magnetostriction associated with the ferromagnetic ordering. In addition it is evident from these curves that the magnetostriction is also the dominant term in the thermal expansion behaviour below 179°K (T_N).

The experimentally observed changes in dimension arising from the applied field are usually written as $\lambda(ij)$ and refer to the change in length in the *i*-direction caused by the application of the magnetic field in the *j*-direction. Normally the directions used are the crystallographic *a*-, *b*- and *c*-axes and two perpendicular directions (*d* and *e*) which lie in the *ac* plane such as to bisect the angle between the *a* and *c* axes.

Above the Néel point Clark *et al* [115] found all the magnetostriction terms to be proportional to H^2 in agreement with the prediction of the Maxwell relation $[\partial\lambda_i/\partial H]_{\sigma_i} = -[\partial M/\partial\sigma_i]_H$, with M the magnetic moment per unit volume, which in this temperature region is proportional to H. In this relation λ_i is the magnetostrictive strain and σ_i the stress corresponding to λ_i. Similar results have also been obtained for holmium [115] and terbium [118]. Below the Néel temperature the magnetostriction is highly sensitive to the applied field strength because of the changes in spin structure which arise as a result of the application of a magnetic field (Section 4.2.6).

The temperature dependence of the magnetostriction has been given in the general treatment of Callen and Callen [124] which uses a single ion model.

This was extended to describe the strain dependence of the exchange energy and dipole–dipole energy by the inclusion of two-ion interactions. An outline of this approach is given here [122].

The Hamiltonian for the specimen is taken as

$$\mathscr{H} = \mathscr{H}_{\mathrm{m}} + \mathscr{H}_{\mathrm{e}} + \mathscr{H}_{\mathrm{me}} + \mathscr{H}_{\mathrm{a}} \tag{4.27}$$

where $\mathscr{H}_{\mathrm{m}}$ relates to the spin system and consists of the isotropic exchange and Zeeman terms. $\mathscr{H}_{\mathrm{e}}$ is the elastic energy associated with homogeneous strain components ε_{xx}, ε_{yy}, ε_{zz}, ε_{xy}, ε_{yz} and ε_{zx}. $\mathscr{H}_{\mathrm{me}}$ is the magnetoelastic contribution coupling the spin system to the strains and $\mathscr{H}_{\mathrm{a}}$ is the magnetocrystalline anisotropy energy for the unstrained lattice.

The strain-dependent energy density leading to the magnetostriction is then

$$E_{\mathrm{ms}} = E_{\mathrm{e}} + E_{\mathrm{me}}. \tag{4.28}$$

The magnetoelastic contribution is of the form

$$\begin{aligned} -E_{\mathrm{me}} = {} & B_1^{\alpha 0}\varepsilon^{\alpha 1} + B_2^{\alpha 0}\varepsilon^{\alpha 2} + B_1^{\alpha 2}\varepsilon^{\alpha 1}(\alpha_z^2 - \tfrac{1}{3}) + B_2^{\alpha 2}\varepsilon^{\alpha 2}(\alpha_z^2 - \tfrac{1}{3}) \\ & + B^{\gamma 2}\varepsilon_1^{\gamma}\tfrac{1}{2}(\alpha_x^2 - \alpha_y^2) + B^{\gamma 2}\varepsilon_2^{\gamma}\alpha_x\alpha_y + B^{\varepsilon 2}\varepsilon_1^{\varepsilon}\alpha_y\alpha_z + B^{\varepsilon 2}\varepsilon_2^{\varepsilon}\alpha_z\alpha_x, \end{aligned} \tag{4.29}$$

where the B's are empirical magnetoelastic constants, and include only the lowest order magnetoelastic effects. The α_j's are the direction cosines of the magnetization with respect to the crystallographic axes. The strain functions $\varepsilon_i^{\mu j}$ are the irreducible strains having the hexagonal close packed symmetry and their relation to the homogeneous strain components is

$$\begin{aligned} \varepsilon^{\alpha 1} &= \varepsilon_{xx} + \varepsilon_{yy} + \varepsilon_{zz}, & \varepsilon^{\alpha 2} &= \tfrac{1}{3}(2\varepsilon_{zz} - \varepsilon_{xx} - \varepsilon_{yy}), \\ \varepsilon_1^{\gamma} &= \tfrac{1}{2}(\varepsilon_{xx} - \varepsilon_{yy}), & \varepsilon_2^{\gamma} &= \varepsilon_{xy}, \\ \varepsilon_1^{\varepsilon} &= \varepsilon_{yz} & \varepsilon_2^{\varepsilon} &= \varepsilon_{zx}. \end{aligned} \tag{4.30}$$

The equilibrium values of the strains resulting from magnetization are obtained by minimizing equation (4.28) with respect to each of the strain components. Once this has been done the hexagonal strains may be transformed to the Cartesian coordinates giving for the saturation magnetostriction

$$\begin{aligned} \lambda = \delta l/l = {} & \tfrac{1}{3}\lambda_{11}^{\alpha}(T) + \tfrac{1}{2}\sqrt{3}\lambda_{12}^{\alpha}(T)[\alpha_z^2 - \tfrac{1}{3}] + 2\lambda_{21}^{\alpha}(T)[\beta_z^2 - \tfrac{1}{3}] \\ & + \sqrt{3}\lambda_{22}^{\alpha}(T)[\alpha_z^2 - \tfrac{1}{3}][\beta_z^2 - \tfrac{1}{3}] + 2\lambda^{\gamma}(T)\{\tfrac{1}{4}[\alpha_x^2 - \alpha_y^2][\beta_x^2 - \beta_y^2] + \alpha_x\alpha_y\beta_x\beta_y\} \\ & + 2\lambda^{\varepsilon}(T)[\alpha_y\alpha_z\beta_y\beta_z + \alpha_x\alpha_z\beta_x\beta_z], \end{aligned} \tag{4.31}$$

with the β_i's the direction cosines of the measurement direction with respect to the crystal axes.

Some differences exist concerning the definitions of the λ^{α}'s in the literature, and while those we have given are consistent with the work of Callen and Callen, we feel it desirable to give the following relations between these coefficients and those used, for example, by Clark *et al.* These are

$$\begin{aligned} \lambda_{11}^{\alpha} &= 2\lambda_1^{\alpha,0} + \lambda_2^{\alpha,0} + 2\lambda_1^{\alpha,2} + \lambda_2^{\alpha,2} \\ (\sqrt{3}/2)\lambda_{12}^{\alpha} &= 2\lambda_1^{\alpha,2} + \lambda_2^{\alpha,2} \\ 2\lambda_{21}^{\alpha} &= -\lambda_1^{\alpha,0} + \lambda_2^{\alpha,0} \\ \sqrt{3}\lambda_{22} &= -\lambda_1^{\alpha,2} + \lambda_2^{\alpha,2}. \end{aligned} \tag{4.32}$$

The two terms $\lambda_1^{\alpha,0}$ and $\lambda_2^{\alpha,0}$ arise from the two-ion interaction term and are independent of the direction of magnetization, varying only with the magnitude of the magnetization. The remaining terms $\lambda_1^{\alpha,2}$ and $\lambda_2^{\alpha,2}$ refer to volume and c/a changes respectively. The coefficients λ^{γ} and λ^{ε} are the same in the different notations, λ^{γ} representing the distortion of the circular symmetry of the basal plane by the rotation of the magnetization and λ^{ε} the distortion of the 90° angle between the basal plane and c-axis directions. The nature of the direction dependent modes are shown in Fig. 4.14.

The temperature (and field) variation of the magnetostriction coefficients occurs through their dependence on the magnetization and has been shown by Callen and Callen to be given by

$$\lambda(T, H) = \lambda(0, 0)\hat{I}_{(2l+1)/2}[\mathscr{L}^{-1}(\sigma)]. \tag{4.33}$$

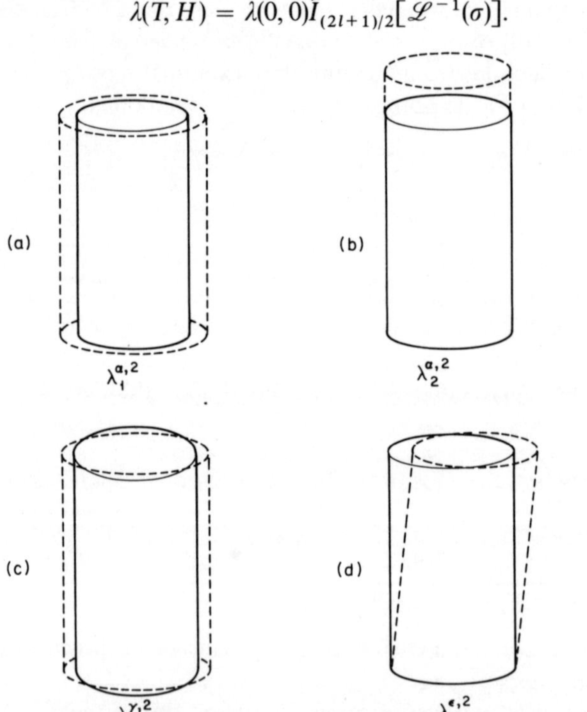

Fig. 4.14 The four possible magnetostriction modes.

In the paramagnetic limit this reduces to $(\frac{3}{5})\lambda(00)\sigma^2$ and at low temperatures (4.33) predicts a variation of $\sigma^{l(l+1)/2}$. The agreement of the experimental values with these theoretical predictions is remarkably good as may be seen from Fig. 4.15(a, b) where the λ^{γ} and λ^{ε} terms are for dysprosium.

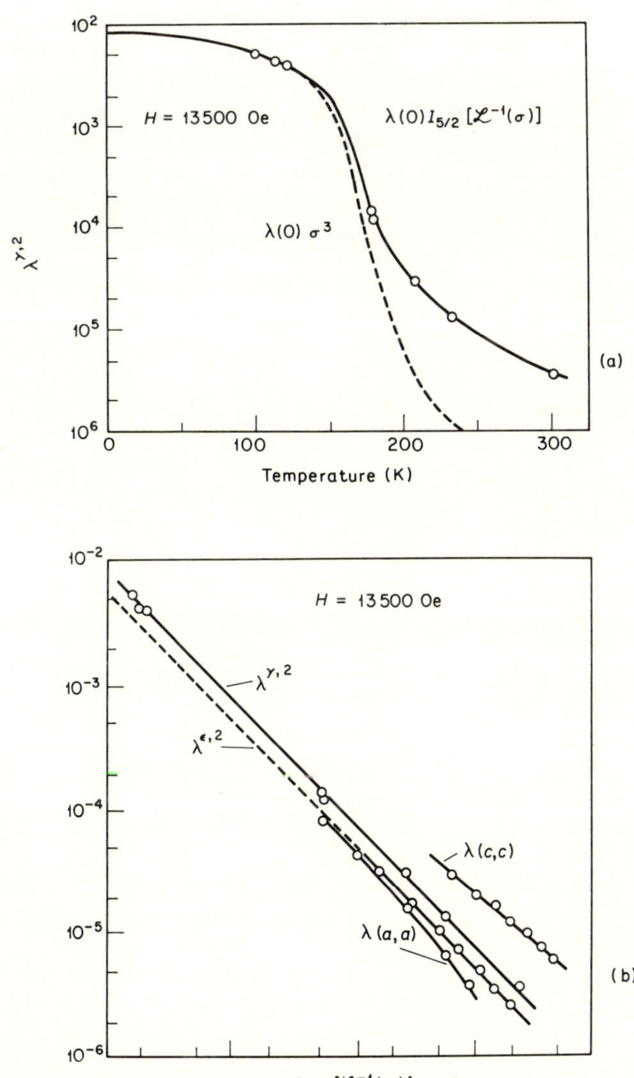

Fig. 4.15 The magnetostriction coefficients of dysprosium [116] compared with the theory of Callen and Callen [121]. (a) $\lambda^{\gamma,2}$ as function of temperature, (b) $\lambda^{\gamma,2}$, $\lambda^{\varepsilon,2}$, $\lambda(a,a)$ and $\lambda(c,c)$ versus $\hat{I}_{\frac{5}{2}}\{\mathscr{L}^{-1}(\sigma)\}$.

4.2.6 *Spin structures and neutron diffraction*
The early neutron diffraction studies carried out at Oak Ridge by Wilkinson, Koehler and their co-workers [123–125] showed that in the ordered state ($T < T_N$) the spin structures of the heavy rare earth metals were much more complex than those of the classical ferromagnetic and antiferromagnetic materials. Neutron scattering from a ferromagnetically aligned spin system shows directly the magnetic order since the peaks in the scattering coincide with those for the nuclear scattering. In the rare earths however, in the temperature range $T_c < T < T_N$ the magnetic scattering in general appears as satellites of the nuclear reflections. The positions of the diffraction peaks for holmium [126] at 77°K are shown in Fig. 4.16(a) where these additional peaks are seen to occur only in the b_3 (equivalent hexagonal) direction. The existence of these satellites was recognized as arising from a sinusoidal modulation of the magnetic scattering amplitudes along the c-axis direction of the holmium single crystal. Analysis showed that the ionic spins in the

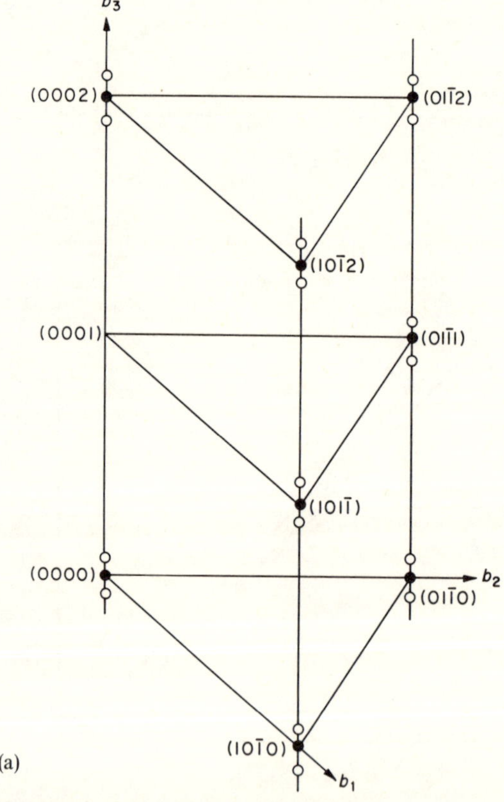

Fig. 4.16 (a) The location of the observed diffraction reflections of holmium at 77°K (helical spin structure) [126]. Full circles represent nuclear scattering from the lattice sites and the open circles represent the magnetic satellites. (b) The helical spin structure.

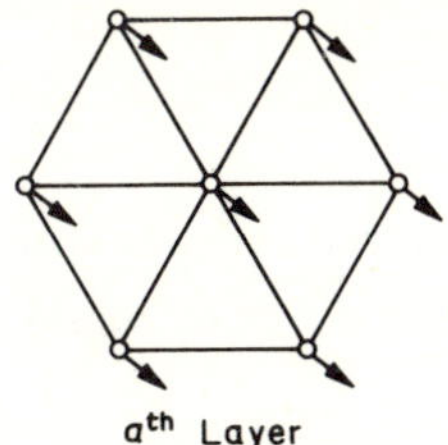

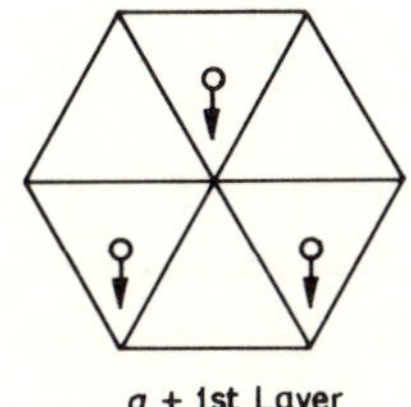

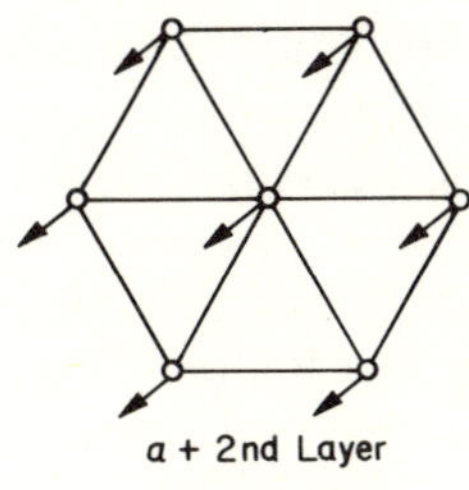

Fig. 4.16 (b)

antiferromagnetic state are arranged in a helical structure with the axis of the helix being the crystallographic c-axis. This structure, shown in Fig. 4.16(b), corresponds to the parallel alignment of spins within any plane perpendicular to the c-axis, but between adjacent atomic planes the spin directions differ by an angle ω, known as the 'turn angle' of the helical spin structure.

This satellite structure may be understood if we insert a term representing the helical spin configuration into the normal long-range order neutron scattering term:

$$\frac{d\sigma}{d\Omega} \propto \exp(-2W(K))|f(K)|^2 \sum_{i,j} \exp[i\mathbf{K}\cdot(\mathbf{R}_i - \mathbf{R}_j)] \times \sum_{\alpha,\beta}\left(\delta_{\alpha,\beta} - \frac{K_\alpha K_\beta}{K^2}\right)\langle S_i^\alpha\rangle\langle S_j^\beta\rangle. \tag{4.34}$$

Here $\exp(-2W(K))$ is the Debye–Waller factor, $f(K)$ the form factor, $\mathbf{K} = (\mathbf{k}_0 - \mathbf{k})$ is the scattering vector and α, β represent a Cartesian co-ordinate system. Writing the helical spin structure such that at any lattice site ($\mathbf{R}_i$) the spin is given by

$$S_i = \mathbf{a}_i S_i^z + \mathbf{b}_i S_i^x, \tag{4.35}$$

where $\mathbf{b}_i$ is the vector (cos ($\mathbf{q}\cdot\mathbf{R}_i$), sin ($\mathbf{q}\cdot\mathbf{R}_i$), 0) and $\mathbf{a}_i$ is a vector parallel to S_i^z. $\mathbf{q}$ is the wavevector of the helix and consequently finite so that $\langle S_\perp \rangle = 0$ and $\langle S_z \rangle \neq 0$.

Using these two equations gives for the total scattering cross-section that

$$\frac{d\sigma}{d\Omega} = \langle S \rangle^2 \sum_{\mathbf{K}_n} \delta(\mathbf{K}-(\mathbf{K}_n \pm \mathbf{q}))|f(K)|^2\left(1+\frac{K_z^2}{K^2}\right)\exp(-2W(K)), \tag{4.36}$$

i.e. diffraction maxima occur for $\mathbf{K} = \mathbf{K}_n \pm \mathbf{q}$, i.e. at each side of the nuclear scattering peak, with an intensity proportional to $\langle S \rangle^2$. Observation of the spacing of a pair of satellites gives a value of $|\mathbf{q}|$ the wavevector of the helical spin structure. The turn angle is then given by $\omega = qc$. Any deviation of the spin structure away from the ideal helix results in the appearance of secondary satellites which may be treated in a similar manner and used to give the Fourier components of the distribution of the moment. The intensities of the different satellite orders give the amplitudes of these Fourier coefficients.

Analysis of the results for the heavy rare earth elements have shown that a helical spin structure, of the type discussed above for holmium, is also observed in the antiferromagnetic range for terbium [124] and dysprosium [127]. In the cases of erbium and thulium however, the ordered moments are in general found to lie parallel to the c-axis. In erbium [128] a magnetic structure change is observed at 53°K, in agreement with specific heat measurements [97] which report an anomaly at this temperature as well as at the Curie and Néel points. Above 53°K the moments are confined to the c-axis direction and their magnitudes are taken to be sinusoidally modulated because of the absence of higher order satellites. The moment distribution is given by $S_i^x = S_i^y = 0$, $S_i^z = \langle S_z \rangle \sin(\mathbf{q}\cdot\mathbf{R}_i + \phi)$, where $\mathbf{q}$ is the modulation vector and ϕ a phase angle. Below 53°K a helical basal plane component of the moment appears together with higher order satellites indicating a departure from the simple sinusoidal modulation. The ordering in thulium [129] is again complex, changing with decreasing temperature from a sinusoidally modulated structure (comparable to erbium above 53°K) which exists between 56°K (T_N) and 40°K to a moment configuration known as the antiphase domain structure below 22°K. In this latter structure, shown in Fig. 4.17, as we go along the c-axis the magnitude and direction of the moments can be represented by a rectangular wave, four adjacent layers of spins pointing in one direction parallel to this axis followed by three layers oppositely directed and so on. Also shown in this figure are the known spin structures of the other rare earth elements in the magnetically ordered state.

Below the Curie temperatures, listed in Table 4.1, the five elements under discussion show various types of ferromagnetic spin ordering. Terbium and

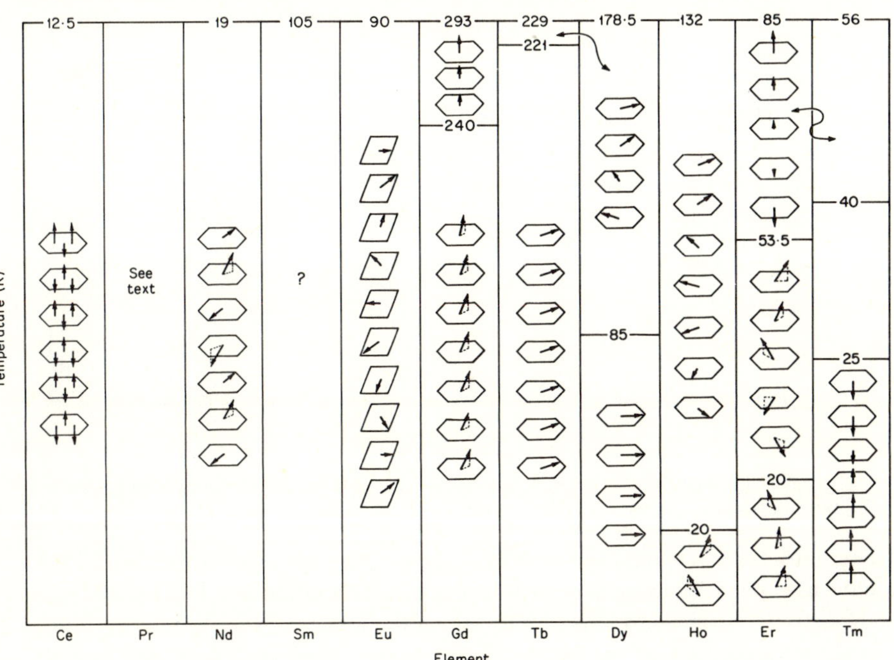

Fig. 4.17 The observed magnetic structures of the rare earth metals in zero applied field.

dysprosium both adopt a parallel alignment of moment in the basal plane with the easy directions (*b*-axis) and (*a*-axis) respectively. In holmium and erbium complete alignment is not observed using neutron diffraction techniques and the structure is a ferromagnetic helical moment arrangement. In this a component of the moment appears parallel to the *c*-axis and is ferromagnetically ordered, while in the basal plane a helical configuration is maintained. A small *c*-axis component has also recently been proposed for dysprosium by Jordan and Lee [72] to explain their magnetization measurements. This component was observed to develop below about 100°K (a 37°K spread was observed between the results using three specimens) but this has not yet been substantiated using neutron diffraction techniques and it may be associated with the presence of other rare earth impurities.

The periodicity of the helical spin structures has been shown to be strongly temperature dependent by Koehler *et al* [124]. It is found that the turn angle of the helical component of the structures decreases as the temperature decreases below the Néel point. The variation of the wave vector describing the structure with the reduced temperature T/T_N is shown in Fig. 4.18.

Gadolinium is truly ferromagnetic over the whole temperature range but nevertheless provides a variety of spin orientations as indicated by the

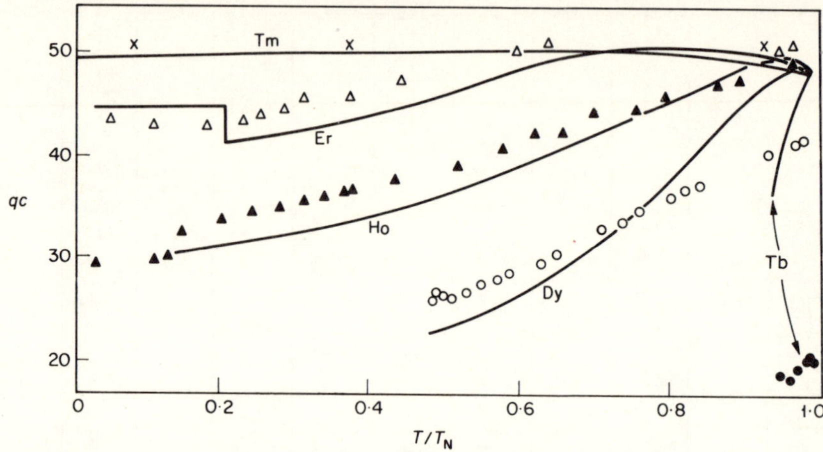

Fig. 4.18 The interlayer turn angle as a function of reduced temperature [198]. Solid curves are theoretical results. Experimental points are shown by the various symbols [125].

magnetocrystalline anisotropy observations. These changes in easy direction have now been confirmed by various neutron diffraction studies [85–87] and show that between 294 and 232°K the moments lie parallel to the crystallographic *c*-axis, but below 232°K the moments move away from this axis to a maximum deviation of 75° near 180°K and then back to within 32° of the *c*-axis at 4·2°K. This is in remarkably good agreement with the results of Corner *et al* [84] obtained using torque measurements.

The spin structures of the light rare earths are not so well known as those of the heavier elements, but a basic picture does exist for all the available metals. The ordering in the double hexagonal phase of cerium is ferrimagnetic within each hexagonal plane [131], but the planes appear to be stacked alternatively, so that the overall structure is antiferromagnetic. Both praseodymium [51, 53] and neodymium [52, 53] show a preferential ordering of the moments on the hexagonal sites over those on cubic sites. In the former element it appears that the cubic sites may never order while the hexagonal sites do so only after application of a magnetic field. In neodymium ordering of the cubic sites occurs several degrees below the temperature at which the moments on the hexagonal sites order. Finally europium [132], which crystallizes in a bcc structure, has a helical antiferromagnetic configuration with the moments lying in the (100) planes with an angle of about 50° between the moments in adjacent planes.

The effects of a magnetic field on the diffraction pattern of holmium have been studied [124, 130] as a function of temperature. At 4·2°K, with the field parallel to either the *a*- or *b*-axis, the satellite intensities are reduced to zero without change in their position. The variation in intensity of the normal lattice reflections with the applied field in the *b*-direction corresponds to the development of the full ferromagnetic moment in the basal plane parallel to this direction, indicating the b-axis as the easy direction. At 13°K, however, there is no indication of an easy axis in the basal plane, the reflection intensities following a $\sin^2\theta$ law, where θ is the angle between the scattering vector and the field direction. The results suggest that the helical spin structure transforms into an intermediate state in which the magnetization vectors of the hexagonal planes oscillate about the field direction. This corresponds to the 'fan' structure predicted by Enz [179] and occurs when the applied field strength exceeds the critical field. Further increase in the field strength causes this fan to collapse to a full ferromagnetic alignment at an applied field H_f.

An analysis of the detailed studies [130] showed that changes of the spin structure occur as a result of the applied field. The nature of these changes depend upon the specimen temperature and can be identified with step-like features in the magnetization–temperature data for $T_c < T < T_N$ [73].

Although little evidence is available from neutron diffraction studies, it is likely that the helical–fan–ferromagnetic spin structure transitions also occur in terbium and dysprosium with increasing applied fields in the basal plane direction.

4.2.7 *Magnetic properties of rare earth-Y, -Sc, and rare earth–rare earth alloys*

Magnetization studies of yttrium-substituted heavy rare earths have been the subject of much interest in the past, the non-magnetic yttrium ions being used as a diluent to create a variation in the average spacing of the magnetic ions in an otherwise constant lattice. Studies of the Gd–Y system [133] showed that antiferromagnetic ordering occurred in samples having an yttrium concentration in excess of 40%. The ferromagnetic Curie temperature and the Néel point variation for this series is shown in Fig. 4.19. Recent neutron diffraction studies [134] have shown that in the antiferromagnetic phase the spins have a helical configuration similar to that in terbium and dysprosium metals. The sudden decrease in T_c at 40% yttrium suggests that the helical spin structure becomes rapidly energetically favourable with respect to a ferromagnetic ordering with increasing yttrium content. This stabilization of the antiferromagnetic phase, and consequent suppression of the antiferro-ferromagnetic transition, is also observed for yttrium substitution into the other heavy rare earths [88, 133, 135, 136], the results for which are given in Fig. 4.20(a) as a function of the mean number of $4f$ electrons. Weinstein *et al* [88] first pointed out that the Néel temperatures vary as a function of the mean value of the de Gennes function $(\overline{G})$ given by

$$T_N \propto \overline{G}^{\frac{2}{3}},$$

although several authors had previously established a general function dependence of T_N on $\overline{G}$ which could be used to represent all the alloy systems.

In the antiferromagnetic phase of these alloys the values of the initial turn angle ω_i (i.e. ω at T_N) also lie on a single curve as a function of $\overline{G}$. Although

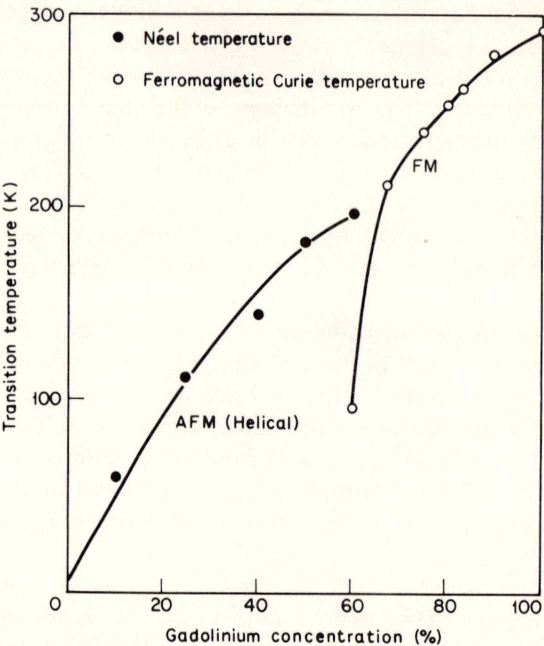

Fig. 4.19 Variation of the transition temperatures in Gd–Y alloys as a function of gadolinium concentration [133].

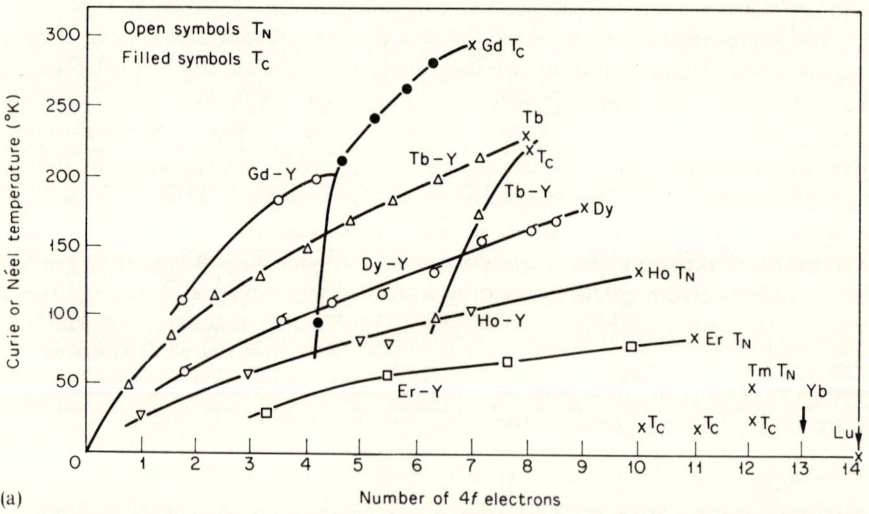

Fig. 4.20 (a) Variation of T_N and T_C for various rare earth-yttrium alloys as a function of the mean number of 4*f* electrons [88, 133, 135, 136]. (b) Variation of T_N, T_C, and other transformation temperatures, for heavy–heavy rare earth alloys as a function of mean number of 4*f* electrons [148]. × – pure elements

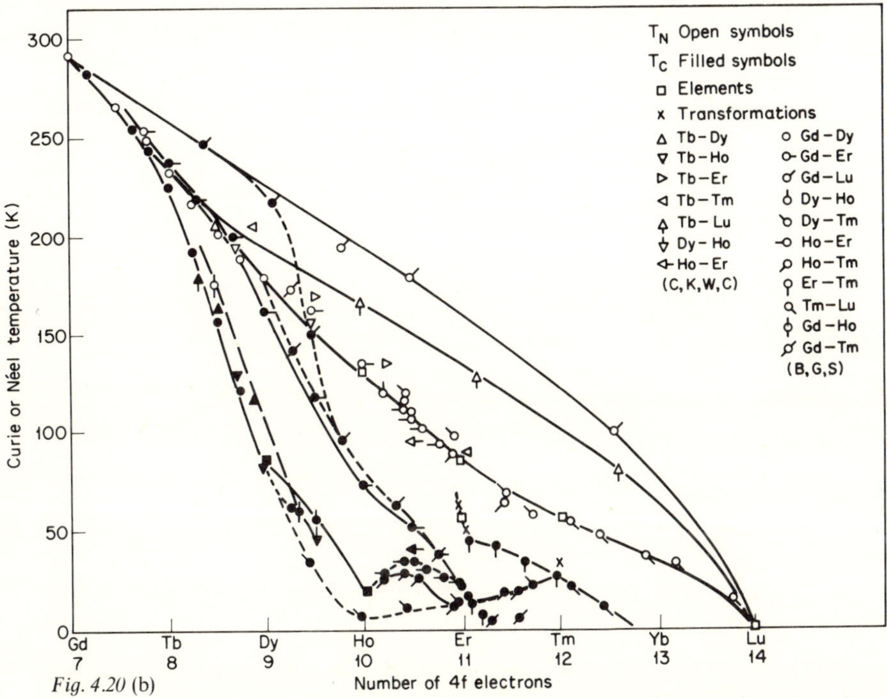

Fig. 4.20 (b)

there is considerable scatter about this curve (Fig. 4.21) it is evident that ω_i saturates at approximately 50° for $\bar{G} \rightarrow 0$. The disappearance of ω_i at $\bar{G} > 11$ is not a continuous process but appears to occur abruptly when ω_i becomes less than 20°, as evidenced by the results of Gd–Y and Gd–Sc [134]. Further indications that there is a lower limit to the possible turn angles of about 20° appears in the observed variation of ω with temperature in both pure terbium and dysprosium and some of their alloys. The nature of the $\omega - T$ curve, however, is also influenced by the value of the de Gennes factor for the specimen under consideration. Regardless of the rare earth involved in an R–Y alloy, ω shows only a small dependence on temperature for small $\bar{G}$, but with increasing $\bar{G}$ the variation becomes more marked and for $\bar{G}$ greater than about 7, ω can show a sudden decrease to zero at some finite temperature. Some selected results from the work of Child *et al* [136] are shown in Fig. 4.22.

For the case of scandium substitution the behaviour of the transition temperatures and the turn angles with $\bar{G}$ does not fit that for the yttrium alloys. The Néel and Curie points decay faster than $\bar{G}^{\frac{2}{3}}$ and Chatterjee [137] has recently pointed out that for the scandium alloys $T_N \propto \bar{G}^{4/3}$, and $\theta \propto \bar{G}^2$, although as with the yttrium-substituted alloys there appears to be no good theoretical reason for this variation.

Neutron diffraction studies of alloys of terbium, holmium and erbium [138, 139], and gadolinium [134], with scandium show that again the

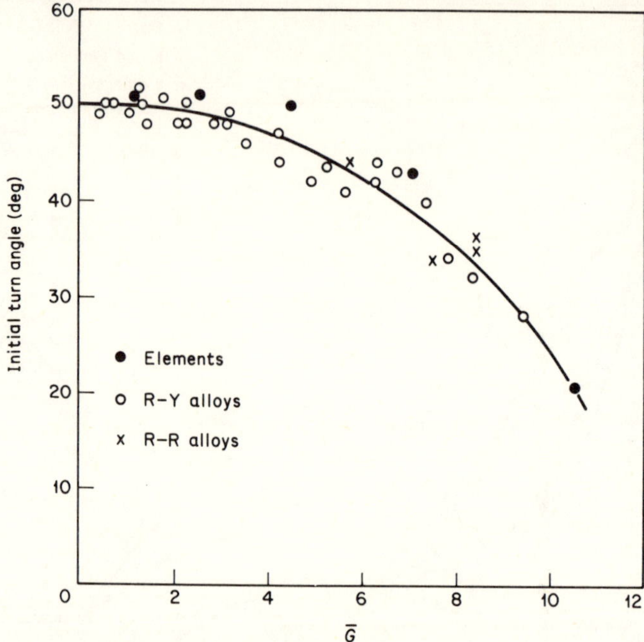

Fig. 4.21 The initial turn angle of the rare earth elements, rare earth-yttrium and rare earth-rare earth alloys as a function of the de Gennes factor.

addition of the non-magnetic constituent stabilizes the periodic magnetic structure at the expense of any ferromagnetic component that may be present.

In rare earth alloys formed with scandium magnetic ordering ceases to be observable at quite high rare earth concentrations, T_N becoming less than 1·3°K at atomic concentrations of approximately 15% gadolinium [134], 25% terbium, 18% holmium, and 39% erbium [138]. This result has been taken as an indication that the magnetic interaction present in the pure metals is drastically modified by the addition of scandium. Wollan [140, 141] has discussed this modification in terms of a size effect. Using the parameter $T_c/\bar{G}$ to represent the coupling energy per magnetic ion, he points out that in the Gd–Y system this remains constant, whereas in the Gd–Sc alloys $T_c/\bar{G}$ decreases linearly with gadolinium concentration. These differences are attributed to the fact that for the Gd–Y specimens we have a constant atomic volume while for the scandium alloys the volume decreases towards the scandium-rich end of the series. The lowering of the transition temperatures with high applied pressures are compared with this decrease and it is suggested that they may be of the same origin.

In very dilute alloys of terbium in yttrium ($<$ 3·71 at.% Tb) Nagasawa and Sugawara [142] have shown that here too there appears to be a critical concentration for the occurrence of magnetic ordering. Below approximately

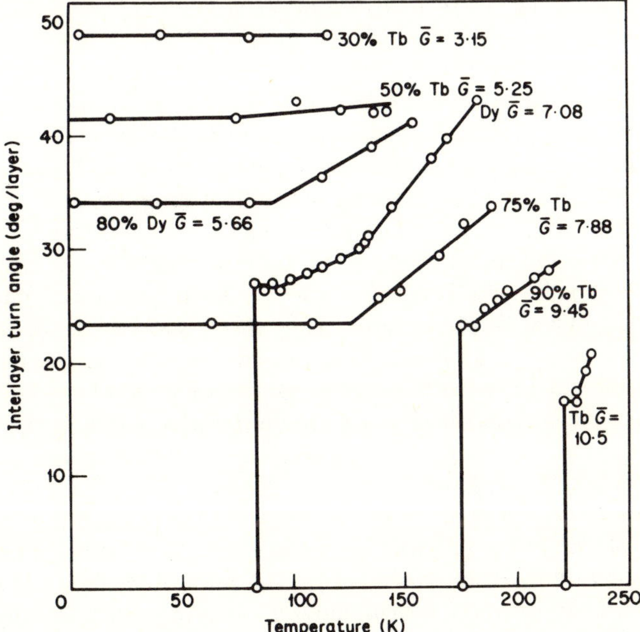

Fig. 4.22 The temperature variation of the turn angle for some rare earth–yttrium alloys and dysprosium and terbium metals. The mean de Gennes factor ($\bar{G}$) is shown [136].

10% terbium concentration the $T_N \propto \bar{G}^{\frac{2}{3}}$ relation breaks down, the transition temperatures falling more rapidly than this relation would suggest and appearing to go to zero at about 1% Tb. A similar limiting concentration also appears likely for Dy–Y alloys as observations on a dilute Dy specimen [143] also exhibit appreciable deviations from the $\bar{G}^{\frac{2}{3}}$ law but do agree well with the results of reference [11] plotted as T_N versus $\bar{G}$. This critical concentration is discussed in terms of both the number of nearest neighbour terbium ions about a terbium ion and effects of the finite mean free paths of the conduction electrons.

At these low concentrations of the magnetic rare earths in a non-magnetic matrix, several workers have found anomalously large values of the effective moment in the paramagnetic region [142, 144, 145]. This effect has usually been attributed to the polarization of the conduction electrons by the magnetic impurity causing an enhanced moment. Using this model Nagasawa and Sugawara obtained a value of +0·15 eV for the exchange interaction parameter in yttrium–terbium alloys.

Observations of intra-rare earth alloys have mainly been limited to those in which the constituents are heavy elements. For the various systems studied it is found that the Néel temperatures follow the same universal curve when plotted against the de Gennes factor, as do the pure elements and yttrium substituted alloys. The Curie temperature variation is, however, different for each system as may be seen from Fig. 4.20(b) which shows the experimental results of various workers [135–151] for a wide variety of these alloy systems. The interesting features of this figure are the maxima in T_c which occur in the Dy–Er, Ho–Er and Ho–Tm systems near to a mean $4f$ electron concentration of 10·5. A trend of this type is also visible for the Gd–Er and Gd–Tm specimens. The elements on opposite sides of this $4f$ electron concentration have different magnetic character, holmium having its easy direction in the basal plane [73] while in erbium the easy direction is parallel to the *c*-axis [74]. The magnetostriction also reverses sign from one element to the other [152] and, like the magnetocrystalline anisotropy effect, is related to the change in sign of the factor α_J multiplying the equivalent operator [21] in the crystal field (see Section 4.3.1) from negative in the metals Tb to Ho to positive for Er and Tm. The sudden break in the T_c-composition curve for Dy–Er has been attributed to the composition at which the magnetic structure changes abruptly from the Dy-type to the Er-type. Neutron diffraction data in these alloys has been discussed at length by Koehler [154, 155] and it is evident that many of the results serve to supplement the information obtained for the pure metals and yttrium-based alloys. The work of Millhouse *et al* [156] on the erbium–dysprosium system showed somewhat unexpectedly that a 4·2°K the dysprosium and erbium moments are aligned parallel to one another in the basal plane. This was taken to imply that multimoment anisotropies [157] dominate in this system and that one major contribution to the observed anisotropy in the rare earth metals is an anisotropic exchange interaction.

4.2.8 *Pressure effects*

The effects of pressure on the magnetic properties of the lanthanides have been studied over pressure ranges of up to about 85 kbar. The transition temperatures are the most readily obtained features of the magnetic behaviour and these have usually been deduced from electrical resistivity measurements (cf. Chapter 5) or alternatively from measurement of the permeability using a toroidal transformer. The core of this transformer is the metal under investigation. In this method the changes in permeability near to the ordering temperatures leads to an increase in the measured secondary voltage. The changes usually appear as a cusp in the vicinity of the Néel point and as a large broad maximum at the transition from non-ferromagnetic to ferromagnetic ordering. The experimental observations are further affected by the occurrence of crystal structure transitions in some specimens, which bring with them other types of magnetic behaviour. The appearance of these other high pressure phases has been discussed in Chapter 2. In discussing the effects of pressure on the magnetic transitions it is convenient to consider initially those results applying to the structure observed at atmospheric pressure and then to examine the magnetic behaviour of the high pressure phases of those metals for which they have been reported.

In the low pressure ranges, with only one exception [164], it has been found that the magnetic transition temperatures are decreased by application of hydrostatic pressure to the sample. The values of dT_N/dp and dT_c/dp obtained from several workers are shown in Table 4.3. The identification

Table 4.3 The variation with pressure of the magnetic ordering temperatures of the heavy rare earth metals [159, 167, 168]. Temperatures are in °K and dT/dp values are in °K/kbar.

	Structure	Transition temperatures (°K)			$\frac{dT_N}{dp}$ (°K/kbar)	$\frac{dT^*}{dp}$ (°K/kbar)	$\frac{dT_c}{dp}$ (°K/kbar)
		T_N	T^*	T_c			
Gadolinium	hcp			293			−1·4
	sm	247			−1·46		
Terbium	hcp	229·3		220	−0·84		−1·24
	sm	196			−0·83		
Dysprosium	hcp	180		88·5	−0·41		−1·27
	sm	166			−0·67		
Holmium	hcp	133			−0·48		
Erbium	hcp	85	53·5	19·6	−0·26	−1·3	

of the different transitions is not always easy and in the case of terbium it has not been possible to resolve T_N and T_c [158–160]. The numbers given in the table relate to two points on the resistance–temperature curves which were identified with the magnetic transitions [160]. The complex nature of some of the magnetic observations may be seen from Fig. 4.23 where the results of McWhan and Stevens [159] are shown for gadolinium. The three peaks appearing at 34 kbar and above are almost certainly associated with the high pressure phase of this metal although unambiguous identification has not been possible.

Assuming that the magnetic transitions are second order, dT/dp is calculated from thermodynamics using Ehrenfest's equation,

$$dT/dp = TV\Delta\alpha/\Delta C_p = \Delta\beta/\Delta\alpha, \tag{4.37}$$

where $\Delta\alpha$, $\Delta\beta$ and ΔC_p are the anomalies in the expansion, compressibility and heat capacity at the transition temperature assuming constant pressure. This relation has been used by McWhan [159], Okamoto [161], Landry [162] and others, and reasonable agreement is found between the experimental and calculated values of dT/dp. Landry, however, points out the danger of using equation (4.37) at such transitions when the detailed behaviour of β and C_p are not reliably known. The variation of transition temperatures with pressure has also been analyzed by considering changes in exchange energies which arise from changes in the interatomic spacing and c/a ratio

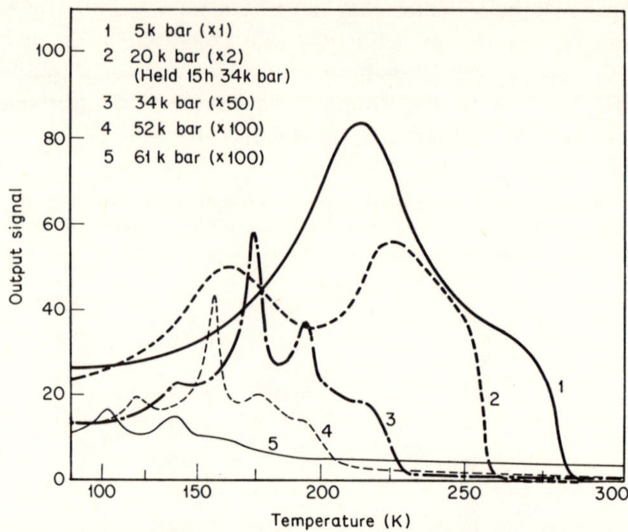

Fig. 4.23 Secondary voltage versus temperature for gadolinium at various pressures [159].

[159, 160, 163, 164]. The uncertainties in evaluating the exchange constants using an elementary approach were discussed by Darby *et al* [165] who showed that the detailed values of the summation in the RKKY interaction were strongly dependent on the Fermi wavevector assumed in the calculation. It has been pointed out, however, that this sum is dependent upon the c/a ratio [160] through the changes caused in the density of states by the anisotropic compressibility. These latter authors display the dependence of $T_i/(g_J-1)^2J(J+1)$ on c/a, where T_i may be any of the usual magnetic transition temperatures, as a means of showing the value of the variation of the $(A_0^2/E_f)\sum\phi(2k_fR)$ term in the indirect interaction with axial ratio. Their results are shown in Fig. 4.24, from which it is evident that the curves for T_c (or T_p) and T_N have opposite dependence on the c/a ratio. If the observed changes in these transition temperatures do occur through a changing c/a ratio then it is perhaps surprising that all the curves of Fig. 4.24 do not have at least a similar form since presumably $d(c/a)/dp$ does not change dramatically on passing through the magnetically ordered regions.

McWhan and Stevens [166] have evaluated the exchange integral $\mathscr{J}(q)$ and its dependence on a dimensionless parameter $\langle r_{4f}^2\rangle^{\frac{1}{2}}/\Omega^{\frac{1}{3}}$ for the pure metals and several alloys. The derived variations for all the systems studied are quite similar and the results have been used to give a value for the change in exchange integral with lattice parameter (a) as $K^{-1}d\mathscr{J}(q)/da = 42 \pm 6°K/Å$. These authors also conclude that it is the variation in the average de Gennes function that is responsible for the observed differences in dT_N/dp, and that the relation $T_N \propto \bar{G}^{\frac{2}{3}}$ is not a universal one.

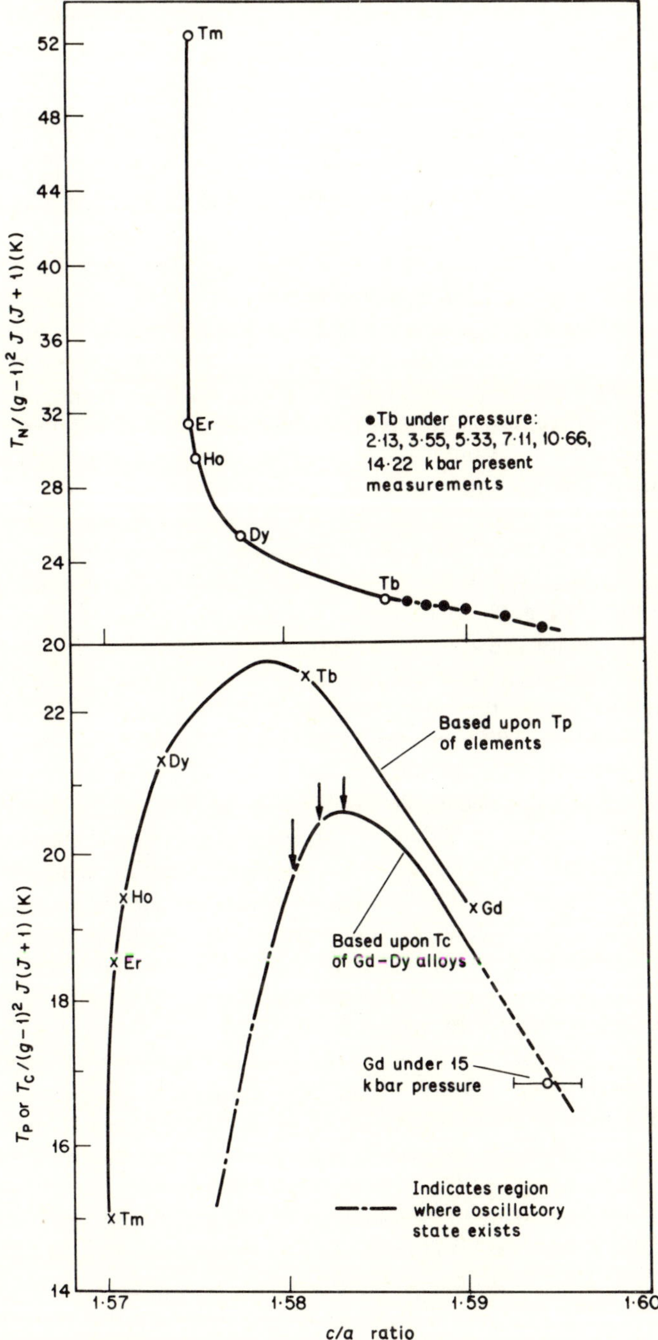

Fig. 4.24 The variation T/G with c/a ratio for the rare earth metals.

The first order crystal structure transitions which occur in some of the metals at high pressures are observable through changes in their magnetic properties as has been indicated earlier. In gadolinium, the high pressure samarium-type phase appears to be antiferromagnetic [159] with an ordering temperature around 240°K, and it is perhaps worth speculating whether there is any relation between this behaviour and the low field magnetization studies of Belov *et al* [167, 168]. The differences in the low and high pressure ordering temperatures near to the phase transitions are always negative (high pressure transition temperature less than low pressure transition temperature) and have been reported to be -37, -21, -12 and -9°K for gadolinium, terbium, dysprosium and holmium respectively. In addition $\mathrm{d}T/\mathrm{d}p$ in the new phases are also negative and have values close to those at low pressure except in gadolinium, for which the type of transition being considered is different. It should be mentioned, however, that in the quenched high pressure phase of samarium Jayaraman and Sherwood [169] observed an increase in the Néel point of 12°K.

The effects of pressure on the Néel and Curie temperatures of a limited number of solid solutions show that the basic features observed for the pure metals are repeated on alloying the metals together or on diluting with the non-magnetic elements yttrium and lanthanum. In the terbium–yttrium system [170] the low pressure phase shows a continuous variation of T_N with composition, the percentage change at 30 kbar decreasing smoothly with increasing yttrium content from a value of 14% in pure terbium to 7% for alloys with less than 70% terbium. This result is taken to indicate that the two-thirds power law for the dependence of the Néel temperature on the de Gennes function is not obeyed at this pressure. Milstein and Robinson [164] have observed a positive $\mathrm{d}T_c/\mathrm{d}p$ at pressures less than 2 kbar in a 42% Gd–58% Dy alloy. This composition has two ordering temperatures at atmospheric pressure, passing through an oscillatory spin state to the ferromagnetic state with decreasing temperature, the range of this antiferromagnetic region being approximately 20°K. For pressures greater than 2 kbar $\mathrm{d}T_c/\mathrm{d}p$ becomes negative. With increasing gadolinium content the antiferromagnetic region disappears at about 50% Gd and at this composition $\mathrm{d}T_c/\mathrm{d}p$ is zero over the lower pressure range but again goes negative for pressures greater than 2 kbar. In the gadolinium-rich specimens in which only a ferromagnetic state is observed $\mathrm{d}T_c/\mathrm{d}p$ is always negative and approximately constant. For all the specimens for which an antiferromagnetic state exists $\mathrm{d}T_c/\mathrm{d}p$ is negative. The appearance of the maximum in T_c for these alloys is associated with the maxima observed for the elements shown in Fig. 4.24, and is taken as an indication of a corresponding change in the c/a ratio for these alloys with increasing pressure.

A neutron diffraction study of terbium and holmium has been made by Umebayashi *et al* [171] and the variation of the turn angle with pressure up to 6 kbar obtained. Using Miwa's theory [172] the pressure dependence of

ω was shown to be

$$\frac{1}{\omega}\frac{\partial\omega}{\partial p} = -\frac{Kf}{\omega f_\omega}\left(\frac{\partial \ln (m^* A_0)}{\partial \ln \Omega}+\frac{2}{3}\right), \tag{4.38}$$

where

$$f(\omega, S, M) = A_0/E_f \propto m^*\Omega^{\frac{2}{3}}A_0,$$

and

$$f_\omega = \partial f/\partial\omega.$$

Here M is the thermal average of $S(=(g_J-1)J)$, the spin quantum number, and A_0 is an averaged matrix element of the s–f exchange interactions. Using this expression, along with the expression for the pressure dependence of the spin disorder resistivity, [173] calculate $\omega^{-1}\partial\omega/\partial p$. Four other experimental properties show reasonable agreement with the neutron diffraction results, as may be seen from Table 4.4. The large difference between the

Table 4.4 The change in interlayer turn angle with pressure for terbium and holmium [171].

	Tb		Ho	
$T-T_N$ (°K)	−4·0	−9	−7	−36·5
ω at 1 atmos.	20·0	17·7	50·0	45·0
$\omega^{-1}(\mathrm{d}\omega/\mathrm{d}p)\times 10^3$/kbar				
obs.	20±2	23±2	1·2±0·2	2·3±0·4
calc.	19	25	0·6	0·9

results for the two elements are thought to be due to the differences in the wavelength of the periodic spin structure, the longer wavelengths being more sensitive to changes in the conduction band characteristics caused by the applied pressure.

4.3 Theory of the ordered states

4.3.1 *Classical theory of the spin structures*

The early theories of the magnetic ordering in the rare earth metals concentrated predominantly on the heavy elements and were concerned with the variation of the observed spin structure from gadolinium to thulium. Various workers [174–177] adopted a molecular field approach and Elliott [174] considered phenomenologically the interactions of the localized $4f$ electrons with the crystal lattice. The crystal field for the hexagonal lattice has been written by Elliott and Stevens [178] in the operator equivalent form

$$V(\mathbf{r}) = A_2^0\langle r^2\rangle\alpha_J O_2^0 + A_4^0\langle r^4\rangle\beta_J O_4^0 + A_6^0\langle r^6\rangle\gamma_J O_6^0 + A_6^6\langle r^6\rangle\gamma_J O_6^6 \tag{4.39}$$

in the notation of Section 1.31. For tripositive ions, the point charge model gives $A_2^0 = -300$, $A_4^0 = -60$, $A_6^0 = +15$ and $A_6^0 = -90$ (cm^{-1}). The effects of this crystal field appear mainly in the dominantly axial anisotropy, with the sign of the coefficient α_J giving the relation of the quadrupole of the f electrons to the direction of **J**, i.e.

$$\sum_i (3z_i^2 - r_i^2) = \alpha_J[3J_z^2 - J(J+1)].$$

For Tb, Dy and Ho α_J is negative while for Er and Tm it is positive. Elliott then takes the known directions as an indication that in all cases the charge cloud orientations are those to be expected from negative α_J and consequently in erbium and thulium the higher order crystal field terms must be such as to favour alignment parallel to the c-axis. Depending on the relative magnitudes and signs of these components the observed spin orientations can be expected to be either perpendicular or parallel to the c-axis or in some intermediate direction. The tendencies produced by these crystal fields were listed by Elliott [174] and are reproduced here in Table 4.5. These spin orientations

Table 4.5 The effect of the various contributions to the crystal field, $\alpha_J A_2^0$, etc., upon the easy direction of magnetization [174]. The directions shown are with respect to the c-axis. The quantities $2\alpha_J J^2$, etc., give the relative magnitudes of the crystal field terms.

Element	Tb	Dy	Ho	Er	Tm
$\alpha_J A_2^0$	$\perp$	$\perp$	$\perp$	$\parallel$	$\parallel$
$\beta_J A_4^0$	$\parallel$	$\angle$	$\angle$	$\parallel$	$\parallel$
$\gamma_J A_6^0$	$\parallel$	$\angle$	$\parallel$	$\angle$	$\parallel$
$\gamma_J A_6^6$	30°	0°	30°	0°	30°
$2\alpha_J J^2$	0·73	0·71	0·28	0·29	0·73
$8\beta_J J^4$	1·27	4·48	2·16	3·36	1·71
$16\gamma_J J^6$	0·84	2·9	4·7	5·9	4·2

are all based on a dominant A_2^0 contribution, but only in thulium do all crystal field effects favour a single direction, this being parallel to the c-axis. In the remaining elements the changes in the relative magnitudes of A_2^0 will result in a tendency to find the moment with some component out of the plane which increases in the direction Dy, Ho, Tb.

The original attempt to account for the helical spin structures was made for dysprosium by Enz [179] following the work by Herpin, Meriel and Villain [180, 181] on $MnAu_2$. In this treatment the spins (**S**) within each lattice plane perpendicular to the c-axis are assumed to be ferromagnetically coupled. The exchange constant between a spin on a selected plane (n) and a spin on the adjacent planes is denoted by $\mathscr{J}_1$; the coupling between the next nearest planes is denoted by $\mathscr{J}_2$, and so on. If θ_n is the direction of the spins in the nth plane with respect to some fixed direction then the total exchange energy of the system is

$$E_{ex} = -S^2 \sum_m \mathscr{J}_m \sum_n \cos(\theta_{n+m} - \theta_n),$$

and this only has a minimum if $(\theta_{n+1} - \theta_n)$ is a constant, equal to the turn angle $\omega = qc$. The exchange energy for N spins is therefore

$$E_{ex} = -NS^2 \sum_m \mathscr{J}_m \cos(mqc) = -\frac{N}{2}\mathscr{J}(q)S^2, \quad \text{say,} \tag{4.40}$$

where $\mathscr{J}(q)$ is clearly the Fourier transform of the exchange interaction between two spins. Considering $\mathscr{J}_1$ and $\mathscr{J}_2$ only, minimization of E_{ex} shows that a helical spin structure is stable, with turn angle $\omega = \cos^{-1}(-\mathscr{J}_1/4\mathscr{J}_2)$, provided $|\mathscr{J}_1/4\mathscr{J}_2| \leq 1$. Within this approximation $E_{ex}/N = -S^2(\mathscr{J}_1 \cos\omega + \mathscr{J}_2 \cos 2\omega)$.

In order for ω to vary with temperature then $\mathscr{J}_1$ and $\mathscr{J}_2$ must also be variable. One might expect these constants to vary with temperature simply through the effects of thermal expansion on the lattice parameters. This possibility seems to have been first proposed by Kittel through a paper by Jarrett and Elliott [183] and led to the addition of the terms $(\partial\mathscr{J}_1/\partial c)(c - c_0)$ and $(\partial\mathscr{J}_2/\partial c)(c - c_0)$ to the expression for the turn angle. Belov *et al* [188] obtained the temperature dependence of $\mathscr{J}_1$ and $\mathscr{J}_2$ directly by combining the experimental values of the turn angle with those of the critical field. This was possible as Herpin and Meriel [181] had showed that the critical field of a helicoidal structure is given by $H_c = -7{\cdot}76\ \mu_s\ \mathscr{J}_2 \sin^4(\omega/2)$, where μ_s is the spontaneous magnetic moment per atom. Consequently along with $\cos\omega = -\mathscr{J}_1/4\mathscr{J}_2$ both $\mathscr{J}_1$ and $\mathscr{J}_2$ may be determined. Combined with the expansion coefficient data of Banister *et al* [189] the dependence of the exchange constants on c are those shown in Fig. 4.25. These results presuppose that all the variation in $\mathscr{J}_1$ and $\mathscr{J}_2$ comes from their dependence on interplanar spacing, and the effects of anisotropy, magnetization and magnetostriction have been neglected in the total energy. This is also the case in much of the other elementary work concerning the temperature variation of the helical spin structures.

After the transition from the helical to the fan spin structure at H_c, further increase in magnetic field strength serves to decrease the deviation of the planar magnetization vectors from the field direction. This collapse of the fan structure is complete, and true ferromagnetic ordering established at a limiting field $H_f = -16\ \mu_s\ \mathscr{J}_2 \sin^4(\omega/2) = H_c/0{\cdot}485$. Herpin and Meriel also derived expressions for the susceptibility of the helical spin system and the magnetization in the region $H < H_c$.

Using these expressions, along with those for H_c and H_f, to calculate the magnetization curves for dysprosium and terbium, Belov *et al* [188] were able to obtain reasonable agreement with experimental data.

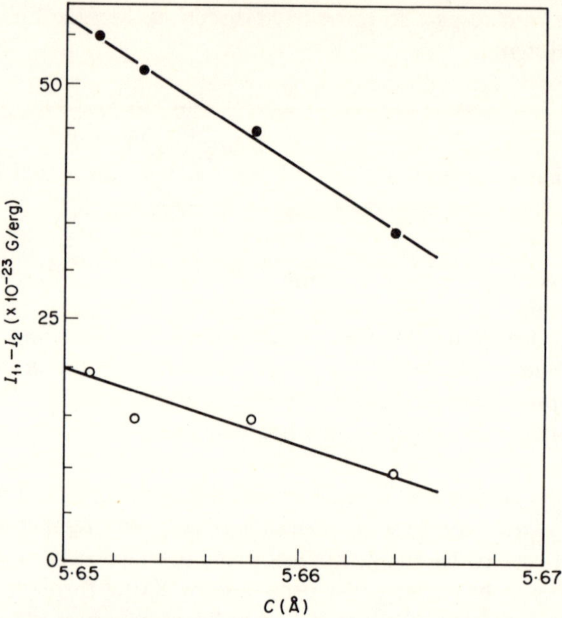

Fig. 4.25 The dependence of $\mathcal{J}_1$ and $\mathcal{J}_2$ of dysprosium upon the axial lattice parameter.

4.3.2 *Theory of the turn angle*

The long range exchange interaction between spins which gives rise to the observed periodic structures is the RKKY interaction discussed at the beginning of the chapter. If the free electron results (4.14) and (4.15) are used for $\mathcal{J}(\mathbf{R}_i)$, then the exchange energy is proportional to

$$\mathcal{J}(\mathbf{q}) = \sum_i \mathcal{J}(\mathbf{R}_i)e^{i\mathbf{q}\cdot\mathbf{R}_i} = A_0^2\chi(\mathbf{q}), \tag{4.41}$$

where a self-energy contribution arising from $\mathbf{R}_i = 0$ has been neglected. This self-energy in fact diverges, but this does not affect the long range order. Since the function $\chi(\mathbf{q})$ in (4.41) has a maximum value at $\mathbf{q} = 0$, so does $\mathcal{J}(\mathbf{q})$, and this model therefore predicts that there can be no oscillatory spin order of any kind. Allowing A to depend upon $\mathbf{q}$ does not alter this situation since $A(\mathbf{q})$ may be assumed in general to have a maximum at $\mathbf{q} = 0$. In order to produce a maximum in $\mathcal{J}(\mathbf{q})$ for $\mathbf{q} \neq 0$ it is therefore necessary to consider a more realistic electronic band structure.

The initial onset of order is determined by the electronic properties in the paramagnetic state, and only this will be considered here. In the ordered state the spin structures modify the energy bands giving rise to extra planes of energy discontinuity which are known as superzones. These effects are described in the next section. Further it will be assumed that only $\mathbf{q}$ parallel to the c-axis will be important since the periodicity of the spin structures is only observed in that direction.

The simplest extension of the free electron model is to introduce the symmetry of the hcp lattice, but still consider the electrons to be free. It is now convenient to use the reduced zone scheme. If $\mathbf{k}$ is restricted to the first zone the interaction energy between spins $\mathbf{S}_i$ and $\mathbf{S}_j$ is then a generalization of expression (4.12), namely, for a Bravais lattice,

$$\mathscr{H}_{ij} = -\sum_{\mathbf{K}_n, q, k} |\langle \mathbf{k}|\mathscr{H}_{\mathrm{sf}}|\mathbf{k}+\mathbf{q}+\mathbf{K}_n\rangle|^2 \left\{\frac{f(\mathbf{k})-f(\mathbf{k}+\mathbf{q}+\mathbf{K}_n)}{E(\mathbf{k}+\mathbf{q}+\mathbf{K}_n)-E(\mathbf{k})}\right\}. \tag{4.42}$$

For the hcp lattice, with two atoms per unit cell, the vectors $\mathbf{R}_i$ in (4.41) are replaced by $\mathbf{R}_i+\boldsymbol{\rho}/2$ and $\mathbf{R}_i-\boldsymbol{\rho}/2$, where $\boldsymbol{\rho}$ joins the atoms. The exchange energy is then proportional to

$$\mathscr{J}(\mathbf{q}) = \sum_{\mathbf{K}_n} |F(\mathbf{K}_n)| A^2(\mathbf{q}+\mathbf{K}_n)\chi(\mathbf{q}+\mathbf{K}_n), \tag{4.43}$$

in which $F(\mathbf{K}_n)$ is the structure factor. Yosida and Watabe [190] have evaluated a sum similar to this, for $\mathbf{q}$ parallel to the c-axis, using the free electron susceptibility function χ, equation (4.10), and assuming $A(\mathbf{q}+\mathbf{K}_n) = A_0$. Their result, which is essentially the curve $\Delta/2k_{\mathrm{f}} = 0$ shown in Fig. 4.28, indicates that $\mathscr{J}(q)$ has a maximum when $qc = 48°$. This value is very close to $\omega_i = qc = 51°$ which is observed in Tm, Er and Ho, but such good agreement must be fortuitous, because of the use of the free electron model. In any case the method is incapable of explaining the variation of turn angle through the series.

Recent calculations of the band structures of the rare earths (cf. Chapter 3) have prompted their use in evaluating more realistic susceptibility functions. However, since these must necessarily be achieved numerically, the results may be in substantial error, and the preliminary results available at the present time, and which are now discussed, must be viewed with caution.

Evenson and Liu [191, 192] have employed the band structures obtained by Keeton and Loucks [193] in the calculation of $\mathscr{J}(\mathbf{q})$ for Gd, Dy, Er and Lu. The use of the double zone representation by them is valid provided the energy band splittings on the zone face AHL, due to relativistic effects, are negligible. In that case, with one atom per unit cell, so that $F(\mathbf{K}_n) = 1$, and assuming $A(\mathbf{q})$ is constant, $\mathscr{J}(\mathbf{q})$ is proportional to the susceptibility function

$$\frac{1}{N}\sum_{\mathbf{k},n,n'} \frac{f_n(\mathbf{k})[1-f_{n'}(\mathbf{k}+\mathbf{q}+\mathbf{K}_0)]}{E_{n'}(\mathbf{k})-E_n(\mathbf{k}+\mathbf{q}+\mathbf{K}_0)}, \tag{4.44}$$

where the bands are labelled by n. Here $\mathbf{K}_0$ reduces $\mathbf{k}+\mathbf{q}$ to the double zone and, restricting attention to the c-axis, it should be observed that the single zone reciprocal lattice vectors, e.g. $\mathbf{K}_1 = (0, 0, 1)$ are not required. Since the conventional structure factor vanishes for these lattice vectors, so that states $\mathbf{k}$ and $\mathbf{k}+\mathbf{q}+\mathbf{K}_1$ do not mix, they will give no contribution to $\mathscr{J}(\mathbf{q})$. Only the bands which determine the Fermi surface were included by Evenson and Liu in expression (4.44). Other bands were found either to contribute only

q-independent terms, or else to be very small because of large energy denominators.

The result of these computations for Dy are shown in Fig. 4.26(a). As observed in Section 4.1.5 a maximum in the susceptibility is expected at a value **q** whenever there are regions of Fermi surface of sufficient area separated by wavevector **q**. Such areas are said to nest together. Within the double zone scheme the most important **q** vectors are those between the two bands, as illustrated in Fig. 4.26(a). Evenson and Liu find that the dominant contribution arises from the vector which calipers the webbing of the whole Fermi surface near *L*. This vector determines the position of the peak in the susceptibility, and is therefore identified with the turn angle.

Keeton and Loucks [193] first observed that the thickness of the webbing (cf. Fig. 3.14) has the same order of magnitude as the wavevector associated

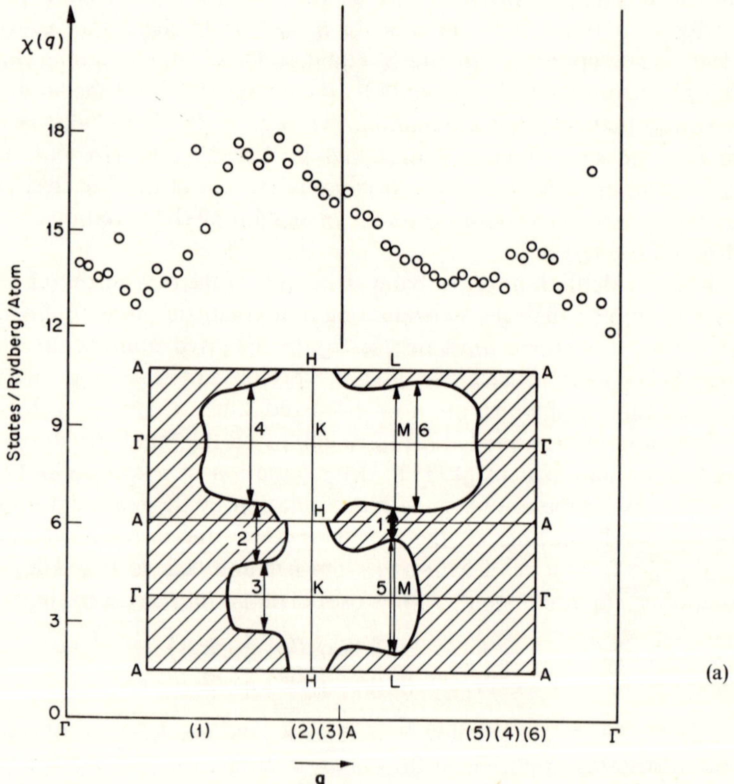

Fig. 4.26 Results of numerical calculations for the susceptibility function for **q** parallel to the *c*-axis. (a) Dysprosium [191]; shaded areas of the inset are regions occupied by holes in the double zone representation. The most important wavevectors which caliper the Fermi surface are shown, and their approximation magnitudes are shown in parentheses on the *q*-axis. (b) Terbium [196].

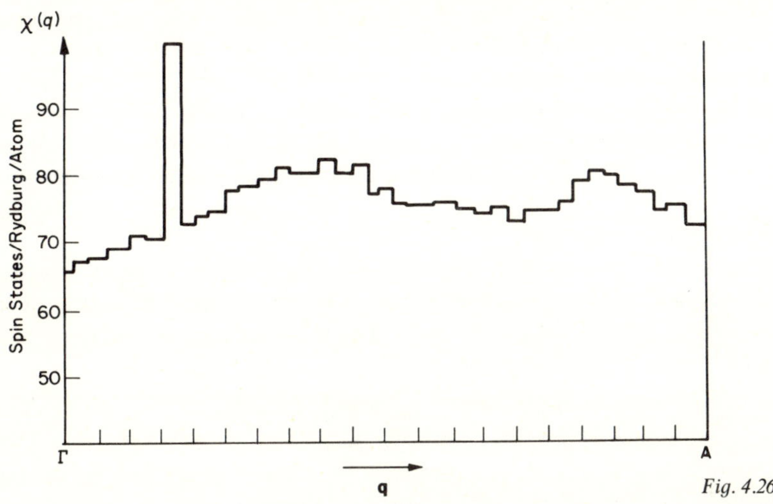

Fig. 4.26 (b)

with the spin structure. Further, it shows the correct trend through the series, and in particular suggests why Gd shows no periodic order, since it has no webbing. However, detailed calculation indicates that the situation is more complex. For Gd Evenson and Liu find that $\chi(\mathbf{q})$ has a maximum at a non-zero **q** vector, and to account for the order a more realistic exchange integral $A(\mathbf{q})$ has been introduced. This was obtained by using an empirical fit of $\mathscr{J}(\mathbf{q})$ to the experimentally determined function for Tb–10% Ho [194] (see Fig. 4.35). The effect of $A(\mathbf{q})$ upon $\chi(\mathbf{q})$ is to reduce the magnitude of $\mathscr{J}(\mathbf{q})$ at large **q**, and to shift its maximum to slightly smaller **q** values. The latter shift results in an improved agreement with experiment, for the elements considered, and appears to be essential in stabilizing the maximum in $\mathscr{J}(\mathbf{q})$ at $\mathbf{q} = 0$ for Gd.

The use of the double zone scheme assumes that the spin–orbit splitting on zone faces AHL is negligible. If the splitting is not negligible, as for the rare earths, there are energy discontinuities on that face, and its associated structure factor is no longer zero. (In any case it is expected on general grounds that the structure factor is never identically zero [195]; see Section 3.2 prior to equation (3.11).) It is then necessary to use the single zone scheme and to include the single zone reciprocal lattice vectors ($\mathbf{K}_1$) in expression (4.44). This has been done by Jackson and Doniach [196] for Tb, and results are shown in Fig. 4.26(b). In addition to the broad maximum found by Evenson and Liu they find a very sharp spike. This is due to a **q**-vector which nests the flat part of the webbing near point L with itself, in the repeated zone scheme, across the faces AHL. Jackson and Doniach conclude that it is this **q**-vector, rather than that of the broad maximum, which determines the turn angle. In fact these same authors have shown that the broad maximum is quite insensitive to the position of the Fermi energy, and this is discussed further in Section 4.3.5.

It must be emphasized that the numerical results available at the present time may not be reliable. However, it is clear that the properties of the paramagnetic Fermi surface play a fundamental role in determining the turn angle at the onset of order. The variation of the periodicity with temperature cannot be explained in terms of the paramagnetic band structure, and it is necessary to consider the effect of the magnetic order upon the bands.

4.3.3 *Superzone boundaries*

As discussed in Section 4.2.6 the spin orderings which occur in rare earth metals are basically of two types. There is a longitudinal oscillation of spin amplitude along the *c*-axis, as in Er or Tm at higher temperatures but below T_N, which may be described by

$$\langle S_n \rangle = (g_J - 1)\langle J_z \rangle \sin(\mathbf{q}\cdot\mathbf{R}_n + \phi), \tag{4.45}$$

and there is also a helical structure with the moments confined to the basal plane as in Tb, Dy and Ho, which may be described by

$$\begin{aligned} \langle S^x_\mathrm{n} \rangle &= (g_J - 1)\langle J_\perp \rangle \cos(\mathbf{q}\cdot\mathbf{R}_\mathrm{n}) \\ \langle S^y_\mathrm{n} \rangle &= (g_J - 1)\langle J_\perp \rangle \sin(\mathbf{q}\cdot\mathbf{R}_\mathrm{n}). \end{aligned} \tag{4.46}$$

Here n labels a lattice site. More complex structures are represented by a combination of these two types, i.e. equation (4.35).

Acting through $\mathscr{H}_\mathrm{sf}$ the ordered moments give rise to a periodic potential

$$\mathscr{H} = \sum_{\mathbf{R}_\mathrm{n}} A(\mathbf{r} - \mathbf{R}_\mathrm{n})\boldsymbol{\sigma}\cdot\mathbf{S}_\mathrm{n} \tag{4.47}$$

which influences the conduction electrons. Since in general the periodicity of (4.47) is different from the lattice its effect is to introduce new boundaries of energy discontinuity in reciprocal space, which are known as *superzone* boundaries. Elliott [197–199] has shown that they are important in determining many properties, particularly the turn angle, as described in the following section, and also electrical resistivity (see Chapter 5).

The effect of any periodic potential of characteristic wave number $\mathbf{q}$ is to mix electrons in states $\mathbf{k}$ and $\mathbf{k}+\mathbf{q}+\mathbf{K}_\mathrm{n}$ where the $\mathbf{k}$'s are restricted to the first Brillouin zone. In the case of $\mathscr{H}$ given by (4.47) the mixing also depends upon the spin of the electrons, and it therefore differs for the two types of order.

For longitudinal order only electron states of like spin are mixed by $\mathscr{H}$ and the matrix elements are $\pm\frac{1}{2}A(K_\mathrm{n}+q)(g_J-1)\langle J_z \rangle F(\mathbf{K}_\mathrm{n})$. The mixing is strongest when states $\mathbf{k}$ and $\mathbf{k}'$ are degenerate in energy, i.e. on planes such that $z = \pm q/2$, and here $\mathscr{H}$ lifts this degeneracy and produces energy gaps of magnitude $A(g_J-1)\langle J_z \rangle F(\mathbf{K}_\mathrm{n})$. Note that for certain $\mathbf{K}_\mathrm{n}$ the conventional structure factor vanishes, and there are no gaps, but see the discussion in the previous subsection.

In the case of the helical spin structure only states which have opposite spin are mixed and then the (off diagonal) matrix elements are $A(g_J-1)\langle J_\perp\rangle F(\mathbf{K}_n)$. As a result a single new plane of energy discontinuity appears in each separate spin sub-band, at $z = \frac{1}{2}q$ for spin up electrons and at $z = -\frac{1}{2}q$ for spin down electrons.

In general, when both types of spin order are present, the electron energies are modified by a single zone boundary to be

$$E'(\mathbf{k}) = \tfrac{1}{2}\{E(\mathbf{k})+E(\mathbf{k}+\mathbf{K}_n\pm\mathbf{q})\pm[(E(\mathbf{k})-E(\mathbf{k}+\mathbf{K}_n\pm\mathbf{q}))^2+\Delta^2|F(\mathbf{K}_n)|^2]^{\frac{1}{2}}\}, \quad (4.48)$$

where the energy gaps are given by

$$\Delta|F(\mathbf{K}_n)|, \text{ and} = A(g_J-1)\{\langle J_z\rangle^2+2\langle J_\perp\rangle^2\pm2\langle J_z\rangle[\langle J_z\rangle^2+\langle J_\perp\rangle^2]^{\frac{1}{2}}\}. \quad (4.49)$$

It is usually more convenient to describe the superzone boundaries in the extended zone scheme when the free electron approximation is employed, since then the unperturbed Fermi surface is a sphere. In this case the superzone boundaries are the planes bisecting the vectors $(\mathbf{K}_n\pm\mathbf{q})$. Of course, as has been repeatedly emphasized, the true Fermi surfaces for the rare earths are nothing like spheres, but in order to make any progress it is often necessary to resort to the simplest approximation.

4.3.4 *The temperature dependence of the turn angle*

The superzone boundaries are important in determining the variation of turn angle $\omega = qc$ with temperature, for they modify the Fermi surface so that their positions ($\mathbf{q}$) are such that the total free energy of the re-distributed electrons is a minimum. The new Fermi surface, which is one of constant energy given by (4.48) depends on the energy gaps $\Delta|F(\mathbf{K}_n)|$ and hence may vary appreciably with temperature through the moment $(g_J-1)\langle J\rangle$. In this way $\mathbf{q}$ depends upon the temperature.

Detailed calculations have only been made using the free electron theory and assuming $A(\mathbf{q}) = A_0$. Since $\mathbf{q}$ has been identified with the vector which calipers the Fermi surface in the c-direction, it would be expected that the largest effects of the superzones would be from the boundaries which lie perpendicular to the c-axis and are close to the old Fermi surface, so that the situation illustrated in Fig. 4.27(a) obtains. This situation would always be expected to be more energetically favourable than those illustrated in Figs. 4.27(b, c), because the occupied states near the boundary have their energy lowered while the unoccupied states have their energy raised. It would also be expected that the larger the energy gaps, i.e. the lower the temperature, the smaller would be the optimum value of $|\mathbf{q}|$, and indeed this is found to be so experimentally.

In general the effects of many zones must be balanced in order to obtain the value of q which minimizes the total energy. In calculating the energy it is more convenient to compare the average energy per electron $E'(q, \Delta)$ with that of the ferromagnetic state for which $q = 0$. This energy difference may

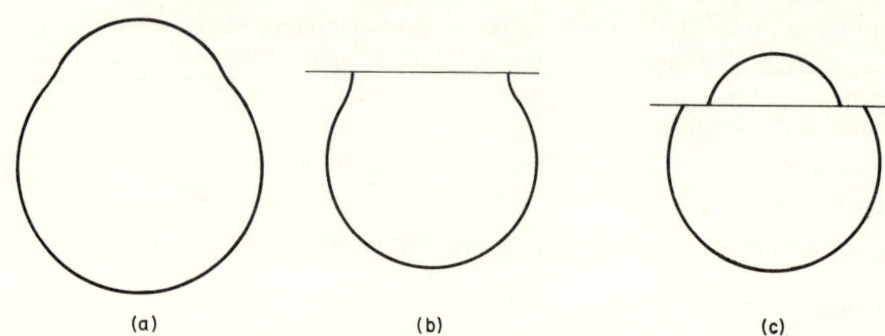

Fig. 4.27 The perturbation of the spherical free electron Fermi surface by a single superzone boundary.

be written, after Elliott and Wedgewood [198], as

$$E'(q,\Delta)-E'(0,\Delta) = -\frac{3\Delta^2}{16E_f}\sum_{\mathbf{K}_n}|F(\mathbf{K}_n)|^2\{\chi(|\mathbf{q}+\mathbf{K}_n|,\Delta F(\mathbf{K}_n)) + \chi(|\mathbf{K}_n-\mathbf{q}|,\Delta F(\mathbf{K}_n))-2\chi(|\mathbf{K}_n|,\Delta F(\mathbf{K}_n))\}, \quad (4.50)$$

where $\chi(q, \Delta F(\mathbf{K}_n))$ is a complicated function which reduces to the free electron susceptibility when $\Delta \to 0$. In particular it is found that deviations from the free electron $\chi(q)$ are negligible for $\mathbf{K}_n$ beyond the four smallest vectors. A graph of equation (4.50) for various values of $\Delta/2E_f$ is shown in Fig. 4.28. The curve $\Delta/2E_f = 0$ corresponds to that obtained by Yosida and Watabe [190] and has a maximum at $qc = 48°$. As the energy gaps increase in size so the q value at maximum energy decreases until at $\Delta/2E_f \sim 0{\cdot}175$ eV it suddenly jumps to $q = 0$ indicating a transition to a ferromagnetic state. Such a transition is a common feature of the heavy rare earths, and more will be said of it in Section 4.3.5.

Values for the turn angles predicted by Elliott and Wedgewood [198] with the choice $A_0 = 0{\cdot}55$ eV are compared with experimental results in Fig. 4.18 and are seen to be in good agreement for Tm, Er and Ho. In the cases of Tm and Er the variations are small because the moments are small, and because both exhibit longitudinal spin order. For this type of order the superzone boundaries occur in pairs so that less freedom is possible in the variation of q than is so for the helical structure [200]. The predicted discontinuity in qc for Er arises from a change in spin structure at $T/T_N = 0{\cdot}22$. The theory also successfully accounts for the variation of the turn angle in Tb–Y alloys [198].

It is clear from Fig. 4.18 that the agreement is not good for Dy and Tb metals, particularly near T_N, where the theory always predicts $qc = 48°$C. Further, the value of A_0 used is considerably higher than values obtained from other data. Thus while the effects of the superzones must be important in

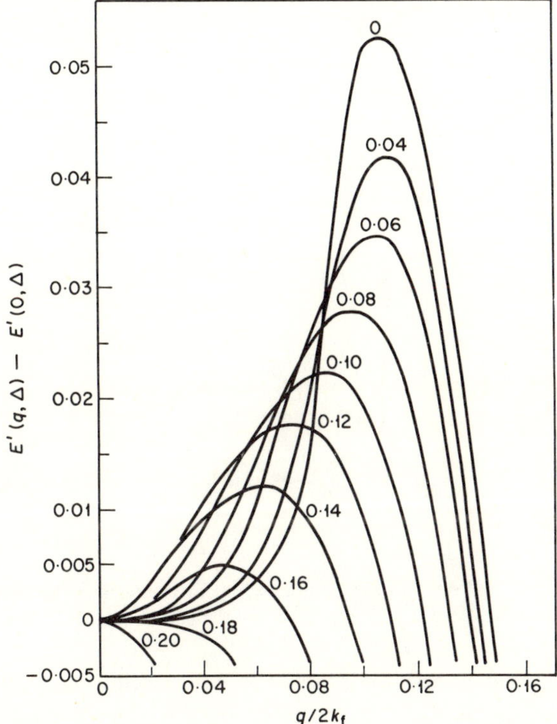

Fig. 4.28 The variation of the energy difference, equation (4.50), between the helical and ferromagnetic spin structures as a function of $q/2k_f$ for various values of $\Delta/2E_f$ [198].

determining q, it is apparent that the use of free electron theory is inadequate, and that some other mechanism dominates near T_N.

The effect of the periodic potential (4.47) upon a realistic energy band structure has been considered by Watson *et al* [201]. They primarily considered Tm and investigated how the paramagnetic Fermi surface obtained by a non-relativistic APW calculation would be modified by the energy gaps of superzones; see Fig. 4.29. Unfortunately, because of uncertainties in the positions of the bands due to the effect of exchange, it was not possible to tell whether the gaps coincided with the Fermi energy, and the repopulation of the electronic states could therefore not be calculated.

Near T_N de Gennes [202] has suggested that another mechanism is important in determining q. At this temperature electron scattering due to spin disorder is appreciable, and in consequence the electrons have a finite mean free path l as discussed in Section 4.1.5. The effect of this is to modify the susceptibility $\chi(q)$ and thereby the value of q which maximizes $\mathscr{J}(q)$. De Gennes and St-James [182] have used values of l obtained from electrical resistivity measurements [203] which vary from 5 Å for Gd to 40 Å for Tm, to calculate the turn angle of the heavy rare earth metals at T_N. They find

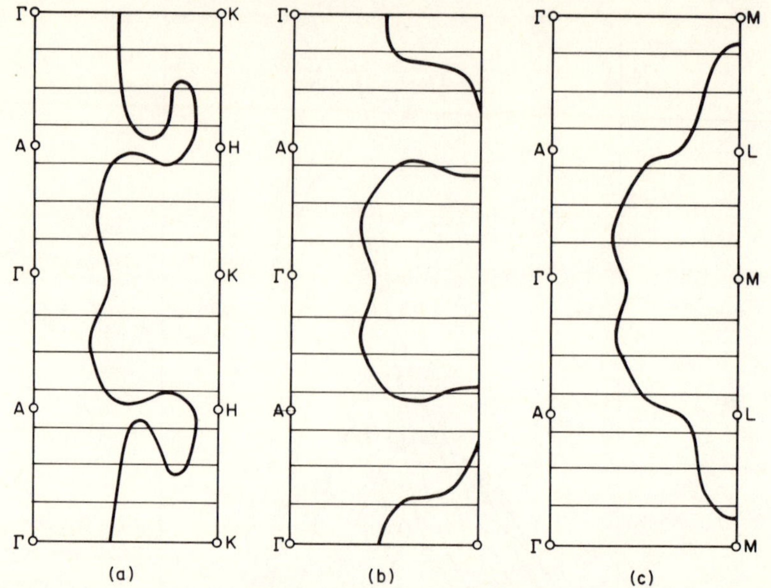

Fig. 4.29 A cross-section of the Fermi surface of thulium. The horizontal lines show the positions of superzone boundaries at $k = \pm n(2\pi/7c)$ introduced by the helical magnetic structure [201].

that the optimum value, assumed along the c-direction, decreases as the scattering increases, i.e. as the ionic moment increases, as is shown in Fig. 4.30. The good agreement with experiment indicates that this is the dominant mechanism at T_N. Of course, as the temperature is lowered the scattering due to spin disorder decreases, but then the superzone boundaries become important.

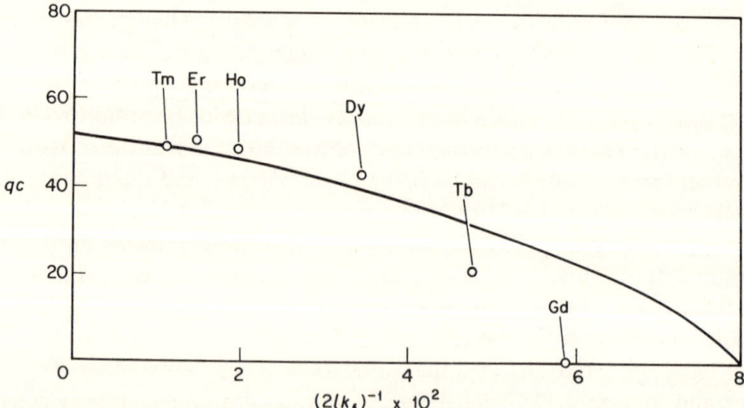

Fig. 4.30 The theoretical variation of the turn angle with electron mean free path (l) near the Néel temperature. The solid circles represent observed values [202].

The effects of superzones and mean free path have been combined in a single theory of Miwa [172]. His results are essentially those discussed above. However, a problem arises in that no consistent value of A_0 exists which will give fits to all the basic experimental data. Again, this difficulty is assumed to stem from the use of the free electron approximation.

4.3.5 *The low temperature transition to ferromagnetic order*

While the existence of the periodic spin structures are maintained by the long range exchange interaction, their exact forms are determined by anisotropy and magnetostriction, both of which favour ferromagnetic order. At lower temperatures the energies due to anisotropy and magnetostriction become increasingly larger, thereby contributing to the reduction in turn angle with temperature, and they must be important in determining the eventual transition to ferromagnetism at the lower ordering temperature T_c. It has been suggested by Cooper [122, 204, 205] that magnetostriction is the dominant process which stabilizes the ferromagnetic order, and his arguments will be discussed below. However, the real situation is more complex and there is evidence [196] that the effects of changes in band structure due to exchange splitting is equally, if not more, important.

The effects of magnetostriction have been considered in detail by Cooper [122]. Here attention is focused upon Dy because this metal has the simplest, helical, spin structure. It is known that the transition from helical [71] to ferromagnetic states for Dy may be induced at $T > T_c$ by applying a magnetic field of sufficient size H_c. The energies of both states, with H_c applied, have been calculated by Elliott [174] using Enz's model [179], and are

$$-[\mathscr{J}_1+(\mathscr{J}_2-CM^2)]M^2-E_d-\mu H_c M, \qquad \text{(ferro)}, \qquad (4.51)$$

$$-[\mathscr{J}_1\cos(qc)+(\mathscr{J}_2-CM^2)\cos(2qc)]M^2, \qquad \text{(helical)}, \qquad (4.52)$$

where E_d is the free energy of the anisotropy plus magnetostriction which gives rise to the transition, and $(\mathscr{J}_2 - CM^2)$ replaces $\mathscr{J}_2$ to allow for the temperature dependence of the turn angle qc. From equation (4.51) and (4.52) an expression for E_d is readily obtained in terms of quantities $\mathscr{J}_1$, $cq(T)$, $H_c(T)$ and $M(T)$ which are known experimentally. In this way Elliott was able to deduce that E_d was proportional to M^6. Since the basal plane anisotropy energy at low temperatures behaves like M^{21}, it is concluded that anisotropy is not dominant in producing the transition at H_c, and therefore is also not dominant at T_c when $H_c = 0$.

The theory of magnetostriction proposed by Callen and Callen [121] has been described previously in Section 4.3.5. The total strain-dependent energy E_{ms}, equation (4.28), is the sum of elastic (E_e) and magnetoelastic (E_{me}) terms and the forms of both are determined by symmetry. With the magnetization restricted to the basal plane in dysprosium, it is sufficient as a first approximation to consider only the strains (ε^{γ}'s) in that plane, and then

$$E_e = \tfrac{1}{2}C^\gamma[(\varepsilon_1^\gamma)^2 + (\varepsilon_2^\gamma)^2], \tag{4.53}$$

$$E_{me} = -B^{\gamma,2}[\varepsilon_2^\gamma \alpha_x \alpha_y + \tfrac{1}{2}\varepsilon_1^\gamma(\alpha_x^2 - \alpha_y^2)], \tag{4.54}$$

where the quantities are as defined after (4.29). A more general treatment has been given by Evenson and Liu [192] but their conclusions are the same as those of Cooper which are given here. The equilibrium strains ($\bar{\varepsilon}^\gamma$) arising from a given net magnetization are those which minimize E_{ms}. If the moments are aligned ferromagnetically along an easy axis E_{ms} is obtained by substitution of the corresponding $\bar{\varepsilon}^\gamma$ into (4.53) and (4.54) with the result

$$E_{ms} = -\tfrac{1}{2}C^\gamma(\bar{\varepsilon}_1^\gamma)^2 = -\tfrac{1}{8}C^\gamma(\lambda^\gamma)^2, \tag{4.55}$$

where λ^γ is defined in Section 4.2.5. It is this term in the magnetostrictive energy which favours ferromagnetic alignment and aids the applied field in producing the transition at T_c. Since at low temperatures λ^γ is the order of M^3, the difference in E_{ms}, given by (4.55), between helical and ferromagnetic configurations varies as M^6, and therefore one concludes that this energy contribution dominates the transition. A similar analysis may be made for terbium.

It was first observed by Elliott and Wedgewood [198] that the effect of exchange upon the electronic band structure leads to the prediction that ferromagnetic order may be energetically more favourable than a helical structure at sufficiently low temperatures. Their work employed the free electron theory and has been briefly described in the previous two sections. The calculated energy differences between the helical and ferromagnetic states are illustrated in Fig. 4.28. As the temperature is lowered the splittings of the energy bands at the superzone boundaries, essentially Δ, increases as the thermally averaged moment $\langle J \rangle$. As Δ increases the turn angle wave-vector q, which determines the positions of the maxima in $E'(q, \Delta) - E'(0, \Delta)$, becomes smaller until at $\Delta/2kf \sim 0{\cdot}013$ Ry the maximum suddenly jumps to $q = 0$, indicating a transition to ferromagnetic order.

The use of free electron theory has severe restrictions and recently Jackson and Doniach [196] have employed realistic RAPW bands [206] to investigate how exchange splitting modifies the susceptibility function given by equation (4.44). To do this they introduced a ferromagnetic exchange splitting Δ_f which rigidly shifts the bands of electrons with opposite spin. For the favoured (up) spins the bands are then $E_n(\mathbf{k}) - \frac{1}{2}\Delta_f$, whilst for down spins the bands become $E_n(\mathbf{k}) + \frac{1}{2}\Delta_f$, and expression (4.44) now explicitly involves a summation over spins. The sums were achieved numerically. The results indicate that the spike feature, Fig. 4.26(b), moves to smaller q values as Δ_f increases until at $\Delta_f \sim 0{\cdot}005$ Ry it suddenly disappears. On the other hand the broad maximum in $\chi(q)$ is still present at even higher Δ_f values, and it is found to be insensitive to Δ_f. Consequently Jackson and Doniach suggest that the spike alone determines the turn angle, and its disappearance indicates a transition to ferromagnetism. That the broad maximum in the susceptibility is unimportant is also substantiated by the results of Evenson and Liu [191] for Gd. As remarked earlier, they find that the maximum exists at a finite q vector, and they have to assume that the exchange integral $A(\mathbf{q})$ is important in stabilizing the ferromagnetic state.

The wavevector of the spike is associated with the thickness of the flat portion of the Fermi surface near L. Its disappearance is due to the exchange splitting causing this feature of the Fermi surface to disappear. Earlier Moller *et al* [194] had suggested a phenomenon of this kind to explain the inelastic neutron scattering results for a Tb–10% Ho alloy; cf. Section 4.4.3.

4.4 Spin waves

The groundstate of a magnetically ordered material at $T = 0°\mathrm{K}$ is one in which all the atomic spins are rigidly fixed with respect to one another. If the system is excited, thermally or otherwise, the low lying excited states are those in which the spin deviations travel through the solid as spin waves, and these determine many of the bulk properties of the material, as will be discussed in subsequent sections. The theory of spin waves in the rare earth metals has been comprehensively reviewed by Cooper [122] and the following discussion draws heavily upon that work. The reader is referred to it for further details.

4.4.1 *Theory of spin waves in planar ferromagnets*

For the planar ferromagnets Dy and Tb, below T_c, and neglecting magnetostriction, the Hamiltonian for the spin system is the sum of exchange and anisotropy terms:

$$\mathscr{H} = -\sum_{i \neq j} \mathscr{J}(\mathbf{R}_i - \mathbf{R}_j)\mathbf{S}_i \cdot \mathbf{S}_j - \sum_i (\mathrm{B}_2^0 S_{iz}^2 - \tfrac{1}{2}\mathrm{B}_6^6[(S_{ix} + iS_{iy})^6 + (S_{ix} - iS_{iy})^6]), \quad (4.56)$$

where the contributions of B_4^0 and B_6^0 are neglected for convenience. For the rare earths it is necessary to project $\mathbf{S}$ on to $\mathbf{J}$ and it will be implicitly assumed here that this has to be done. In the standard treatments of spin waves anisotropy is neglected, but it is large for the rare earths, and therefore must be included. There is a further complication in obtaining the spin wave energies in that it is strictly necessary [207] to subdivide the hcp lattice into two simple hexagonal Bravais lattices. However, the correct result for the *acoustic* spin waves may be obtained by considering the hcp lattice as a whole, and at $T = 0$ the spin wave energy is [208, 209]

$$\hbar\omega(\mathbf{k}) = S\{[-2\mathrm{B}_2^0 + 2\mathscr{J}(0) - 2\mathscr{J}(\mathbf{k}) - 6\mathrm{B}_6^6 S^4] \times [2\mathscr{J}(0) - 2\mathscr{J}(\mathbf{k}) - 36\mathrm{B}_6^6 S^4]\}^{\frac{1}{2}}, \quad (4.57)$$

where $\mathbf{k}$ is the wavenumber. This exhibits an energy gap at $\mathbf{k} = 0$, of approximate magnitude $S^3(\frac{9}{4}\mathrm{B}_6^6\mathrm{B}_2^0)^{\frac{1}{2}}$, because of the anisotropy which always tends to oppose deviations of the spins from equilibrium. For Dy the predicted

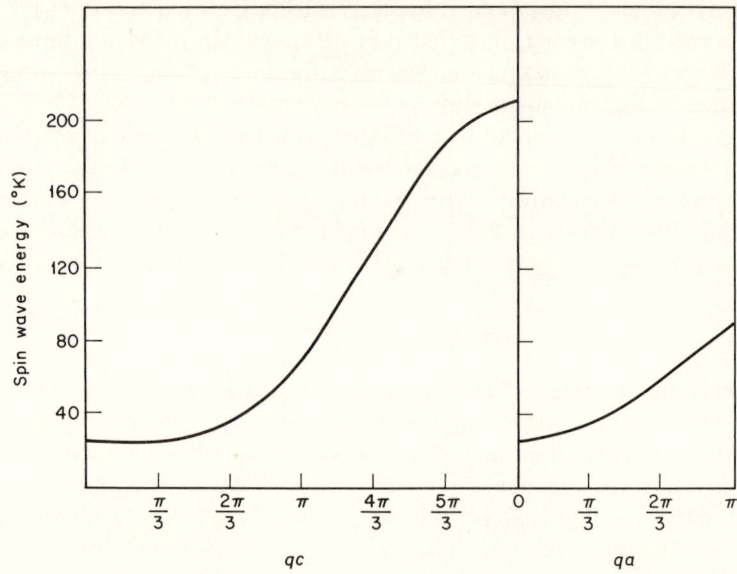

Fig. 4.31 The predicted low temperature spin wave spectrum for dysprosium [122].

energy gap is about 24°K. The dispersion law (4.57) is illustrated in Fig. 4.31. At $T \neq 0$ the temperature dependence of the anisotropy may be introduced [194, 21] by multiplying the constants $\mathrm{B}_n^m(T=0)$ by the appropriate $\hat{I}_{(2l+1)/2}$ factor, and dividing by the reduced magnetization. It is usually assumed [194] that $\mathscr{J}(\mathbf{k})$ is simply scaled by the reduced magnetization, although experiment indicates that the **k**-dependence of this quantity does in fact alter with temperature. Attempts [211] to put the temperature dependence on a firmer theoretical basis have had limited success.

The effect of an applied field upon the spin wave energies has been described by Cooper *et al* [212] and may be summarized in the dispersion relation.

$$\hbar\omega(\mathbf{k}) = \{[2S(\mathscr{J}(0)-\mathscr{J}(\mathbf{k})-\mathrm{B}_2^0-3\mathrm{B}_6^6S^4\cos 6\delta)+g\mu_\mathrm{B}H_\parallel\cos\delta+g\mu_\mathrm{B}H_\perp\sin\delta]$$
$$\times[2S(\mathscr{J}(0)-\mathscr{J}(\mathbf{k})-18\mathrm{B}_6^6S^4\cos 6\delta)+g\mu_\mathrm{B}H_\parallel\cos\delta+g\mu_\mathrm{B}H_\perp\sin\delta]\}^{\frac{1}{2}}. \tag{4.58}$$

Here $H_\parallel$ and $H_\perp$ are respectively the components of the *internal* field parallel and perpendicular to the easy direction of magnetization, and δ is the angle between the equilibrium direction of the magnetization and the easy direction. It is often convenient to define an hexagonal anisotropy field H_h by $g_J\mu_\mathrm{B}H_\mathrm{h} = \mathrm{B}_6^6S^5$. For an internal field of magnitude $36H_\mathrm{h}$ along a hard direction it may be shown that the moments are entirely pulled into that direction ($\delta = \pi/3$), and that $\omega(0) = 0$, as is illustrated in Fig. 4.32. This later

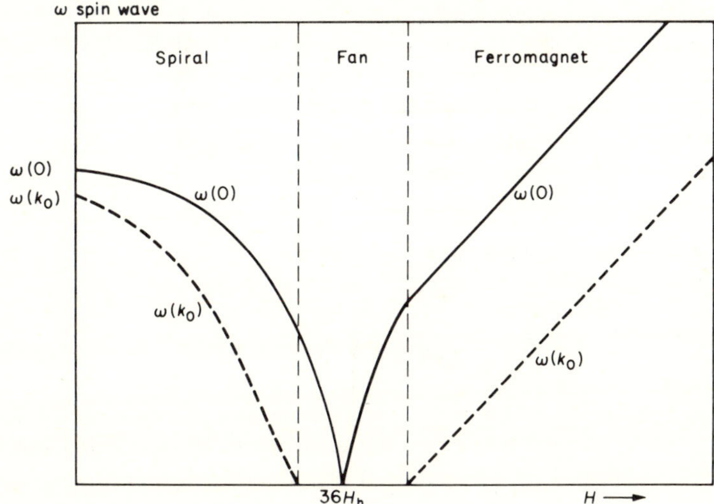

Fig. 4.32 The predicted variation of $\omega(0)$ and $\omega(\mathbf{k})$ for a planar ferromagnet with the magnetic field applied in the hard basal plane direction [212].

fact is important in ferromagnetic resonance studies, as is described in Section 4.4.4.

It has been seen in Section 4.3.5 that magnetoelastic effects are comparable to those of anisotropy. It would therefore be expected that they are important in determining spin wave energies and this has been considered in detail by Cooper [204, 205]. There are two limiting approximations.

In the *frozen lattice* approximation it is assumed that for small $\mathbf{q}$ the lattice strains are frozen so that they have their static, equilibrium values. Since the magnetization is restricted to the basal plane, the equilibrium strains to be considered are $\bar{\varepsilon}_1^\gamma$ and $\bar{\varepsilon}_2^\gamma$; compare equations (4.53) and (4.54). The spin wave energies are then

$$\hbar\omega(\mathbf{k}) = S\{[2\mathscr{J}(0)-2\mathscr{J}(\mathbf{k})-2\mathrm{B}_2^0-6\mathrm{B}_6^6S^4+\bar{\mathrm{B}}_2^\gamma\lambda^\gamma/2] \times[2\mathscr{J}(0)-2\mathscr{J}(\mathbf{k})-2\mathrm{B}_2^0-36\mathrm{B}_6^6S^4+\bar{\mathrm{B}}^\gamma\lambda^\gamma]\}^{\frac{1}{2}}, \quad (4.59)$$

in which the temperature dependence of the magnetostatic constant $\bar{\mathrm{B}}^\gamma$ may be obtained in terms of λ^γ and c^γ through the relation $\bar{\mathrm{B}}^\gamma = 3c^\gamma\lambda^\gamma/S(2S+1)$. Because the magnetoelastic energy has cylindrical symmetry, it contributes to $\omega(0)$ even at $H = 36H_\mathrm{h}$ when an internal field $\mathbf{H}$ acts along a hard axis.

The other extreme approximation assumes that for long wavelength modes the strain is able to follow the oscillations of the moments. In this case the magnetoelastic contribution to $\hbar\omega(\mathbf{k})$ considered above vanishes, for the energy has cylindrical symmetry, and is therefore independent of the orientation of the strains, and therefore of the moments. Of course, when the strains are frozen the moments are able to have different orientations with them, and there is a magnetoelastic contribution, as in (4.59). To obtain a contribution in the unfrozen case it is necessary to include second order magnetostriction effects, i.e. to include fourth order terms in S in the magnetoelastic energy so that [213]

$$E_{\mathrm{me}} = \varepsilon_1^\gamma[2C_1(\alpha_x^2+\alpha_y^2)+C_2(8\alpha_x^2\alpha_y^2-1)] + 4\varepsilon_2^\gamma[C_1\alpha_x\alpha_y+C_2\alpha_x\alpha_y(\alpha_x^2-\alpha_y^2)], \tag{4.60}$$

where the α's are direction cosines of $\mathbf{S}$. Minimizing the sum, E_{ms}, of this and E_{e} given by (4.53), to obtain the strains, and substituting back into E_{ms}, gives

$$E_{\mathrm{ms}} = \frac{C(\lambda^\gamma)^2}{4}(\alpha_x^2+\alpha_y^2)+\frac{c^\gamma\lambda^\gamma}{4}\cos 6\phi. \tag{4.61}$$

Here $C = \lambda^\gamma/2 = -C_1/c^\gamma$ and $A = -2C_2/c^\gamma$, and are respectively the first and second order magnetostriction parameters of Rhyne and Legvold which may be obtained experimentally by expressing the a-axis magnetostriction as

$$\frac{\delta l}{l} = -2C\sin^2\phi + A\sin 2\phi, \tag{4.62}$$

where ϕ is the angle made by the magnetization with that axis. The first order term in (4.61) is cylindrically symmetric and is that previously considered. The second order term, however, has hexagonal symmetry and therefore can contribute to the spin wave energy even when the strain follows the magnetization. It has the same effect as the basal plane anisotropy and both terms may be compounded into $(\mathrm{B}_6^6)' = \mathrm{B}_6^6 + c^\gamma\lambda^\gamma A/4$, say. The spin wave dispersion relation is then given by (4.57) with B_6^6 replaced by $(\mathrm{B}_6^6)'$ but it should be noted that while B_6^6 varies with temperature as $I_{13/2}$ the contribution $\lambda^\gamma A$ varies as $I_{5/2}I_{9/2}$. In an applied field along a hard planar direction the energy $\hbar\omega(0)$ is zero at $H = 36(\mathrm{B}_6^6)'S^5$. Hence if an energy gap is observed experimentally at this field, it must arise from the frozen lattice mechanism. In fact ferromagnetic resonance experiments (Section 4.4.4) indicate that there is no gap.

4.4.2 *Theory of spin waves for the periodic spin structures*

Here the theory is much more complex than that for the planar ferromagnets and no attempt is made to give anything but the basic results. For details reference should be made to Cooper *et al* [212], Cooper and Elliott [214], or Cooper [122].

Attention is focused upon the simplest helical structures, such as are observed in Dy and Tb. With a field applied in the basal plane, and ignoring anisotropy in that plane, the appropriate Hamiltonian for the spin system is

$$\mathscr{H} = \sum_{i\neq j}\mathscr{J}(\mathbf{R}_i-\mathbf{R}_j)\mathbf{S}_i\cdot\mathbf{S}_j - \mathrm{B}_2^0\sum_i S_{iz}^2 - g\mu_{\mathrm{B}}H\sum_i S_{ix}. \tag{4.63}$$

For zero field this gives for the spin wave energies

$$\hbar\omega(\mathbf{k}) = S\{[2\mathscr{J}(\mathbf{q})-\mathscr{J}(\mathbf{q}+\mathbf{k})-\mathscr{J}(\mathbf{q}-\mathbf{k})][-2\mathscr{J}(\mathbf{k})+2\mathscr{J}(\mathbf{q})-2B_2^0]\}^{\frac{1}{2}}, \quad (4.64)$$

where $\mathbf{q}$ is the wavevector of the spin structure. It should be noted that in general $\hbar\omega(\mathbf{k}) \propto \mathbf{k}$ for $\mathbf{k} \to 0$, and that the energies are largest for $\mathbf{k} \sim \mathbf{q}$. The application of a magnetic field tends to align the moments, thereby distorting the helix. Above the critical value H_f (Section 4.3.1) the field produces a complete transition to the ferromagnetic state, and the spin wave energies are then given by equation (4.58), with B_6^6 assumed zero here. In particular, for $\mathbf{k} = 0$,

$$\hbar\omega(0) = \{[-2B_2^0S+g\mu_BH]\mu_BH\}^{\frac{1}{2}}. \quad (4.65)$$

At small magnetic field strengths it has been shown [215] that $\hbar\omega(0) = 0$ independent of field. The behaviour of $\hbar\omega(q)$ to order H^2 has also been considered and in this case the effect of H is to lift the degeneracy of the $\pm q$ states and give rise to two modes, a symmetric one denoted by $\hbar\omega(\cos q)$ and an antisymmetric one $\hbar\omega(\sin q)$. These are found to be not very sensitive to H so that extrapolation of the low field results to $H = H_c$ should be reasonably accurate.

For fields just less than H_f the $\mathbf{k} = 0$ excitations of the fan structure may be shown to be given by

$$\hbar\omega(0) = \{g\mu_BH[g\mu_BH-2B_2^0S]-8S[\mathscr{J}(\mathbf{q})-\mathscr{J}(0)]\beta^2(-2B_2^0S)\}^{\frac{1}{2}}, \quad (4.66)$$

where β is the angular amplitude of the fan. In this case also, the degeneracy of the $\mathbf{k} = \pm q$ states is lifted, and Nagamiya [216] has shown that

$$\hbar\omega(\cos q) = 0, \text{ independent of } H,$$

and (4.67)

$$\hbar\omega(\sin q) = \beta\{-16B_2^0[2S^2(\mathscr{J}(\mathbf{q})-\mathscr{J}(0))+S^2(\mathscr{J}(\mathbf{q})-\mathscr{J}(2\mathbf{q}))\}^{\frac{1}{2}}.$$

It is assumed that these results may be extrapolated to $H = H_c$, at which field there must clearly be a discontinuity in spin wave energies. This is illustrated in Fig. 4.33.

The effect of including basal plane anisotropy is to shift the critical fields H_c and H_f, to $H_c - H_h$ and $H_f - 36H_h$ respectively, when H is applied along an easy axis. For very large anisotropy it is therefore possible to completely eliminate the fan structure.

Again ignoring basal plane anisotropy, the behaviour of the *conical* spin structure may be accounted for by including the higher axial anisotropy terms B_4^0 and B_6^0 in the Hamiltonian (4.63). The spin wave dispersion relation is still such that $\hbar\omega(0) = 0$, but it is now asymmetric with respect to $\mathbf{k} = 0$, i.e. $\hbar\omega(\mathbf{k}) \neq \hbar\omega(-\mathbf{k})$. This might be expected since the ferromagnetic component of the cone defines a unique direction in the crystal. In particular, as

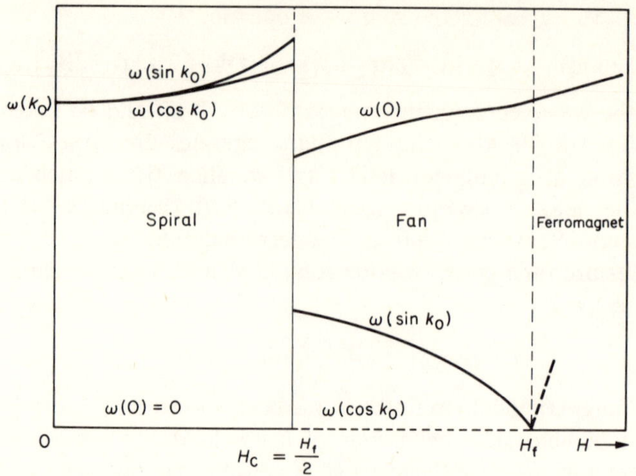

Fig. 4.33 The predicted variation of $\omega(\mathbf{k})$ with applied field assuming no basal plane anisotropy [122].

the cone angle $\theta \to 0$, the spectrum becomes symmetric about $\mathbf{k} = \mathbf{q}$ so that $\hbar\omega(2\mathbf{q}) \to \hbar\omega(0) = 0$.

The excitation of the *longitudinal* spin structures are not adequately described by spin waves. In such a structure there is much disorder, as only the z-component of the spins varies periodically and the resulting large entropy is important for its stability [174].

4.4.3 *Inelastic neutron scattering experiments*

The most direct experimental method of obtaining spin wave dispersion curves is to excite the spin system by causing thermal neutrons to be inelastically scattered from it. In this way spin wave energies may be studied for all values of $\mathbf{k}$ except near $\mathbf{k} = 0$. As yet no experiments have been reported in which an external magnetic field is applied.

Of the rare earths Tb has the lowest cross-section for thermal neutron capture, and the first inelastic scattering experiments were made upon this metal by Moller *et al* [217, 194]. Results for the dispersion curves at a temperature (90°K) in the planar ferromagnetic region are shown in Fig. 4.34 and it should be noted that optical branches are present in the a and b directions. The degeneracy at K is significant in that Brinkman and Elliott [218] have shown that it implies that the exchange interaction is isotropic. The degeneracy is removed by anisotropic exchange.

By fitting an expression like (4.57) to the dispersion curves, a function $\mathscr{J}(\mathbf{k}) - \mathscr{J}(0)$ may be obtained [194, 219], and for Tb below T_c this shows a maximum at $\mathbf{k} = 0$. Moller *et al* [194, 219] have also studied the temperature dependence of the spin wave energies and found that $\mathscr{J}(\mathbf{k}) - \mathscr{J}(0)$ scales

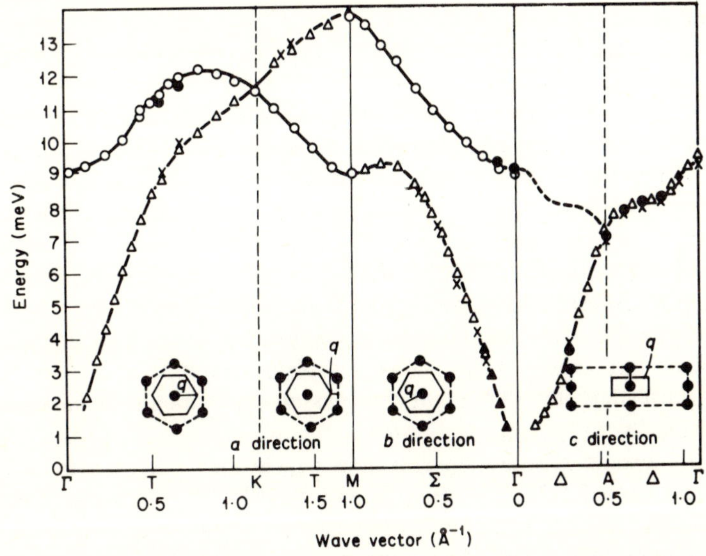

Fig. 4.34 Experimental spin wave dispersion curves for terbium at 90°K.

approximately as the ordered moment, as was previously mentioned in Section 4.4.1.

The addition of Ho to Tb increases the temperature range of the helical spin structure and enables the spin wave energies to be studied in this region, Moller *et al* have studied the alloy Tb–10% Ho, for which $T_N = 221°K$ and

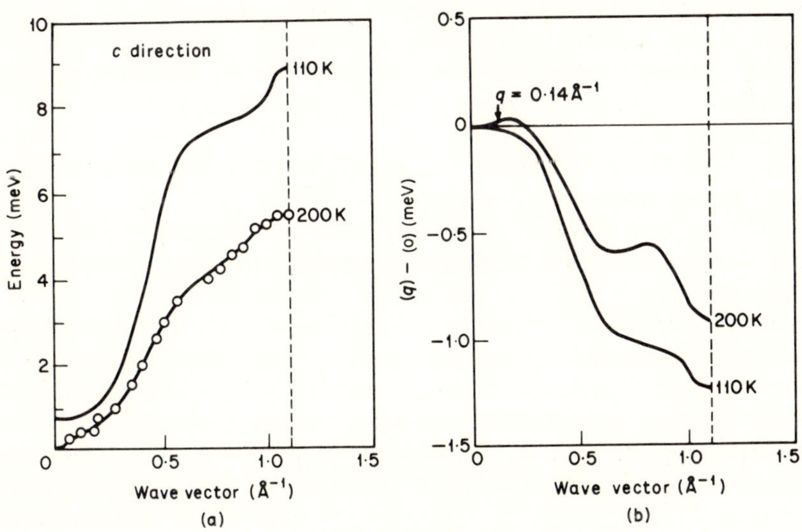

Fig. 4.35 (a) Spin wave dispersion curves for Tb–10% Ho in the ferromagnetic (110°K) and helical (200°K) phases. (b) Values of $\mathscr{J}(\mathbf{q}) - \mathscr{J}(0)$ deduced from (a) [194].

$T_c = 195°K$, and results are shown in Fig. 4.35 for both the ferromagnetic and helical regions. The functions $\mathscr{J}(\mathbf{k}) - \mathscr{J}(0)$ deducted by fitting expressions like (4.57) and (4.64) to the dispersion curves are also shown. The results explicitly illustrate the characteristic features previously discussed in Sections 4.4.2 and 4.3.2. For example in the helical region, at $\mathbf{k} = 0$ there is no energy gap, and the energy is proportional to $\mathbf{k}$. Further, $\mathscr{J}(\mathbf{k}) - \mathscr{J}(0)$ exhibits a maximum at $\mathbf{k} = 0{\cdot}14$ Å^{-1}, which corresponds to the period of the helix. On the other hand, the deduced, best fit values of the anisotropy parameter B_2^0 from neutron intensities [220] and B_6^0 from the magnitude of the $\mathbf{k} = 0$ energy gap [219] differ considerably from those obtained by torque measurements [101] for Tb. The reason for this discrepancy is not clear at the present time. The function $\mathscr{J}(\mathbf{k}) - \mathscr{J}(0)$ also has a second maximum at $\mathbf{k} \simeq 0{\cdot}8$ Å^{-1}, and it has been suggested [122] that this is associated with a spin wave Kohn type anomaly.

Neutron inelastic scattering experiments have also been performed on Er [221] and Ho [222, 223]. Compared with the Tb–10% Ho alloy, pure Ho shows a much stronger tendency to helical order, and in consequence the observed maximum in $\mathscr{J}(\mathbf{k}) - \mathscr{J}(0)$ is more pronounced. Again discrepancies appear between the deduced anisotropy constants [223] and those obtained from torque measurements.

4.4.4 *Magnetic resonance experiments*

Neutron inelastic scattering experiments cannot determine the value of the spin wave dispersion curve at $\mathbf{k} = 0$. In principle magnetic resonance experiments can fill this gap, but in practice their interpretation is complicated by the very large width of the absorption line, which is of the order of a few kilo-Oersteds. Most work has been done on the planar ferromagnets Dy and Tb with the d.c. field applied in the basal plane. Only results for these two elements will be mentioned here. However, resonant absorption has also been observed [79, 224] in Er and Ho.

Several authors [79, 225–227] have reported results of experiments on Dy in the helical region in which both r.f. and d.c. magnetic fields are applied parallel to an easy axis of magnetization. There is a single broad line which has a sharp rise and slow fall, and the temperature dependence of the position of the rise corresponds very closely to that of the critical field H_c.

The most interesting experiments have been those in which the d.c. field has been applied along a hard magnetization direction. In the ferromagnetic state resonance is expected when the magnitude of the applied field is comparable with the hexagonal anisotropy field $36H_h$, because then $\omega(\mathbf{k} \to 0) \to 0$, as discussed in Section 4.4.1. In Fig. 4.36 are shown results for Dy obtained by Rossol and Jones [228]. The curve at 100·8°K is typical of the behaviour for the helical state ($T > 85°K$). The rise at approximately 1·6 kOe corresponds to the transition from helix to fan, and the peak at about 6 kOe occurs when the applied field just cancels $36H_h$. At higher temperatures these two

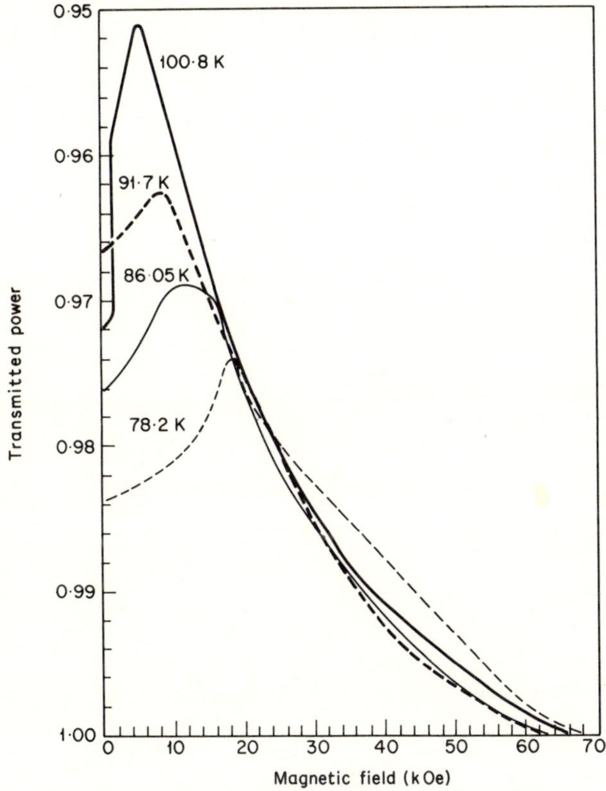

Fig. 4.36 Field dependence of the transmitted power at 37·7 GHz for dysprosium with applied field along a hard axis in the basal plane [228].

features move together, and the resonance due to the effect of H_h disappears when they coalesce at 104°K. Similar results have been observed for Tb by Bagguley and Liesegang [79].

For the helix, theory predicts two absorption lines ~ 1 kOe apart, but in experiments these cannot be resolved. Cooper [122] has shown that with realistic parameters for Dy substituted, equation (4.58) predicts the temperature dependence of an average resonance field, corresponding to $\hbar\omega(0) = 0$, which agrees well with experiment, as can be seen in Fig. 4.37. The same author also shows that the use of the frozen lattice approximation, equation (4.59), for Dy predicts that no resonance should be observed at all, so that this approximation cannot be valid. In making theoretical calculations for Tb there is uncertainty in what to take for the basal plane anisotropy constant. As discussed previously, the value obtained by fitting the neutron inelastic scattering experimental results differs from the value obtained from static measurements. Cooper has used the neutron experiment value in equation (4.58) to predict the variation of the resonance field with temperature. While

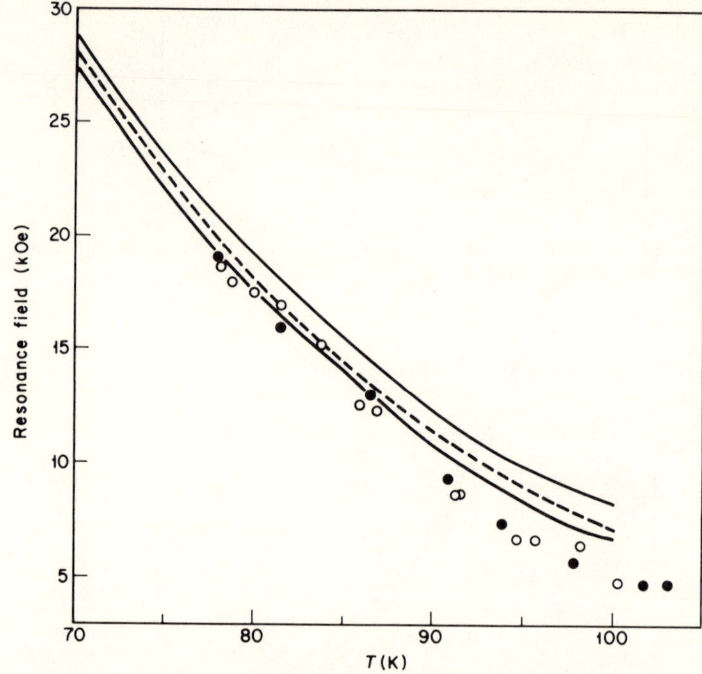

Fig. 4.37 The calculated temperature dependence of the resonance field of dysprosium at 37·7 GHz for applied field along a hard axis in the basal plane. The broken curve assumes $\omega(0) = 0$ while the solid curves assume that $\omega(0) = 1{\cdot}81°K$. The open circles and squares are experimental data [122].

theory correctly gives a field of ~6 kOe at 130°K, the curve falls off more quickly with temperature than the experimental one, going to zero at 210°K, whereas the experimental value is ~2 kOe at that temperature. It is possible [122] that magnetostriction may be important for Tb, but its effect upon the basal plane anisotropy does not improve the agreement with the resonance results. Some experimental results by Stanford and Young [229] at the much higher frequency 100 GHz show several anomalies, and are not understood.

4.4.5 *Magnetic specific heat*

Spin wave effects are also in evidence in the temperature dependence of the heat capacity of the rare earth metals and at low temperatures the following relations are predicted [230] for that component of the heat capacity with the magnetically ordered phase.

$C_M \propto T^{\frac{3}{2}}$	(Simple ferromagnetic spin wave theory)
$C_M \propto T^3$	(antiferromagnet at low temperatures with linear magnon dispersion relation)

$$C_M \propto T^{\frac{3}{2}} \exp(-\Delta E/kT) \quad \text{(ferromagnet with a spin wave activation energy } \Delta E = (K_2 K_6)^{\frac{1}{2}}) \tag{4.68}$$

$$C_M \propto T^3 \exp(-\Delta E/kT) \quad \text{(ferromagnetic spiral structure with spin wave activation energy).}$$

Cooper has also given the relations:

$$\begin{aligned} C_M &\propto (\Delta E/kT + 2 + 2kT/\Delta E) \exp(-\Delta E/kT), && (kT \ll \Delta E) \\ C_M &\propto T && (kT \geq \Delta E) \end{aligned} \tag{4.69}$$

for dysprosium, but these have not been tested experimentally.

Experimental measurements are complicated by the necessity of extracting the other terms of the total specific heat from the observed values in order to isolate C_M, and published results show considerable variation in their analysis.

Table 4.6 The form of the observed temperature variation of the magnetic contribution to the specific heat of the heavy rare earth metals [232].

	Temperature dependence	Temperature range (°K)	Spin wave energy gap (°K)
Gadolinium	$C_M = 0{\cdot}19\, T^{2{\cdot}7}$	13–25	
	$= 24\, T^{3/2} \exp(-26/T)$	17–25	−26
Terbium	$C_M = 36\, T^{3/2} \exp(-23{\cdot}5/T)$	8–25	−23·5
Dysprosium	$C_M = 107\, T^{3/2} \exp(-31/T)$	9–20	−31
Holmium	$C_M = 1{\cdot}5\, T^{3{\cdot}2}$	3–9	0
Erbium			
Thulium	$C_M = 8{\cdot}3\, T^{2{\cdot}3}$	3–20	0

Table 4.6 gives the possible fits to experimental data obtained by Lounasmaa and Sündstrom [232] in their investigation of the heavy rare earth metals. The relations are the ones finally chosen as a result of extensive examination of the temperature dependence of other properties (magnetization and spin disorder resistivity) and comparison of the known structures with equations (4.68). The best fit is often ambiguous, however, and much work needs to be done before we can be satisfied that we are observing and interpreting this contribution to the specific heat accurately.

4.5 Hyperfine interactions

The parameters of the hyperfine interactions in the rare earth elements have been the subject of considerable experimental and theoretical work in recent years. Experimentally, the investigations have been carried out using nuclear magnetic resonance and Mössbauer effect techniques, specific heat measurements and neutron-transmission measurements. The first two of these methods provide a direct measure of the splitting of the nuclear Zeeman levels and hence of the hyperfine fields. The latter techniques rely for their interpretation on a knowledge of the Hamiltonian describing the interaction, since one is looking at effects arising from the thermal distribution over all the levels.

Since nmr and Mössbauer measurements are limited by the problem of obtaining sensitivity over a wide frequency range of examination and a shortage of suitable nuclei, most attention has been concentrated on specific heat measurements. These, of course, suffer from the disadvantages of extracting the various contributions from the observed total specific heat where assumptions have to be made concerning the behaviour of the magnetic component.

It is well known that the hyperfine interaction constants of the rare earths are reasonably independent of the environment in which the nucleus is situated, a fact which has been used to associate the dominant contribution to the hyperfine field with the electrons of the unfilled $4f$ shell of the ion in which the nucleus is located. This has been discussed in Section 1.1.5. The effects of the other electrons and nuclei in the solid appear only as a perturbation on this contribution.

The interaction Hamiltonian proposed by Bleaney and Hill [233] includes quadrupole terms and is written

$$\mathscr{H} = a_0' I_z + P\{I_z^2 - \tfrac{1}{3}I(I+1)\} + P'(I_x^2 - I_y^2), \tag{4.70}$$

where

$$a_0' = a_J \langle J \rangle,$$

and

$$P = -\frac{3e^2 Q\langle r^{-3}\rangle \alpha_J}{4I(2I-1)}\langle 3J_z^2 - J(J+1)\rangle,$$

and

$$P'/P = -1/(2J-1), \qquad \text{for integral } J,$$
$$= -1/2J, \qquad \text{for(half-integral } J,$$

in the usual notation. The derivation of these terms in P' assumes that the moment is confined entirely to the basal plane, whereas in fact there is a small component perpendicular to this plane [234]. Additional second order correction effects associated with inter- and intra-ion interactions were considered by Bleaney [235] and shown to be generally small. Further the P' contribution is also usually small since J for these elements is large. This treatment omits environmental effects due to conduction electron polarization and the crystal lattice. However, crystal field calculations show that for the latter $\langle J \rangle = J$ (to $\pm 0{\cdot}1\%$) so that neglecting this term should not lead to appreciable difficulties.

4.5.1 *Nuclear specific heat*

Since one only needs to consider the two leading terms in the Hamiltonian, only the coefficients a_0' and P are required from the experimental data. Consequently analysis of specific heat observations at very low temperatures ($< 1°$K) is possible with reasonable accuracy.

Associated with the heat capacity of the $(2I+1)$ nuclear levels is the Schottky anomaly which appears as a maximum in C_N at a temperature $T \simeq \langle \Delta\omega \rangle / k$ $(= \mu_I H_{\text{eff}}/I)$ where $\langle \Delta\omega \rangle$ is the mean spacing of the energy levels. To the high temperature side of this anomaly the Boltzmann exponential may be expanded as a power series and the nuclear heat capacity becomes

$$C_N/R = \sum_{n=2}^{\infty} C_n (kT)^{-n} \tag{4.71}$$

The coefficients in this expansion are given by

$$C_2 = \frac{(I+1)}{3I}\mu_I H_{\text{eff}} + \frac{1}{45}[P^2 + 3(P')^2] I(I+1)(2I-1)(2I+3)$$

$$C_3 = -\frac{1}{15} P \frac{(I+1)}{I} (2I-1)(2I+3)\mu_I^2 H_{\text{eff}}^2,$$

and

$$C_4 = -\frac{1}{30}\frac{(I+1)}{I^3}(2I^2+2I+1)\mu_I^4 H_{\text{eff}}^2, \quad \text{etc.}$$

However, for strong hyperfine interactions the exact formula must still be retained.

The interaction constant a_0' is found to be of the order of one or two tenths of a degree in the rare earth metals and consequently the Schottky anomaly is centred about this temperature. The spin entropy associated with the peak is $S_s = R \ln (2I+1)$.

Most of the accurate, low temperature, heat capacity observations [236–250] have been performed for $T > 0{\cdot}4°$K and consequently include only the tail of the anomaly. While in principle both a_0' and P may be obtained from these measurements, their extraction is frequently limited by the uncertain behaviour of the other heat capacity contributions in this temperature range, namely, the electronic ($C_e = AT$), lattice ($C_L = BT^3$) and magnetic (C_M) heat capacities. This last term is the total heat capacity associated with the magnetic ordering and has been briefly discussed earlier.

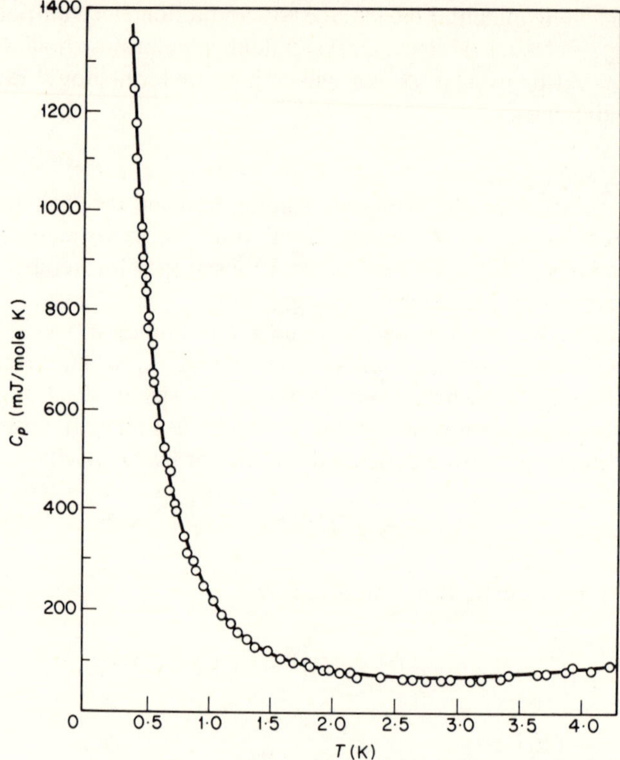

Fig. 4.38 The specific heat of terbium metal between 0·4 and 4·3°K obtained by various workers [239].

The results for Tb are shown in Fig. 4.38. However, it should be mentioned that differences have often been reported above 1°K which are due largely to the effects of impurities, particularly metal oxides. Below 1°K, however, the spread of results is rather small and reasonably consistent values of both a_0' and P have been obtained. The results of Lounasmaa for many of the metals between 0·4° and 4°K are shown in Fig. 4.39. The continuous decrease of C below 1°K in both ytterbium and gadolinium is a consequence of the full and half-full $4f$ shell associated with these elements. The values of the interaction constants derived from these and other measurements are given in Table 4.7 along with the theoretical estimates of Bleaney. As may be seen the results are in good agreement with these predictions.

Direct observation of the Schottky anomaly in both holmium and terbium has recently been made by Krusius *et al* [251] in a series of measurements going down to 0·03°K. Their results for terbium are given in Fig. 4.40 and the results of Lounasmaa [239] and Van Kempen [246] are also included for completeness. The entropy change associated with the anomaly in both cases is within 1% of the theoretical value and supports the evidence obtained in

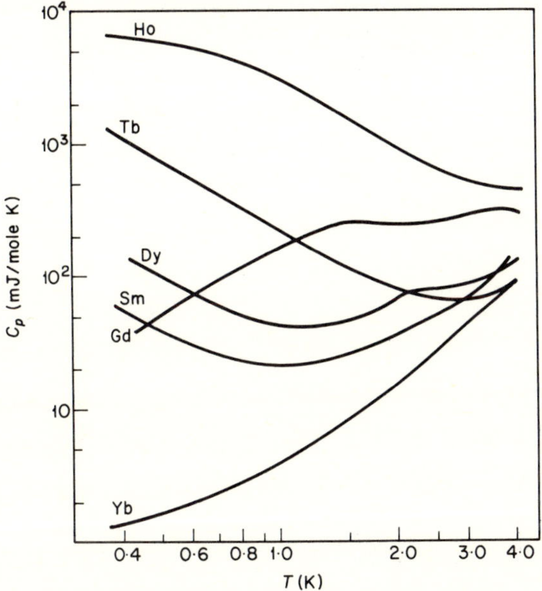

Fig. 4.39 The low temperature specific heat of samarium, gadolinium, terbium, dysprosium, holmium and ytterbium [237–241].

earlier specific heat results suggesting that impurity effects are negligible below about 1°K.

4.5.2 *Nuclear spectroscopy*

Direct observation of the nuclear level splitting using nmr has been greatly assisted by the experimental specific heat and Mössbauer effect measurements

Table 4.7 The hyperfine interaction constants of the heavy rare earth metals. D is the coefficient of the leading term in the expansion of the nuclear specific heat in inverse powers of T ($C_N = DT^{-2}$). Constants deduced from specific heat measurements (SH) [231, 237–240, 242, 263] are compared with the theoretical predictions of Bleaney [235]. Results marked with an asterisk are for trivalent ytterbium.

		Sm	Gd	Tb	Dy	Ho	Yb
D (mJ/K/mole)	SH	8·6		238	26·4	4480	0·012
	Bleaney	8·9	0·02	250	26·6	4200	16*
a'_c (°K)	SH	0·026		0·150	0·048	0·320	0·002
	Bleaney	0·027	0·002	0·153	0·048	0·312	0·09*
P (°K)	SH			0·002	0·008	0·007	
	Bleaney	$<0·001$		0·023	0·008	0·001	
H_{eff} (MOe)		3·3		4·1	7·1	9·3	0·14

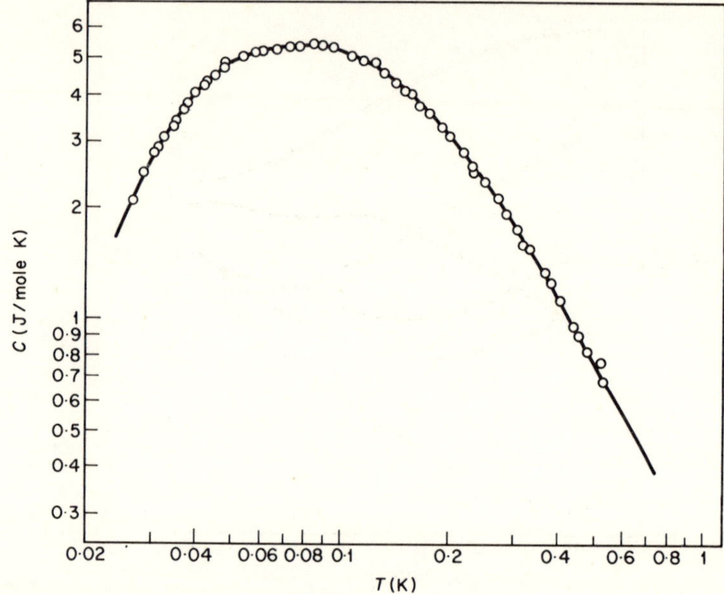

Fig. 4.40 The specific heat of terbium below 1°K, showing the Schottky anomaly [239, 246, 251]. The solid line represents a best fit theoretical curve based on this data and hyperfine parameters determined from nmr measurements.

(where a suitable isotope is available). These have provided reliable estimates of the resonant frequency and so helped to avoid the tedious searches common to nuclear resonance in ferromagnetic materials.

Suitable isotopes for Mössbauer studies are available for nearly all the elements, most work having been carried out on dysprosium and europium. In addition to the nuclear field and quadrupole interactions, which are listed in Table 4.7, the isomer shifts have also been obtained using this technique. The magnitude of this shift depends on the magnitude of the 6*s* electrons wavefunction at the nucleus (i.e. $|\psi_{6s}(0)|^2$) and the fractional difference in radii of the excited and ground states of the nucleus ($\Delta r/r$). In principle it is possible to determine these parameters, but for dysprosium the results conflict with those obtained from other means.

Both the hyperfine field and electric quadrupole interaction have been studied by Kobayashi *et al* in terbium [252] and dysprosium [253] using spin echo techniques. Their results, which agree closely with the calculations of both Kondo [254] and Bleaney [235], are given in Table 4.7.

The hyperfine field at the holmium nucleus has been extensively studied by Mackenzie *et al* [255]. The total splitting of the eight nuclear levels in ^{165}Ho is approximately 2°K, the level separation being given by Bleaney as 6500 MHz. Consequently for measurements at 1°K the nuclear spin system is largely polarized. In the holmium–gadolinium system it is found that the total hyperfine interaction constant increases linearly with $(g_J - 1)J$ (averaged for the alloy) as is also the case in holmium–terbium and holmium–dysprosium alloys. This variation from the free ion (a_0') value has been attributed to the effect of the conduction electron polarization, and it is well known that there are two contributions:

(a) from conduction electron polarization by the neighbouring ions (a_0'');
(b) from the conduction electron polarization by the ions own $4f$ electrons (a_0''').

On this basis we may write

$$a_0 = a_0' + a_0'' + a_0'''. \tag{4.72}$$

Rigorously the second term also includes other neighbour effects, such as those of overlap and covalency, but the separation of these contributions is difficult and they are usually considered together with (*a*) above.

We know that the conduction electron polarization may be given by $\langle S_z \rangle \propto A\overline{(g_J - 1)J}$ (the bar indicating an average value in the case of alloys), and to a first approximation A may be taken as constant for all the heavy rare earths. Assuming that the contribution to the hyperfine interaction is proportional to $\langle S_z \rangle$ then we can expect $a_0'' \propto \overline{(g_J - 1)J}$. The value of a_0''' can be expected to be constant for any ion under investigation.

Experimental results for some holmium alloys are shown in Fig. 4.41, plotted against $\overline{(g_J - 1)J}$, along with the hyperfine field values for the same specimens. Extrapolating to zero spin, i.e. infinite dilution, gives a value of a_0 in excess of the free ion value suggesting that the a_0''' term is positive

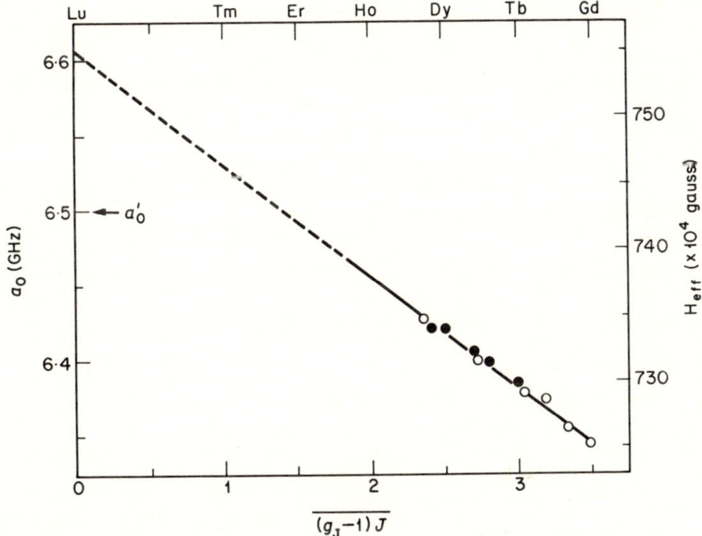

Fig. 4.41 The total hyperfine interaction constant of ^{165}Ho (a_0) as a function of an average $(g_J - 1)J$ for some holmium–rare earth alloys [255].

Table 4.8 Analysis of the various contributions to the hyperfine field in europium, gadolinium and holmium [255, 256]. Magnetic fields are in kOe.

	Europium 151	Gadolinium	Holmium 165
Total hyperfine field (kOe)	−265	−350	7380
Core polarization (kOe)	−340	−340	7430
Conduction electron polarization by neighbour ions, etc. (kOe)	−115	−200	−170
Conduction electron polarization by own 4*f* electrons (kOe)	+190	+240	+120

(since $a_0'' = 0$ for $\overline{(g_J-1)J} = 0$) and of magnitude 105 MHz corresponding to a hyperfine field contribution of +100 kOe. Since the values of a_0 (and the hyperfine field strength B_0) for the alloys is always less than the free ion value the contribution to the hyperfine interaction from the environmental effects must be negative, its value given by the gradient of this curve being $a_0'' = \overline{(g_J-1)J}$; ($H_{\text{eff}}'' = -10\overline{(g_J-1)J} \times 10^4$ Oe).

A similar analysis has also been made for other systems using a variety of techniques [256–261]. The contributions to the hyperfine field using interactions obtained in this way are shown in Table 4.8 for europium, gadolinium and holmium.

References

[1] R. M. Bozorth and C. D. Graham, *I.B.M. Research Report*, RC 1635 (1966).
[2] F. H. Spedding, S. Legvold, A. H. Daane and L. D. Jennings, *Low Temp. Physics*, **2**, 368 (1958).
[3] K. P. Belov, M. A. Belyanchikova, R. Z. Levitin and S. A. Nikitin, *Rare Earth Ferromagnetics and Antiferromagnetics*, Nanka, Moscow (1965).
[4] J. M. Lock, *Proc. Phys. Soc.*, **B70**, 476 (1957).
[5] J. M. Lock, *Proc. Phys. Soc.*, **B70**, 566 (1957).
[6] H. Leipfinger, *Z. für Physik*, **150**, 415 (1958).
[7] S. Arajs and R. V. Colvin, *Rare Earth Research*, ed. E. V. Kleber, MacMillan Co., New York (1961) p. 178.
[8] M. A. Ruderman and C. Kittel, *Phys. Rev.*, **96**, 99 (1954).
[9] T. Kasuya, Prog. *Theoret Phys.* (Kyoto), **16**, 45 (1956).
[10] K. Yosida, *Phys. Rev.*, **106**, 893 (1957).
[11] Y. A. Rocher, *Advan. Phys.*, **11**, 233 (1962).
[12] P. G. de Gennes, *J. Phys. Radium*, **23**, 510 (1962).
[13] T. Kasuya, *Magnetism* ed. G. T. Rado and H. Suhl, Academic Press, London, **2B**, 215 (1966).
[14] C. Kittel, *Solid State Phys.*, 22, 1 (1968).
[15] G. J. Bowden, Ph.D. Thesis, Manchester University (1967).
[16] G. J. Bowden, D. St. P. Bunbury and M. A. H. McCausland, unpublished.
[17] S. H. Liu, *Phys. Rev.*, **121**, 451 (1960); *ibid*, **123**, 470 (1961).

[18] T. A. Kaplan and D. H. Lyons, *Phys. Rev.*, **128**, 2072 (1962).
[19] F. Specht, *Phys. Rev.*, **162**, 389 (1967).
[20] V. Jaccarino, B. T. Matthias, M. Peter, H. Suhl and J. H. Wernick, *Phys. Rev. Letts.*, **5**, 251 (1960).
[21] R. E. Watson, S. Koide, M. Peter and A. J. Freeman, *Phys. Rev.*, **139**, A167 (1965).
[22] R. E. Watson, A. J. Freeman and S. Koide, *Phys. Rev.*, **186**, 625 (1969).
[23] R. E. Watson and A. J. Freeman, *Phys. Rev.*, **152**, 566 (1966).
[24] L. I. Schiff, *Quantum Mechanics*, 2nd Ed., McGraw Hill, New York (1955) p. 161.
[25] J. H. Van Vleck, *Rev. Mod. Phys.*, **34**, 681 (1962).
[26] D. C. Mattis, *Theory of Magnetism*, Harper and Row, New York (1965) pp. 201, 275.
[27] F. Holtzberg, T. R. McGuire, S. Methfessel and J. C. Suits, *Phys. Rev. Letts.*, **13**, 18 (1964).
[28] J. M. Ziman, *Principles of the Theory of Solids*, Cambridge University Press (1964) p. 126.
[29] P. A. Wolf, *Phys. Rev.*, **120**, 814 (1960).
[30] D. R. Harman and A. W. Overhauser, *Phys. Rev.*, **143**, 183 (1966).
[31] B. Giovannini, M. Peter and J. Schrieffer, *Phys. Rev. Letts.*, **12**, 734 (1964).
[32] P. G. de Gennes, *J. Phys. Radium*, **23**, 630 (1962).
[33] A. J. Heeger, A. P. Klein and P. Tu, *Phys. Rev. Letts*, **17**, 803 (1966).
[34] L. Roth, H. J. Zeiger and T. A. Kaplan, *Phys. Rev.*, **149**, 519 (1966).
[35] P. R. P. Silva, *Phys. Rev.*, **166**, 679 (1968).
[36] A. Blandin, *J. Phys. Chem. Solids*, **22**, 507 (1961).
[37] J. H. Van Vleck, *Theory of Electric and Magnetic Susceptibilities*, O.U.P. (1932).
[38] W. Klemm and H. Bommer, *Z. anorg. Chem.*, **231**, 138 (1937).
[39] W. Klemm and H. Bommer, *Z. anorg. Chem.*, **241**, 264 (1939).
[40] H. Bommer, *Electrochem.*, **45**, 357 (1939).
[41] F. Trombe, *Compt. Rend.*, **198**, 1591 (1934).
[42] F. Trombe, *Compt. Rend.*, **201**, 652 (1934).
[43] F. Trombe, *Ann. Phys.*, **7**, 383 (1937).
[44] M. Owen, *Ann. Phys.*, **37**, 657 (1912).
[45] C. H. La Blanchetais, *Compt. Rend.*, **220**, 392 (1945).
[46] C. R. Burr and S. Ehara, *Phys. Rev.*, **149**, 551 (1966).
[47] D. H. Parkinson and L. M. Roberts, *Proc. Phys. Soc.*, **B70**, 476 (1957).
[48] D. H. Parkinson, F. E. Simon and F. H. Spedding, *Proc. Roy. Soc.*, **A207**, 137 (1951).
[49] O. V. Lounasmaa and W. Sundstrom, *Phys. Rev.*, **158**, 591 (1967).
[50] M. K. Wilkinson, H. R. Child, C. J. McHargue, *Phys. Rev.*, **122**, 1409 (1961).
[51] J. W. Cable, R. M. Moon, W. C. Koehler and E. O. Wollan, *Phys. Rev. Letts.*, **12**, 553 (1964).
[52] R. M. Moon, J. W. Cable and W. C. Koehler, *J. Appl. Phys.*, **35**, 1041 (1964).
[53] B. Lebech and B. D. Rainford, *J. de Phys.*, 32, C1-370 (1971).
[54] N. Nagasawa and T. Sugawara, *J. Phys. Soc. Japan*, **23**, 701 (1967).
[55] T. Johansson, K. A. McEwen and P. Tomborg, *J. de Phys.*, 32, C1-372 (1971).
[56] R. M. Bozorth and J. H. Van Vleck, *Phys. Rev.*, **118**, 1493 (1960).
[57] M. Schieber, S. Foner, R. Docho and E. J. McNiff, *J. Appl. Phys.*, **39**, 885 (1968).
[58] J. F. Elliott, S. Legvold and F. H. Spedding, *Phys. Rev.*, **91**, 28 (1953).
[59] H. Leipfinger, *Z. anorg. Allgem. Chem.*, **231**, 138 (1937).
[60] W. C. Thoburn, S. Legvold and F. H. Spedding, *Phys. Rev.*, **112**, 56 (1958).
[61] F. Trombe, *Compt. Rend.*, **221**, 19 (1945).
[62] F. Trombe, *Compt. Rend.*, **236**, 591 (1953).
[63] F. Trombe, *J. Phys. Rad.*, **12**, 22 (1951).
[64] J. F. Elliott, S. Legvold and F. H. Spedding, *Phys. Rev.*, **94**, 1143 (1954).
[65] B. L. Rhodes, S. Legvold and F. H. Spedding, *Phys. Rev.*, **109**, 1544 (1958).
[66] J. F. Elliott, S. Legvold and F. H. Spedding, *Phys. Rev.*, **100**, 1959 (1953).
[67] D. D. Davis and R. M. Bozorth, *Phys. Rev.*, **118**, 1543 (1960).

[68] W. E. Henry, *Rare Earth Research*, ed. E. V. Kleber, MacMillan (1960) p. 165.
[69] H. E. Nigh, S. Legvold and F. H. Spedding, *Phys. Rev.*, **132**, 1092 (1963).
[70] D. E. Hegland, S. Legvold and F. H. Spedding, *Phys. Rev.*, **131**, 158 (1963).
[71] D. R. Behrendt, S. Legvold and F. H. Spedding, *Phys. Rev.*, **109**, 1544 (1958).
[72] R. G. Jordon and E. W. Lee, *Proc. Phys. Soc.*, **92**, 1074 (1967).
[73] D. L. Strandburg, S. Legvold and F. H. Spedding, *Phys. Rev.*, **127**, 2046 (1962).
[74] R. W. Green, S. Legvold and F. H. Spedding, *Phys. Rev.*, **122**, 827 (1961).
[75] D. B. Richards and S. Legvold, *Phys. Rev.*, **186**, 508 (1969).
[76] J. J. Rhyne, S. Foner, E. S. McNiff and R. Doclo, *J. Appl. Phys.*, **39**, 892 (1968).
[77] R. B. Flippen, *J. Appl. Phys.*, **35**, 1047 (1964).
[78] P. H. Bly, Thesis, Durham University (1967).
[79] D. M. S. Bagguley and J. Leisegang, *Proc. Roy. Soc.*, **A300**, 497 (1967).
[80] K. P. Belov, R. Z. Levitin, S. A. Nikitin and A. V. Pedko, *Soviet Phys. JETP*, **13**, 1096 (1961).
[81] K. P. Belov and A. V. Pedko, *Soviet Phys. JETP*, **15**, 62 (1962).
[82] D. B. McWhan and A. L. Stevens, *Phys. Rev.*, **139**, A682 (1965).
[83] C. D. Graham, Jr., *J. Phys. Soc. Japan*, **17**, 1310 (1962).
[84] W. D. Corner, W. Roe and K. N. R. Taylor, *Proc. Phys. Soc.*, **80**, 927 (1962).
[85] J. W. Cable and E. O. Wollan, *Phys. Rev.*, **165**, 733 (1968).
[86] G. Will, R. Nathans and H. A. Alperin, *J. Appl. Phys.*, **35**, 1045 (1964).
[87] V. M. Kuchin, V. A. Semenkov, S. Sh. Shilshtein, *Soviet Phys. JETP*, **28**, 649 (1969).
[88] S. Weinstein, R. S. Craig and W. E. Wallace, *J. Appl. Phys.*, **34**, 1354 (1963).
[89] D. H. Parkinson, F. E. Simon and F. H. Spedding, *Proc. Roy. Soc.*, **A207**, 137 (1951).
[90] D. H. Parkinson and L. M. Roberts, *Proc. Phys. Soc.*, **B70**, 471 (1957).
[91] L. M. Roberts, *Proc. Phys. Soc.*, **B70**, 434 (1957).
[92] L. D. Jennings, E. D. Hill and F. H. Spedding, *J. Chem. Phys.*, **31**, 1240 (1959).
[93] M. Griffel, R. E. Skochdopole and F. H. Spedding, *Phys. Rev.*, **93**, 657 (1954).
[94] L. D. Jennings, R. M. Stanton and F. H. Spedding, *J. Chem. Phys.*, **27**, 909 (1957).
[95] M. Griffel, R. E. Skochdopole and F. H. Spedding, *J. Chem. Phys.*, **25**, 75 (1957).
[96] B. C. Gerstein, M. Griffel, L. D. Jennings, R. E. Miller, R. E. Skochdopole and F. H. Spedding, *J. Chem. Phys.*, **27**, 394 (1957).
[97] R. E. Skochdopole, M. Griffel and F. H. Spedding, *J. Chem. Phys.*, **23**, 2258 (1955).
[98] J. D. Jennings, E. Hill and F. H. Spedding, *J. Chem. Phys.*, **34**, 2082 (1961).
[99] M. I. Darby and K. N. R. Taylor, Proc. Int. Conf. Magnetism, Nottingham (1964) p. 742.
[100] C. D. Graham, *J. Appl. Phys.*, **38**, 1375 (1967).
[101] J. J. Rhyne and A. E. Clark, *J. Appl. Phys.*, **38**, 1379 (1967).
[102] R. Z. Levitin and B. K. Ponomaryov, *Soviet Phys. JETP*, **26**, 1121 (1968).
[103] K. Tajima and S. Chikazumi, *J. Phys. Soc. Japan*, **23**, 1175 (1967).
[104] C. Zener, *Phys. Rev.*, **96**, 1335 (1954).
[105] E. Callan and H. B. Callen, *J. Phys. Chem. Solids*, **27**, 1271 (1966).
[106] P. H. Bly, W. D. Corner, K. N. R. Taylor and M. I. Darby, *J. Appl. Phys.*, **39**, 1336 (1968).
[107] T. T. Jew and S. Legvold, USAEC Report IS-867 (1963).
[108] W. D. Corner and F. Hutchinson, *Proc. Phys. Soc.*, **75**, 781 (1960).
[109] R. M. Bozorth and T. Wakiyama, *Tech. Rep. ISSP*, **A57** (1962).
[110] E. W. Lee and L. Albers, *Proc. Phys. Soc.*, **79**, 977 (1962).
[111] S. Legvold, J. Alstad and J. J. Rhyne, *Phys. Rev. Letts.*, **10**, 509 (1963).
[112] A. E. Clark, R. M. Bozorth and B. de Savage, *Phys. Letts.*, **5**, 100 (1963).
[113] K. P. Belov and S. A. Nikitin, *Soviet. Phys. JETP*, **15**, 279 (1962).
[114] J. J. Rhyne and S. Legvold, *Phys. Rev.*, **138**, A507 (1965).
[115] J. J. Rhyne, S. Legvold and E. T. Rodine, *Phys. Rev.*, **154**, 266 (1967).
[116] A. E. Clark, B. de Savage and R. M. Bozorth, *Phys. Rev.*, **138**, A216(1965).
[117] J. J. Rhyne and S. Legvold, *Phys. Rev.*, **140**, A2143 (1965).
[118] P. de V. Du Plessis, *Phil. Mag.*, **18**, 145 (1968).

[119] P. de V. Du Plessis and L. Albers, *Solid State Comm.*, **3**, 251 (1965).
[120] F. J. Darnell and E. P. Moore, *J. Appl. Phys.*, **34**, 1337 (1963).
[121] E. Callen and H. B. Callen, *Phys. Rev.*, **139**, A455 (1965).
[122] B. R. Cooper, *Solid State Phys.*, **21**, 393 (1969).
[123] W. C. Koehler, E. O. Wollan, M. K. Wilkinson and J. W. Cable, *Rare Earth Research*, ed. E. V. Kleber, MacMillan (1961) p. 149.
[124] W. C. Koehler, J. W. Cable, E. O. Wollan and M. K. Wilkinson, *J. Phys. Soc. Japan*, **17**, Suppl. B.III. 32 (1962).
[125] W. C. Koehler, E. O. Wollan, H. R. Child and J. W. Cable, *Rare Earth Research*, ed. J. S. Vorres, Gordon and Breach (1963) p. 199.
[126] J. W. Cable, H. R. Child, W. C. Koehler and M. K. Wilkinson, *Pile Neutron Research in Physics* (1962) p. 379.
[127] M. K. Wilkinson, W. C. Koehler, E. O. Wollan and J. W. Cable, *J. Appl. Phys.*, **32**, 485 (1961).
[128] J. W. Cable, E. O. Wollan, W. C. Koehler and M. K. Wilkinson, *Phys. Rev.*, **140**, A1896 (1965).
[129] W. C. Koehler, J. W. Cable, E. O. Wollan and M. K. Wilkinson, *Phys. Rev.*, **126**, 1672 (1962).
[130] W. C. Koehler, J. W. Cable, H. R. Child, M. K. Wilkinson and E. O. Wollan, *Phys. Rev.*, **158**, 450 (1967).
[131] H. K. Wilkinson, H. R. Child, C. J. McHargue, W. C. Koehler and E. O. Wollan, *Phys. Rev.*, **122**, 1409 (1961).
[132] N. G. Nereson, C. E. Olsen and G. P. Arnold, *Phys. Rev.*, **135**, A176 (1964).
[133] W. C. Thoburn, S. Legvold and F. H. Spedding, *Phys. Rev.*, **110**, 1298 (1958).
[134] H. R. Child, J. W. Cable, *J. Appl. Phys.*, **40**, 1003 (1969).
[135] S. Weinstein, R. S. Craig and W. E. Wallace, *J. Chem. Phys.*, **39**, 1449 (1963).
[136] H. R. Child, W. C. Koehler, E. O. Wollan and J. W. Cable, *Phys. Rev.*, **138**, A1655 (1965).
[137] D. Chatterjee, Ph.D. Thesis, Durham University (1971).
[138] H. R. Child and W. C. Koehler, *Phys. Rev.*, **174**, 562 (1968).
[139] H. R. Child and W. C. Koehler, *J. Appl. Phys.*, **37**, 1353 (1966).
[140] E. O. Wollan, *Phys. Rev.*, **160**, 369 (1967).
[141] E. O. Wollan, *J. Appl. Phys.*, **38**, 1371 (1967).
[142] H. Nagasawa and T. Sugawara, *J. Phys. Soc. Japan*, **23**, 711 (1967).
[143] D. T. Nelson and J. Legvold, *Phys. Rev.*, **123**, 80 (1961).
[144] J. Popplewell, A. M. Harris and R. S. Tebble, *Proc. Phys. Soc.*, **85**, 347 (1965).
[145] G. A. Shafiquillina, V. I. Chechernikov and I. A. Markova, *Vesta. Mosk. Univ.* Ser. III, **21**, 96 (1966).
[146] H. R. Child, Thesis, University of Tennessee (1965).
[147] H. R. Child, W. C. Koehler, E. O. Wollan and J. W. Cable, *Bull. Am. Phys. Soc.*, **9**, 213 (1964).
[148] R. M. Bozorth and R. J. Gambino, *Phys. Rev.*, **147**, 487 (1966).
[149] R. M. Bozorth, *J. Appl. Phys.*, **38**, 1366 (1967).
[150] F. H. Spedding, R. G. Jordan and R. W. Williams, *J. Chem. Phys.*, **51**, 509 (1969).
[151] L. Tissot and A. Blaise, *J. Appl. Phys.*, **41**, 1180 (1970).
[152] T. Tsuya, A. E. Clark and R. M. Bozorth, *Proc. Int. Conf. Magnetism*, Nottingham (1964) p. 250.
[153] K. W. H. Stevens, *Proc. Phys. Soc.*, **65**, 209 (1952).
[154] W. C. Koehler, Keller Reports KR-132, 156 (1969).
[155] W. C. Koehler, *J. Appl. Phys.*, **36**, 1078 (1965).
[156] A. H. Millhouse, W. C Koehler and H. R. Child, *J. Appl. Phys.*, **40**, 1006 (1969).
[157] R. E. Watson, A. J. Freeman and J. O. Dimmock, *Phys. Rev.*, **167**, 497 (1968).
[158] C. B. Robinson, S. Tan and K. F. Sterrett, *Phys. Rev.*, **141**, 548 (1966).
[159] D. B. McWhan and A. L. Stevens, *Phys. Rev.*, **139**, A682 (1965).

[160] A. R. Wazzen, R. S. Vitt and L. B. Robinson, *Phys. Rev.*, **159**, 400 (1967).
[161] T. Okamoto, H. Fujii, Y. Hidaka and E. Tatsumoto, *J. Phys. Soc. Japan*, **24**, 951 (1968).
[162] P. Landry and R. Stevenson, *Can. J. Phys.*, **41**, 1273 (1963).
[163] L. B. Robinson, F. Milstein and A. Jayaraman, *Phys. Rev.*, **134**, A187 (1964).
[164] F. Milstein and L. B. Robinson, *Phys. Rev.*, **159**, 466 (1967).
[165] M. I. Darby and K. N. R. Taylor, *J. Appl. Phys.*, **37**, 1442 (1966).
[166] D. B. McWhan and A. L. Stevens, *Phys. Rev.*, **154**, 438 (1967).
[167] J. E. Milton and T. A. Scott, *Phys. Rev.*, **160**, 387 (1967).
[168] H. Bartholin and D. Bloch, *J. Phys. Chem. Solids*, **29**, 1063 (1968).
[169] A. Jayaraman and R. C. Sherwood, *Phys. Rev.*, **134**, A691 (1964).
[170] D. B. McWhan, E. Corenzwit and A. L. Stevens, *J. Appl. Phys.*, **37**, 1355 (1966).
[171] H. Umebayashi, G. Shirane, B. C. Frazer, *Phys. Rev.*, **165**, 688 (1968).
[172] H. Miwa, *Proc. Phys. Soc.*, **85**, 1197 (1965).
[173] D. Bloch and R. Pauthenet, *Proc. Int. Conf. Magnetism*, Nottingham (1964) p. 255.
[174] R. J. Elliott, *Phys. Rev.*, **124**, 346 (1961).
[175] T. A. Kaplan, *Phys. Rev.*, **123**, 329 (1961).
[176] T. Nagamiya, K. Nagata and Y. Kitano, *Prog. Theor. Phys.* (Kyoto), **27**, 1253 (1962).
[177] R. J. Elliott, *Magnetism*, Vol. IIA, eds. G. T. Rado and H. Suhl, Academic Press, New York (1965) p. 397.
[178] R. J. Elliott and K. W. H. Stevens, *Proc. Roy. Soc.*, **A218**, 553 (1953).
[179] U. Enz, *Physica*, **26**, 698 (1960).
[180] A. Herpin, P. Meriel and J. Villain, *Compt. Rend.*, **249**, 1334 (1959).
[181] A. Herpin, P. Meriel, *J. Phys. Rad.*, **22**, 337 (1961).
[182] P. G. de Gennes and D. St. James, *S. S. Comm.*, **1**, 62 (1963).
[183] H. S. Jarett and R. J. Elliott, *J. Phys. Soc. Japan*, **17**, Suppl. B–1, 8 (1962).
[184] F. J. Darnell and E. P. Moore, *J. Appl. Phys.*, **34**, 1337 (1963).
[185] P. C. Landry and R. Stevenson, *Can. J. Phys.*, **411**, 1273 (1963).
[186] P. C. Landry, *Phys. Rev.*, **156**, 578 (1967).
[187] E. W. Lee, *Proc. Phys. Soc.*, **84**, 693 (1964).
[188] K. P. Belov, R. Z. Levitin and S. A. Nikitin, *Sov. Phys.-Uspekhi*, **7**, 179 (1964).
[189] J. R. Banister, S. Legvold and F. H. Spedding, *Phys. Rev.*, **94**, 1140 (1954).
[190] K. Yosida and A. Watabe, *Progr. Theoret. Phys.* (Kyoto), **28**, 361 (1962).
[191] W. E. Evenson and S. H. Liu, *Phys. Rev. Letts.*, **21**, 432 (1968).
[192] W. E. Evenson and S. H. Liu, *Phys. Rev.*, **178**, 783 (1969).
[193] S. C. Keeton and T. L. Loucks, *Phys. Rev.*, **168**, 672 (1968).
[194] H. B. Møller, J. C. G. Houmann and A. R. Mackintosh, *Phys. Rev. Letts.*, **19**, 312 (1967).
[195] R. Harris, *Phys. Letts.*, **30A**, 473 (1969).
[196] C. Jackson and S. Doniach, *Phys. Letts.*, **30A**, 328 (1969).
[197] R. J. Elliott and F. A. Wedgewood, *Proc. Phys. Soc.*, **81**, 846 (1963).
[198] R. J. Elliott and F. A. Wedgewood, *Proc. Phys. Soc.*, **84**, 63 (1964).
[199] R. J. Elliott, *Magnetism*, eds. G. T. Rado and H. Suhl, Academic Press, London, **2A**, 409 (1965).
[200] A. Arrott, *ibid*, **2B**, 296 (1966).
[201] D. Smith, *Scr. Met.*, **1**, 157 (1957).
[202] P. G. de Gennes, *J. Phys. Radium*, **23**, 630 (1962).
[203] R. V. Colvin, S. Legvold and F. H. Spedding, *Phys. Rev.*, **120**, 741 (1960).
[204] B. R. Cooper, *Phys. Rev. Letts.*, **19**, 900 (1967).
[205] B. R. Cooper, *Phys. Rev.*, **169**, 281 (1968).
[206] C. Jackson, *Phys. Rev.*, **178**, 949 (1969).
[207] K. Niira, *Phys. Rev.*, **117**, 129 (1960).
[208] K. Yosida and H. Miwa, *J. Appl. Phys.*, **32**, 85 (1961).
[209] K. Yosida and H. Miwa, *Prog. Theoret. Phys.* (Kyoto), **26**, 693 (1961).
[210] J. Liesegang, Thesis, Oxford University (1966).

[211] M. S. S. Brooks, D. A. Goodings and B. I. Ralph, *J. Phys. C.*, **I**, 132 (1968).
[212] B. R. Cooper, R. J. Elliott, S. J. Nettel and H. Suhl, *Phys. Rev.*, **127**, 57 (1962).
[213] W. P. Mason, *Phys. Rev.*, **96**, 302 (1954).
[214] B. R. Cooper and R. J. Elliott, *Phys. Rev.*, **131**, 1043 (1963).
[215] R. J. Elliott and R. V. Lange, *Phys. Rev.*, **152**, 235 (1966).
[216] T. Nagamiya, *5th Rare Earth Res. Conf.*, Ames, Iowa (1965).
[217] H. B. Mølller and J. C. G. Houmann, *Phys. Rev. Letts.*, **16**, 737 (1966).
[218] W. Brinkman and R. J. Elliott, *Proc. Roy. Soc.*, **294A**, 343 (1966).
[219] H. B. Mølller, *J. Appl. Phys.*, **39**, 807 (1968).
[220] P. A. Lindgard, A. Kowalska and E. P. Laut, *J. Phys. Chem. Solids*, **28**, 1357 (1967).
[221] A. D. B. Woods, T. M. Holden and B. M. Powell, *Phys. Rev. Letts.*, **19**, 908 (1967).
[222] M. W. Stringfellow, T. M. Holden, B. M. Powell and A. D. R. Woods, *J. Appl. Phys.*, **40**, 1443 (1969).
[223] R. M. Nicklow, H. A. Mook, H. G. Smith, R. E. Reed and M. K. Wilkinson, *J. Appl. Phys.*, **40**, 1452 (1969).
[224] D. M. S. Bagguley and J. Liesegang, *J. Appl. Phys.*, **37**, 1220 (1966).
[225] F. C. Rossol, Thesis, Harvard University (1966).
[226] D. M. S. Bagguley and J. Liesegang, *Phys. Letts.*, **17**, 96 (1965).
[227] H. A. Blackstead and P. L. Donoho, 6th *Rare Earth Res. Conf.*, Gatlinburg, Tennessee, U.S.A. (1967).
[228] F. C. Rossol and R. V. Jones, *J. Appl. Phys.*, **37**, 1227 (1966).
[229] J. S. Stanford and R. C. Young, *Phys. Rev.*, **157**, 245 (1967).
[230] B. R. Cooper, *Proc. Phys. Soc.*, **80**, 1225 (1962).
[231] O. V. Lounasmaa and L. J. Sundstrom, *Phys. Rev.*, **134**, A1620 (1964).
[232] O. V. Lounasmaa and L. J. Sundstrom, *Phys. Rev.*, **150**, 399 (1966).
[233] B. Bleaney and R. W. Hill, *Proc. Phys. Soc.*, **78**, 313 (1961).
[234] G. J. Bowden, Thesis, Manchester University (1967).
[235] B. Bleaney, *J. Appl. Phys.*, **34**, 1024 (1963).
[236] N. Kurti and R. S. Safrata, *Phil. Mag.*, **3**, 780 (1958).
[237] O. V. Lounasmaa, *Phys. Rev.*, **126**, 1352 (1962).
[238] O. V. Lounasmaa and R. A. Guenther, *Phys. Rev.*, **126**, 1357 (1962).
[239] O. V. Lounasmaa and P. R. Roach, *Phys. Rev.*, **128**, 622 (1962).
[240] O. V. Lounasmaa, *Phys. Rev.*, **128**, 1136 (1962).
[241] *Idem, ibid*, **129**, 2460 (1963).
[242] *Idem, ibid*, **133**, A211 (1964).
[243] *Idem, ibid*, **133**, A219 (1964).
[244] *Idem, ibid*, **133**, A503 (1964).
[245] E. C. Heltemes and C. A. Swenson, *J. Chem. Phys.*, **35**, 1264 (1961).
[246] H. Van Kemper, A. R. Miedema and W. J. Huistamp, *Physica.*, **30**, 229 (1961).
[247] J. E. Gordon, C. W. Dempsey and T. Zoller, *Phys. Rev.*, **124**, 724 (1961).
[248] B. Dreyfus, B. B. Goodman, A. Lacaze and G. Trolliet, *Compt. Rend.*, **253**, 1764 (1961).
[249] J. G. Dash, R. D. Taylor, P. P. Craig, *Proc. 7th Int. Conf. Low Temp. Phys.*, Toronto (1960) p. 705.
[250] B. Dreyfus, B. B. Goodman, G. Trolliet and L. Weil, *Compt. Rend.*, **253**, 1085 (1961).
[251] M. Krusius, A. C. Anderson and B. Holmstrom, *Phys. Rev.*, **177**, 910 (1969).
[252] S. Kobayashi, N. Sano and J. Itoh, *J. Phys. Soc. Japan*, **22**, 676 (1967).
[253] S. Kobayashi, N. Sano and J. Itoh, *J. Phys. Soc. Japan*, **21**, 1456 (1966).
[254] J. Kondo, *J. Phys. Soc. Japan*, **16**, 1690 (1961).
[255] I. S. MacKenzie, M. A. H. McCausland and A. R. Wagg (to be published).
[256] S. Hufner, *Phys. Rev. Letts.*, **19**, 1034 (1967).
[257] M. Kalvius, P. Kienle, H. Heike, W. Wedermann and C. Schule, *Phys.*, **172**, 231 (1964).
[258] R. L. Cohen, *Phys. Rev.*, **134**, A94 (1964).
[259] S. Ofer, M. Rakavy, E. Legel and B. Khurgin, *Phys. Rev.*, **138**, A241 (1965).

[260] F. J. Shore, G. A. Reynolds, V. L. Sailor and G. Brunhard, *Phys. Rev.*, **138**, B1361 (1965).
[261] L. Passell, V. L. Sailor and R. L. Shermer, *Phys. Rev.*, **135**, A1769 (1964).
[262] G. Brunhart, H. Postina and V. L. Sailor, *Phys. Rev.*, **137**, B1484 (1965).
[263] O. V. Lounasmaa, *Proc. Int. Conf. Low Temp. Phys.*, London, Butterworth (1964) p. 223.

5

Transport Properties

It is well known that the transport properties of most materials which pass through a magnetically ordered state exhibit features which may be directly associated with the degree and nature of alignment of the ionic moments. This is particularly true of the rare earth metals in which a direct exchange interaction occurs between the conduction electrons and the $4f$ electrons. Further, the complexity of behaviour is increased by the presence of the helical spin structures and by the highly anisotropic nature of the Fermi surface of these metals. With the possible exception of the numerous studies of the superconducting behaviour of lanthanum and its alloys [1–5] there has been significantly less work done on the transport properties than in studies of the magnetic properties of these elements. Early measurements of the electrical and thermal conductivity of polycrystalline samples revealed the correlation with the magnetic ordering which was to be expected. In addition, the negative slope of the resistivity temperature variation near to the Néel point and the rapid fall in the resistivity at the Curie point were observed. These facts subsequently led to the introduction of the concept of superzone boundaries to explain their behaviour. These features will be discussed in the following, along with the associated changes in the thermal conductivity and thermoelectric properties.

5.1 Electrical resistivity

The variation of the resistivity of ferromagnetic metals with temperature may be considered in terms of Mattheissen's rule, as the sum of various scattering contributions. Thus the total resistivity $\rho(T)$ at any temperature (see for example Fig. 5.1) is given by

$$\rho(T) = \rho_{\text{res}} + \rho_{\text{ph}}(T) + \rho_{\text{m}}(T), \tag{5.1}$$

where ρ_{res} is the residual resistivity given experimentally by the zero temperature intercept, and originates in the crystalline imperfections in the solid. $\rho_{\text{ph}}(T)$ arises from the scattering of conduction electrons by phonons and is given by $\rho_{\text{ph}} \alpha T^n$ at low temperatures and $\rho_{\text{ph}} \alpha T$ above the Debye temperature. For scattering in a single band the Bloch theory shows $n = 5$, whereas if the scattering also involves s–d transitions, as is the case for the rare earths, the situation is less well understood and experimentally n is the

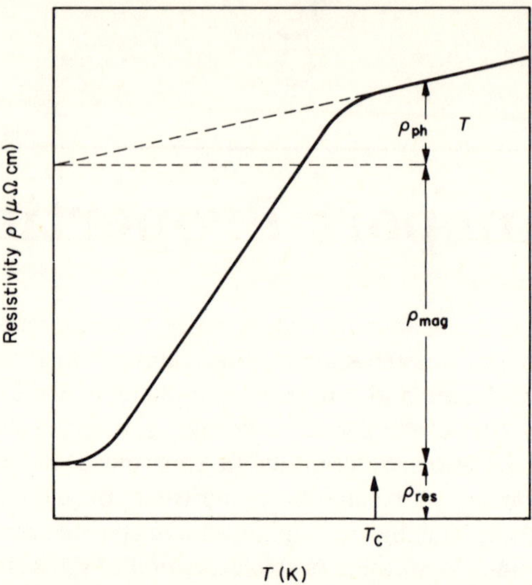

Fig. 5.1 The variation of the resistivity of a ferromagnetic metal with temperature, showing the various contributions to the total resistivity.

order of 3. Finally the magnetic contribution $\rho_m(T)$ caused by the disorder of the spin system can be discussed in terms of both spin wave scattering effects and a molecular field model.

Experimentally one often finds it difficult to separate out the low temperature phonon contribution from the observed resistivity behaviour, especially when it is accompanied by other contributions which are changing rapidly in this temperature region. This is particularly true of holmium, erbium and thulium in which the changes in the resistivity at the magnetic transition temperatures near to 20°K mask the phonon contribution and the separation of this latter term in these metals is somewhat ambiguous. The remaining terms of equation (5.1) however, are rather more readily obtained.

The polycrystalline resistivity behaviour of the rare earths has been reported by several workers [6–14] and data is available for single crystal specimens of most of the heavy metals [15–21]. In addition to the normal increase in resistivity with temperature due to spin-disorder and phonon scattering, two dominant features appear in the *c*-axis resistivity of single crystal specimens of terbium to thulium. These are the occurrence of a pronounced maximum at temperatures just below the Néel point and a sudden, steplike, drop as the temperature is lowered through the Curie point. Similar behaviour is also observed in the polycrystalline resistivity of the light elements and is presumably also associated with the *c*-axis component. As the results for dysprosium show (see Fig. 5.2), the basal plane resistivity is relatively free of these anomalies, having only a continuous rapid increase in the magnetically ordered region followed by slower variation due to

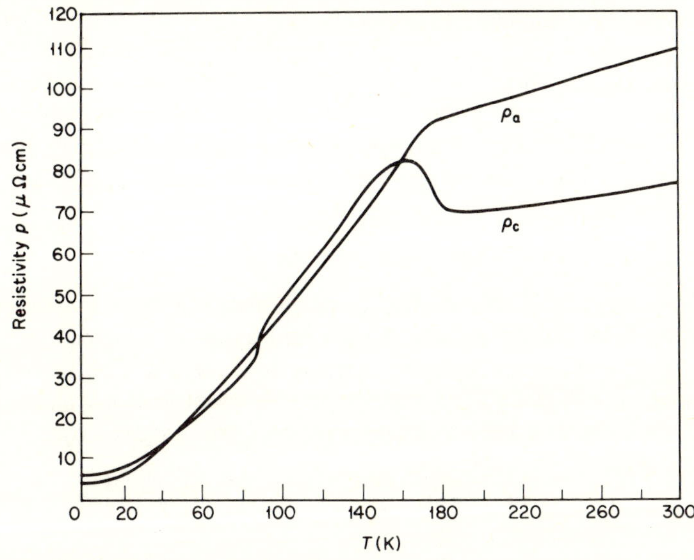

Fig. 5.2 Single and polycrystal resistivity data for dysprosium between 4·2 and 300°K.

phonon scattering when the specimen is paramagnetic. The derived variation of the polycrystalline resistivity ($\rho_{\text{poly}} = \frac{1}{3}(\rho_c + 2\rho_{\text{basal plane}})$) are in good agreement with the earlier results.

Examination of the results for all the metals show that the peak in ρ_c at T_N and the discontinuity at T_c occur only in those metals whose magnetic order passes through an antiferromagnetic moment configuration. The

Table 5.1 Magnitude of the various contributions to the total resistivity for the rare earth metals. * indicates polycrystalline data.

Element	Phonon coefficient C ($\rho = CT$ μohm cm°K^{-1}) basal plane	c-axis	$\rho_{m_{\max}}$ (μohm cm) basal plane	c-axis	$\left[\frac{d\rho_{ph}}{dT}\right]_b \left[\frac{d\rho_{ph}}{dT}\right]_c$
Ce*	0·06–0·09				
Pr*	0·13				
Nd*	0·13				
Sm*	0·21				
Eu*	0·134				
Gd	0·088	0·03	108	108	2·9
Tb	0·188	0·039 ± 0·01	85·1	88·1	2·5
Dy	0·150	0·08	61·9	47·4	1·875
Ho	0·194	0·112	40	24	1·73
Er	0·205	0·107	22·4	11	1·92
Tm	0·209	0·124	22·3	7·4	1·97
Lu	0·259	0·127		—	—

purely ferromagnetic element gadolinium does, however, show a slight maximum in ρ_c at the Curie point which is thought to arise from short range magnetic order effects. These, of course, will also contribute to the anomalies in the other metals. The resistivity of the non-magnetic elements is typical of that to be expected for phonon and residual scattering only. The magnitudes of the various resistivity contributions for all the metals are given in Table 5.1.

5.2 Spin disorder resistivity

The exchange interaction between the localized magnetic moments and the conduction electrons, which has been discussed earlier in connection with the magnetic behaviour, leads to a scattering process which is dependent upon the ionic spin. Taking the scattering cross-section for each scattering centre as C_m leads to a resistivity arising from this process given by

$$\rho_m = (\hbar k_f N / ne^2) C_m = \frac{3\pi N m^*}{8\hbar e^2 E_f} A_0^2 S(S+1)\left[1 - \frac{\langle S \rangle^2}{S(S+1)}\right], \tag{5.2}$$

where N and n are the volume density of centres and conduction electrons respectively and the remaining terms have their usual meaning.

In the high temperature, paramagnetic state, for which all possible moment orientations are possible, it follows that a temperature-independent spin disorder contribution results, whose magnitude is given by [22, 23]

$$\rho_{m_{\max}} = (3\pi N m^* / 8\hbar e^2 E_f) A_0 (g_J - 1)^2 J(J+1). \tag{5.3}$$

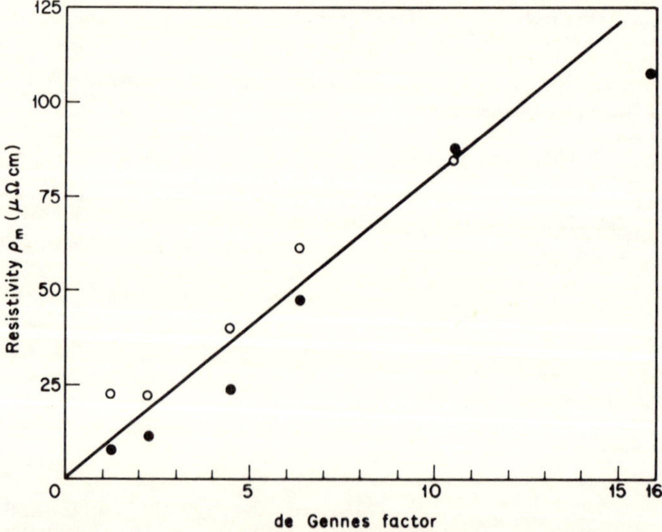

Fig. 5.3 The variation of the *c*-axis and basal plane spin disorder resistivity for the heavy rare earth metals with the de Gennes factor $(g_J - 1)^2 J(J+1)$ and the factor $S(S+1)$.

Table 5.2 Values of A_0 and m^* for the heavy rare earth elements [25]. (a) and (b) assume free electron k_f values. In (c) k_f is from [26].

Element	Gd	Tb	Dy	Ho	Er	Tm
(a) A_0 (eV Å^3)	3·1	3·2	3·0	2·9	2·2	1·9
m^*	2·6	2·8	2·9	2·9	4·2	5·7
(b) A_0	3·0	3·09	2·93	2·73	2·04	1·87
m^*	2·69	2·90	2·97	3·09	4·52	5·81
k_f (Å^{-1})	1·45	1·45	1·44	1·43	1·42	1·41
(c) A_0	5·81	5·01	3·91	3·33	2·27	1·85
m^*	1·36	1·76	2·2	2·52	4·08	5·91

The experimental values of this term are plotted in Fig. 5.3 as a function of the de Gennes factor $(g_J-1)^2J(J+1)$, the agreement being surprisingly good. In view of this correlation, various workers [22–25] have used the expression for the paramagnetic Curie temperature from the RKKY theory (equation 4.16) along with equation (5.3) to evaluate both the exchange constant A_0 and the effective mass m^* values for these elements. The values of the sum appearing in these expressions have been derived for the different lattice constants of the heavy metals and the effects of allowing slight changes in the radius of the Fermi sphere have been examined [25]. The values of m^* and A_0 so obtained are given in Table 5.2 for Fermi momenta corresponding to: (i) the free electron model with an electron density of 3 electrons per ion; and (ii) the k_f value obtained by Yosida [26] from measurements of the helical spin turn angle.

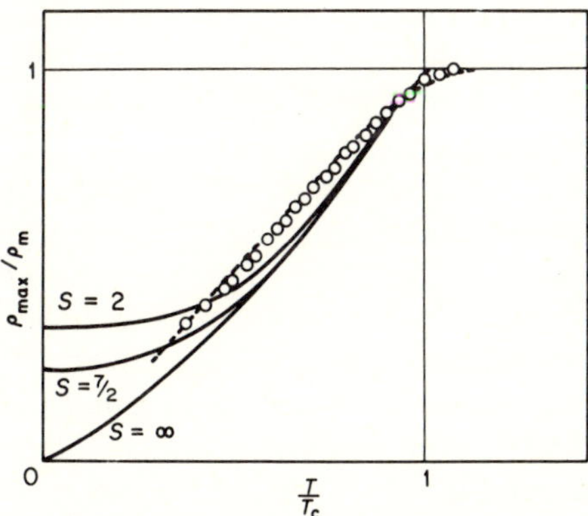

Fig. 5.4 Spin disorder resistivity of gadolinium as a function of reduced temperature. The solid curves are obtained theoretically for various values of $(g_J-1)J$. The broken curve represents the experimental points [27].

For temperatures immediately below the upper ordering temperature, it follows from equation (5.2) that for a general conical structure

$$\rho_{\mathrm{m}} = \rho_{\mathrm{max}}(1 - M^2 - \tfrac{1}{2}M'^2) \tag{5.4}$$

where M and M' are the magnetizations in the axial and basal plane directions respectively. De Gennes and Friedel [27] have successfully applied this approach to gadolinium with the results shown in Fig. 5.4.

5.3 Spin wave scattering

At low temperatures the magnetic scattering is best discussed in terms of a spin wave model. It is now well known that the scattering process in a ferromagnetic metal depends on the spin wave spectrum and in a simple ferromagnet with no anisotropy the spin wave resistivity has been shown to have the form

$$\rho_{\mathrm{m}} = CT^2. \tag{5.5}$$

By analogy with the behaviour of the observed magnetization in anisotropic ferromagnets [29], Mackintosh [30] has predicted that the spin wave contributions to the total resistivity will also depend on the value of the spin wave activation energy ΔE and may be written

$$\rho_{\mathrm{m}} = CT^2 \exp(-\Delta E/kT). \tag{5.6}$$

The experimental results for terbium and dysprosium can be fitted to equation (5.6) with $\Delta E/k \sim 20°\mathrm{K}$ in agreement with theory (Section 4.4.1) for dysprosium. It should be pointed out however, that the fitting accuracy is somewhat insensitive to the value of $\Delta E/k$ and that the only available data for use in this comparison was taken using polycrystalline samples. Consequently, since the more recent single crystal experimental results show appreciable anisotropy of ρ_{m}, this fit is perhaps to be viewed with caution. Nevertheless the overall nature of the results indicate that the spin wave activation energy should be taken into consideration when attempting to account for the observed resistivity of these and other ferromagnetic materials.

In the magnetically ordered spiral structures (e.g. holmium and erbium) the spin wave spectrum is linear at small wave numbers [31, 32] and consequently leads to a spin wave scattering contribution to the resistivity which is proportional to the fourth power of the temperature. Here again an activation energy will be introduced by the magnetocrystalline anisotropy and the resistivity may be expected to have the form

$$\rho_{\mathrm{m}} = CT^4 \exp(-\Delta E/kT). \tag{5.7}$$

As yet however, there is little evidence to support this, and the thin film observations of Lodge [33, 34] give a closer fit to equation (5.6) than to (5.7) and are in keeping with the other elements examined in this investigation in having $\Delta E/k = 20°K$.

5.4 The effect of superzone boundaries

The discussions of Sections 5.2 and 5.3 do not provide any understanding of the rapid changes in resistivity observed in the pure metals at T_N and T_c, and another mechanism must be sought. Between these two ordering temperatures the moments have various periodic forms and, as discussed in Section 4.3.3, a consequence is that superzone boundaries appear within the Brillouin zone. The effect of the superzones is to give rise to an appreciable reduction in Fermi surface area, and Mackintosh [35] and Miwa [36] first suggested that the appearance of such boundaries perpendicular to the c-axis could account for the observed increase in resistivity parallel to that direction. A detailed treatment for free electron bands has been given by Elliott and Wedgewood [37], and the results given below are taken from their work.

The simplest theory is assumed in which the scattering mechanisms discussed above are written in terms of an isotropic total relaxation time τ. The conductivity tensor is then given by

$$\sigma_{ij} = \frac{e^2\tau}{4\pi^3\hbar}\int v_i \mathrm{d}S_j, \tag{5.8}$$

where $v_i = \hbar^{-1}\partial E(\mathbf{k})/\partial k_i$ is a component of the electron's group velocity, and the integration is over the Fermi surface. Elliott and Wedgewood took the unperturbed surfaces of constant energy to be spheroidal. For a single superzone boundary at wavevector $\mathbf{l}$ the perturbed energies are given by an expression like (4.48) of Section 4.3.3 and the Fermi surface may be considered to be distorted in one of the three ways illustrated in Fig. 4.27. The results for the conductivity may conveniently be expanded in the parameter (Δ/E_f), where Δ is the energy gap at the superzone boundary, and they may be summarized in the form

$$\sigma_{ii} = \sigma_{ii}^0(1-\delta_{ii}), \tag{5.9}$$

where σ_{ii}^0 is the unperturbed value. To order $(\Delta/E_f)^2$ it is found that $\delta_{xx} = 0$, and also that $\delta_{zz} = 0$ provided the Fermi surface does not touch the superzone boundary. When the boundary (i) cuts and (ii) touches the Fermi surface then to order Δ/E_f:

$$\text{(i)}\ \delta_{zz} = 3\pi l\Delta/4k_f E_f \tag{5.10}$$

$$\text{(ii)}\ \delta_{zz} = 1-(l/k_f)^3+(3\pi l\Delta/8k_f E_f). \tag{5.11}$$

It can be seen that the effect is largest when the superzone cuts the surface and in this case the complete expression for ρ_{zz} is

$$\rho_{zz} = \frac{\rho_{\text{res}} + \rho_{\text{ph}}(T) + \frac{\rho_{\text{max}}}{J^2}(1 - \frac{1}{2}\langle J_z \rangle^2 - \langle J_\perp \rangle^2)}{1 - \frac{\Gamma}{J^2}(\langle J_z \rangle^2 + \langle J_\perp \rangle^2)}, \qquad (5.12)$$

where

$$\Gamma = \frac{3\pi A_0 J}{4E_f} \sum_{l_i} \frac{l_i}{k_f}$$

the sum being over all superzone boundaries which cut the Fermi surface. The numerator in (5.12) is an explicit expression for ρ given by (5.1) and (5.4).

The form of (5.12) with a best fit constant value of Γ shows qualitative agreement with experimental results for dysprosium and holmium single crystals (Fig. 5.5), and the importance of superzone boundaries is established. In fact since the boundary with largest l will only touch, and not cut, the Fermi surface, its position will depend upon temperature as described in Section 4.3.4 and therefore Γ will also depend upon temperature. Indeed Elliott and Wedgewood find that the required temperature dependence of Γ for a good fit is very similar to that of the temperature variation of the turn angle. However, it must be emphasized again that quantitative agreement of theory with experiment cannot be expected when free electron bands are assumed.

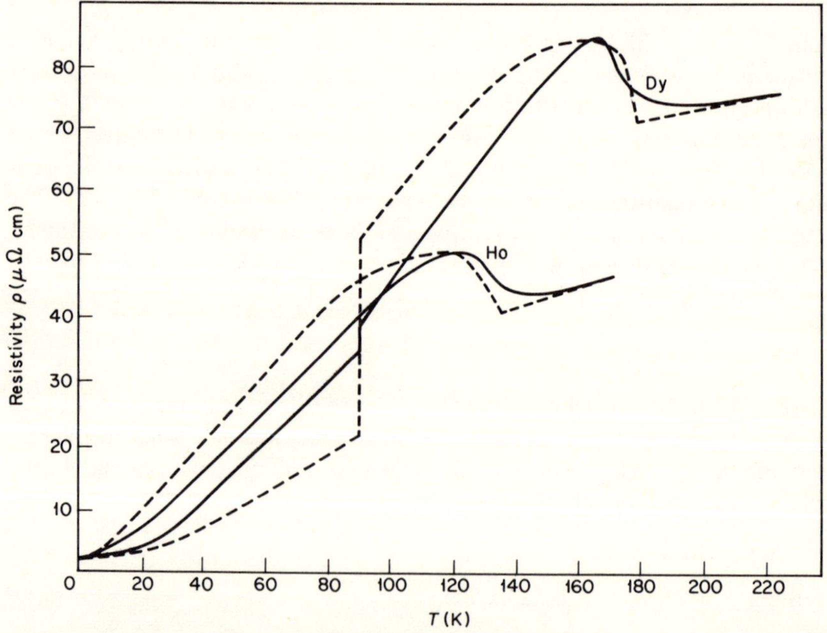

Fig. 5.5 Experimental (full curves) and theoretical (broken curves) values of the *c*-axis resistivities of dysprosium and holmium [15, 19, 37].

5.5 Resistivity above the ordering temperature

Above the highest ordering temperatures, resistivity measurements show few features although small discontinuities occur at temperatures close to those at which crystal structure changes take place [38, 39].

The gradient of the resistivity–temperature plots in this region are given in Table 5.1 and for the heavy rare earths the ratio of the axial to basal plane gradients continuously decreases across the series. Assuming that there is a constant spin disorder contribution at these temperatures then to a first approximation

$$\frac{d\rho_{\text{ph}_i}}{dT} \propto \sum_{\text{Fermi surface}} dS_i \tag{5.13}$$

Values for ΣdS_i have been obtained [40] for all the heavy metals. Comparison of the ratios $\Sigma dS_c/\Sigma S_{\text{basal}}$ with the experimental ratios of the gradients of the resistivity–temperatures curves shows quite good agreement for

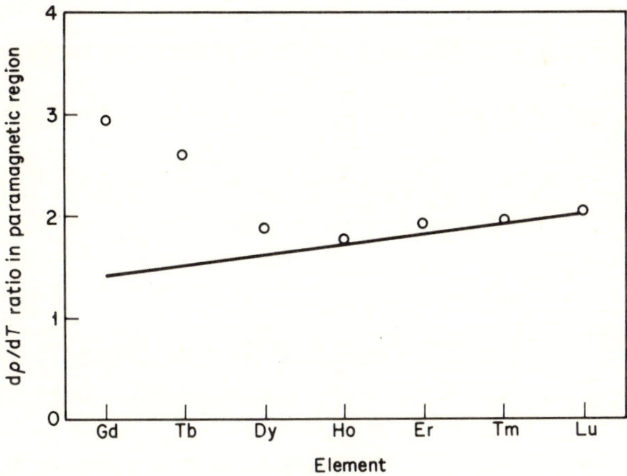

Fig. 5.6 A comparison of the $(d\rho/dT)_c/(d\rho/dT)_{\text{basal}}$ ratios with the theoretical ratios $\sum dS_c/\sum dS_{\text{basal}}$ of the projections of the Fermi surface areas.

the elements dysprosium–lutetium as may be seen from Fig. 5.6. Kasuya [42] has proposed that the spin disorder resistivity will also be temperature-dependent above the Néel points and has tabulated a series of values for this dependence. These are in general small however, and will have little effect on the values given in Fig. 5.6.

5.6 The effects of alloying and pressure

5.6.1 *Alloying behaviour*

As yet, relatively few studies of the resistivity of the rare earth alloys have been made. Those which have show a behaviour similar to that which might be expected from the magnetic behaviour of the alloys. In addition however, Smidt and Daane [43] observed a correlation between the magnitude of the residual resistivity and the difference in angular momentum quantum numbers of the components. This additional contribution was associated with the spin disorder existing at $T = 0°K$ because of the random distribution of the two moment species, in other words a residual magnetic resistivity. A similar behaviour has been found in Gd_x–$Y_{(1-x)}$ by Popplewell *et al* [44] who give the concentration dependence of ρ_{res} as $\rho_{res} - 8{\cdot}4 \simeq 200[x(1-x)]$ in agreement with the predictions of Dekker [45].

In the case of an alloy containing two types of magnetic ion, Dekker [23] has shown that equation (5.3) for the spin disorder resistivity may be written:

$$\rho_m(x) = \frac{3\pi N m^*}{2\hbar e^2 E_f}[xA_{0_a}^2(g_{J_a}-1)^2 J_a(J_a+1)+(1-x)A_{0_b}^2(g_{J_b}-1)^2 J_b(J_b+1)$$

$$-x(1-x)\{A_{0_a}(g_{J_a}-1)J_a - A_{0_b}(g_{J_b}-1)J_b\}^2], \tag{5.14}$$

where a and b refer to the various constants for the constituents of the alloy of composition $A_xB_{(1-x)}$. When one of the components is non-magnetic this reduces to:

$$\rho_m(x) = \frac{3\pi N m^*}{2\hbar e^2 E_f}(g_{J_a}-1)^2 A_{0_a} J_a x(1+J_a x). \tag{5.15}$$

Using this expression Popplewell derived values for A_{0_a} as a function of yttrium composition which increased from approximately 3 to 6 eV Å^3 for pure gadolinium to 2·4 atomic % gadolinium. The Ziman parameter also increased from gadolinium to yttrium and it was assumed that both these variations could be attributed to changes in the Fermi surface on alloying.

Although this method of obtaining A_0 is straightforward it is incapable of resolving the sign of the interaction, and further, since assumptions have to be made concerning E_f and m^*, the resultant values may be no better than order of magnitude estimates. It is not clear from the work of Popplewell what values were used for these terms, and one must assume that they adopted the free electron values used by other workers.

Observation of the Kondo effect in dilute alloys however, provides a means of obtaining both the magnitude and sign of A_0, although again some approximation must be made for E_f. According to Kondo [47] the magnetic contribution to the resistivity of a dilute magnetic alloy is given by

$$\rho'_m = \rho_m[1+2A_0 N(E_f)\log T], \tag{5.16}$$

where ρ_m is the usual spin disorder term calculated in the first Born approximation of the s–f interaction (equation (5.3)), and $N(E_f)$ is the density of states at the Fermi surface. If A_0 is negative then a Kondo minimum appears in the total resistivity, whereas for positive A_0 the resistivity anomaly appears as an increase in the gradient of the resistivity–temperature curve.

Observations by Sugawara [48] on yttrium-based alloys showed that A_0 is positive for all the heavy rare earths studied and cerium, but negative for the remainder. Similar results were also found [49] from a study of the low temperature resistivity and superconducting transition temperatures of lanthanum containing small amounts of rare earth impurity. The depression of the superconducting transition temperature in the presence of an exchange interaction has been evaluated by several workers [50–52] and is given by

$$\Delta T_c = -(\pi^2/16k)cN(E_f)(g_J - 1)^2 J(J+1)A_0^2 \tag{5.17}$$

and again allows $|A_0|$ to be determined. In (5.17) c is the impurity concentration.

The variation of A_0 obtained by these methods is given in Fig. 5.7 for both the yttrium and lanthanum alloys. Sugawara [49] also considers the effects of the mixing of the $4f$ electrons with the conduction electrons on the

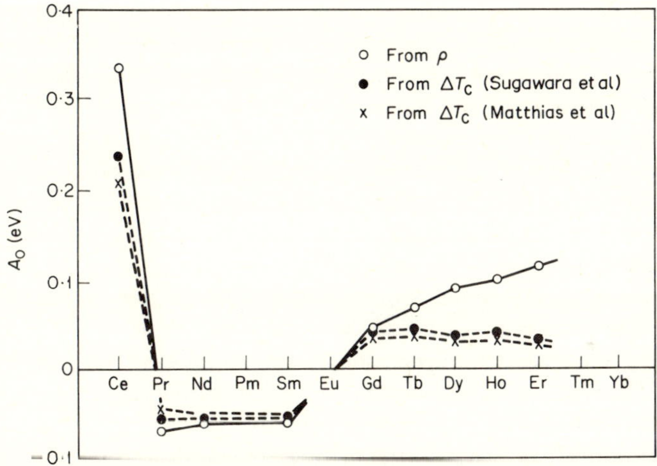

Fig. 5.7 The experimental values of the exchange constant A_0 for all the rare earth elements obtained from measurements of the Kondo effect and the depression of the superconducting transition temperature in the dilute magnetic alloys of lanthanum.

value of A_0 and concludes that the degree of mixing and the nature of the conduction bands is similar in both yttrium and lanthanum. Their results also show that the residual resistivities of the lanthanum alloys are better described by assuming that there is zero spin–orbit coupling in the impurity ions. On this basis, recalculation of A_0 leads to positive values for all the metals except cerium. In a later publication Sugawara used a more complete expression for the resistivity of magnetically dilute alloys [53] to obtain A_0, and again found similar signs for this parameter.

The depression of the superconducting temperature in the lanthanum–cerium system and the values of the superconducting critical fields observed by Sugawara are in excess of those to be expected from the Abrikosov–Gor'kov theory [50]. This deviation, along with observed susceptibility values of both the lanthanum–cerium alloys [55–57] and α- and β-cerium [58], have led Edelstein [58] to conclude that cerium is in a spin-compensated state in the elemental and dilute alloy states, i.e. the conduction electron polarization cancels the effect of the cerium ionic moment. Evidence is also available for spin compensation over most of the alloying range in the yttrium–cerium system [59]. Smith [60] however, has used the results of measurements at high pressures on rare earth doped lanthanum to suggest that these anomalies may originate from the cerium $4f$ electron lying in a virtual bound state [24].

5.6.2 *Pressure effects*

As a matter of convenience in investigating both the structural and magnetic changes caused by the application of high hydrostatic pressures to the rare earth metals, the resistivity behaviour has frequently been the experimental parameter used to determine the transition temperatures and pressures of these changes. Consequently a considerable literature exists relating to the influence of pressure on the resistivity itself. Interpretation of the results however, has never been satisfactorily accomplished as the number and complexity of the different contributions to the observed changes is rather unwieldy. These include the previously discussed structural and magnetic transitions and the related changes in the band structure which must occur with increasing pressure.

Since the early measurements of Bridgman [61, 62], more recent investigations of the resistivity variation have included those of Vereshchagin *et al* [63, 64], Souers and Jura [65], Stager and Drickamer [66] and Wazzan *et al* [67]. The results of these workers are in general agreement and show a wide variety of transitions many of which have been tentatively associated with the electronic behaviour of the metals.

5.7 The Hall effect

It has been known for some considerable time that the Hall effect in ferromagnetic materials contains, in addition to the normal $1/ne$ term, a second component dependent on the magnetization M. The electric current in such a material may be written in component form

$$j_i = \sigma_{ij}E_j + (\sigma_k)_{ijk}B_jE_k + (\sigma'_k)_{ijk}M_jE_k, \tag{5.18}$$

where E is the electric field, and B is the magnetic field. It follows then that the Hall field per unit current density is given by

$$e_H = R_0B + R_1 4\pi M, \tag{5.19}$$

where R_1 is the anomalous (ferromagnetic) Hall coefficient. Experimentally

R_1 is appreciably greater than R_0 and as may be expected is strongly dependent on temperature. In the transition metals this behaviour of R_1 is generally associated with a spin–orbit coupling of the magnetic electrons which also act as charge carriers [68]. In the rare earths however, in which the *f*-electrons are highly localized this cannot apply and both Kondo [69] and Maranzana [70] have considered this problem specifically for these metals. The former considers the case of an intrinsic spin–orbit coupling of the *f*-electrons together with a anisotropic *s–f* interaction leading to the transverse Hall current. This model requires an unquenched angular momentum of the *f*-electrons and would not therefore account for the large anomalous effect in gadolinium [71–73]. Using this model, Maranzana has introduced the asymmetry through an interaction between the *f*-electron spin and the orbital momentum of the charge carriers. While this interaction is inherently of low magnitude, the high mobility of the *s*-electrons keeps its contribution to the observed Hall field in excess of that due to the normal Hall effect.

Experimental observations on polycrystalline materials were first carried out by Kevane [74] using an a.c. technique but were unfortunately limited in general to temperatures in excess of the magnetic transitions. Since then, more complete studies have been made on several of the metals and single crystal results are available for gadolinium [71, 73], terbium [75], dys-

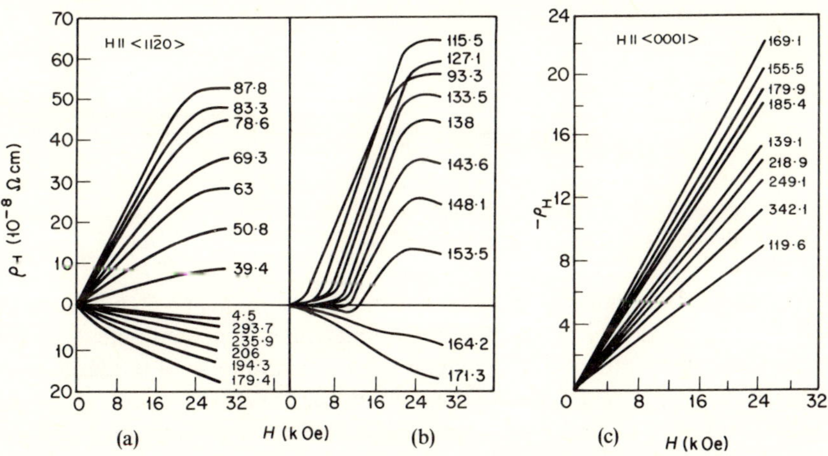

Fig. 5.8 The isothermal dependence of the Hall resistivity of single crystal dysprosium on the applied magnetic field strength. (a) The basal plane results for $T < T_C$, $T > T_N$, and (b) $T_C < T < T_N$. (c) *c* axis for $119·6 < T < 342·1$°K [76].

prosium [76, 77], lutetium [73] and yttrium [73]. Fig. 5.8 shows the observed variation of the Hall resistivity for a basal plane and *c*-axis dysprosium specimen as a function of applied field strength under isothermal conditions at various temperatures. For $T > T_N$ and $T < T_c$ these observations predominantly show negative and positive Hall coefficients respectively. In the helical spin range however, for temperatures well below T_N, a rapid change in sign occurs (from negative to positive) as the applied field increases through the critical field value. Using the relation for the Hall resistivity at an applied field equal to the demagnetizing field

$$(\rho_H)_{H_{appl}} = 4\pi NM_s = (R_0+R_1)4\pi M_s \tag{5.20}$$

and assuming $R_0 \ll R_1$, Rhyne derived values of R_1 corresponding to the spontaneous magnetization M_s which are illustrated in Fig. 5.9 for the basal plane specimen. The small negative value at low temperatures is thought to arise from the neglected R_0 term.

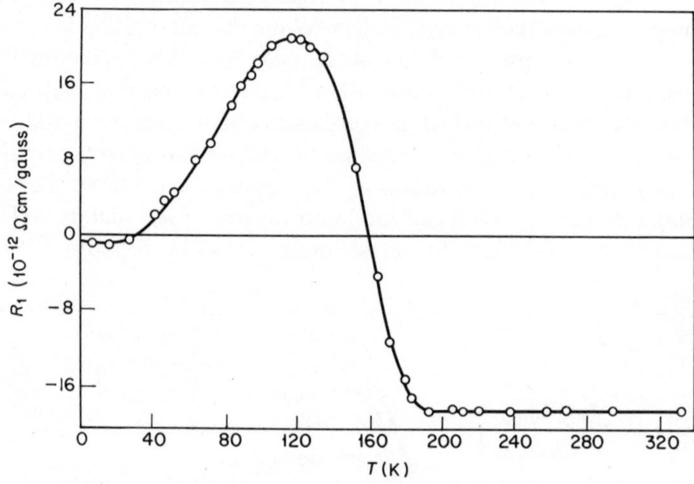

Fig. 5.9 Anomalous Hall coefficient for single crystal dysprosium specimen with $H//a$-axis and current along the b-axis [76].

Although the spontaneous magnetization component of the Hall coefficient vanishes above T_N, an anomalous magnetic component is still observable due to the induced magnetization resulting from the large ionic moment in this material. Under these conditions one then expects that both components will be linear with applied field. The anomalous component is, however, directly proportional to the effective susceptibility $\chi^* = (\chi/1+4\pi N\chi)$. The results shown in Fig. 5.10 confirm this approach and lead to a value for the normal coefficient of $-0{\cdot}3\times10^{12}$ ohm cm gauss^{-1} and for the anomalous coefficient $R_1 = -18\times10^{-12}$. In the c-axis specimen the corresponding values are $-3{\cdot}7\times10^{-12}$ and -23×10^{-12} respectively.

Using a molecular field approximation, Irkhin and Abel'skii [78] have considered R_1 on the basis of both the intrinsic spin–orbit coupling of the magnetic electrons and the 'mixed' s–d coupling. They obtain for both mechanisms

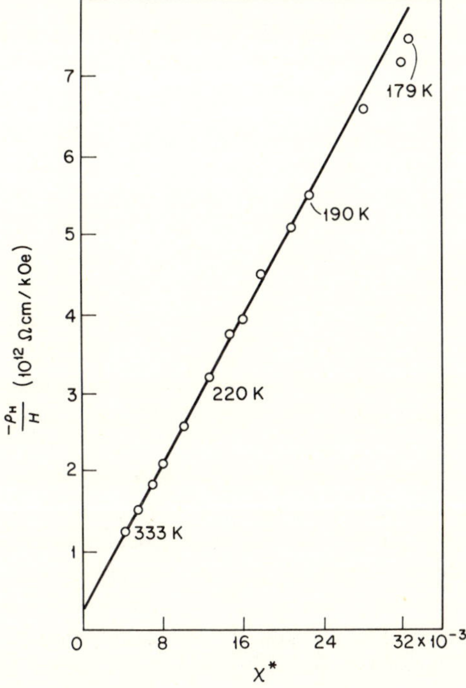

Fig. 5.10 Slope of the Hall resistivity of dysprosium versus applied field curve in the paramagnetic region as a function of the effective magnetic susceptibility [76].

$$R_1 = (9\pi A_0^2/64E_f)(m^*\zeta_{\text{eff}}/e^2N\hbar M_c(0))[\tfrac{1}{4}-\gamma^2], \tag{5.21}$$

where ζ_{eff} is the effective spin–orbit coupling constant, $M_s(0)$ the magnetization at $T=0$ and $\gamma = \frac{1}{2}[M_s(T)/M_s(0)]$ with $M_s(T)$ the spontaneous magnetization at temperature T. This then reduces to

$$R_1 = \pm(3\zeta_{\text{eff}}/16E_f)(\rho_m/M_s(0)), \tag{5.22}$$

by using the expression for ρ_m from the spin wave resistivity theory of Kasuya and Mannari [79, 80]. Equation (5.21) also provides a simple dependence of R_1 on the magnetization, given by

$$R_1 = C[M_s^2(0) - M_s^2(T)], \tag{5.23}$$

a relation which has been proposed by several authors.

The temperature dependence of R_1 has been considered using a spin wave model in conjunction with the expression obtained by Maranzana [70] for the anomalous Hall resistivity. These calculations show a T^4 dependence of R_1 at low temperatures and a T^2 dependence at higher temperatures.

The experimental results of Rhyne and others [72, 81] show that the T^2 prediction of Kagan and Maksimov [104] is obeyed over an appreciable range of temperature. The results for dysprosium [76] are also in excellent agreement with equations (5.22) and (5.23) while both Lee and Legvold [73] and Babushkina [72] find that $R_s \propto \rho_m^2$.

5.8 Thermal conductivity

Experimental investigations [82–94] show that the rare earth metals are poor heat conductors compared with other metals and in conjunction with the electrical resistivity measurements one finds anomalously large values of the Lorentz function (L) which are usually in excess of that obtained for electronic heat conduction only. Since spin-wave contributions to the observed specific heat are known to be significant [95, 96] it is not unlikely that they may also act as heat carriers in the thermal conduction process and several authors have analyzed their results with this in mind.

Single crystal results are available for all of the heavy metals [41, 97, 98], with the exception of ytterbium, and in all of these there is appreciable anisotropy of the c-axis and basal plane conductivity and Lorentz functions at all temperatures. Only in the c-axis lutetium specimens does the observed value of L compare favourably with the 'electron only' value.

The experimental results for dysprosium [91] shown in Fig. 5.11 contain most of the important features found for the other magnetically ordered heavy metals. In all the metals in the paramagnetic range the c-axis conductivity exceeds that in the basal plane, and with the exceptions of terbium and holmium the ratio K_c/K_a becomes less than unity below the magnetic ordering temperature. This arises partially from the sudden decrease in the

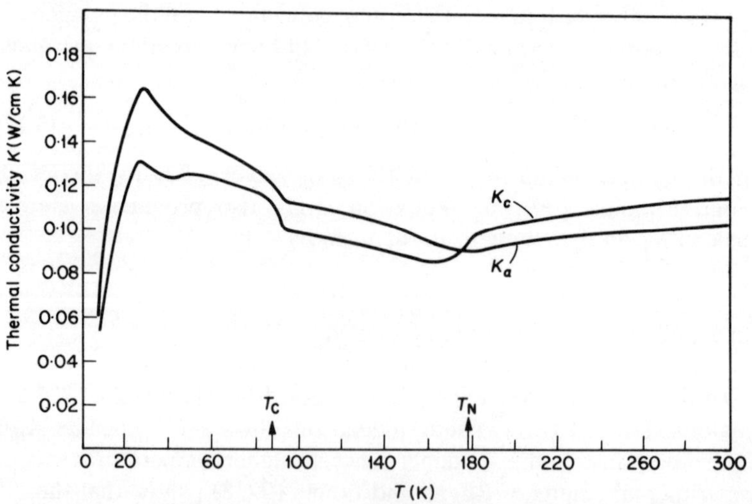

Fig. 5.11 The thermal conductivity of single crystal dysprosium as a function of temperature.

c-axis component at the Néel point due to the appearance of the superzone boundaries but as this inversion of the anisotropy ratio also occurs in gadolinium other contributing factors should be considered. Whether or not these are related to the magnetic properties of the elements still remains to be seen.

The lowering of the thermal conductivity in the antiferromagnetic region is entirely analogous to the behaviour of the electrical resistivity discussed earlier and arises because, with the appearance of the superzone boundary structure, the integral over the Fermi surface in the expression for the thermal conductivity

$$K = (L_0 T e^2 / 4\pi^3 \hbar) \int v_i \mathrm{d}S_j = L_0 T \rho_i^{-1} \tag{5.24}$$

decreases. This reduction of the Fermi surface occurs predominantly for those regions parallel to the c-axis and consequently, as with the resistivity, the effects are most easily seen in the c-axis component of K. The magnitudes of the changes in the effective surface area (ΔA) have been calculated by Loucks and Liu [40] for erbium, the results being

$$\Delta A_c = 6\%, \quad \text{and} \quad \Delta A_b \simeq 0{\cdot}6\%.$$

Experimentally, $\Delta K_c / \Delta K_b$ for dysprosium is found to be in close agreement with the value predicted using these changes in Fermi surface area.

In the other metals however, the comparison is much less convincing and experimental results show only that $\Delta K_c > \Delta K_b$. Further, in many of these metals any sudden changes in the thermal conductivity at the transition temperatures may be masked by the rapid variation in K in their vicinity.

In the paramagnetic state, equation (5.24) indicates that the anisotropy of the thermal conductivity will be dominated by the Fermi surface integral $\int v \mathrm{d}S$ and one may expect the value of the ratio of the two projected areas to give the magnitude of this anisotropy. The value estimated by Loucks and Liu of $A_c / A_b = 2{\cdot}1$ is then to be compared with K_c / K_b values ranging from 1 to 1·7 for terbium and thulium, respectively.

The anomalously large Lorentz numbers observed in all the magnetic metals suggest that other conduction mechanisms must be considered in addition to the electron contribution. Attempts have been made to isolate the various scattering mechanisms from the low temperature thermal conductivity data in terms of impurity, phonon and magnon effects. While as yet it is inconclusive the data appear to show some support for a magnon component in the observed conductivity at low temperatures.

5.9 Thermoelectric properties

phenomena, is sensitive to both the electron scattering mechanisms and the detailed electronic structure of the material. The measurements of Legvold and others [13, 100] on the heavy rare earth metals have shown an interesting dependence on the magnetic properties and an appreciable anisotropy of S both above and below the upper magnetic ordering temperature [101]. With increasing temperature the general behaviour (illustrated in Fig. 5.12

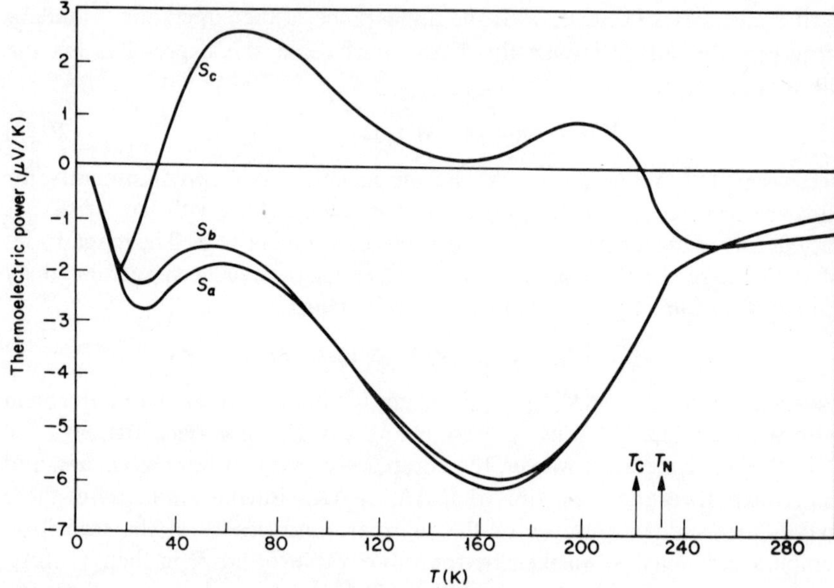

Fig. 5.12 The temperature dependence of the Seebeck coefficients of single crystal terbium.

for terbium) shows a maximum negative value of S at about 20°K followed by a region of rapidly varying S for $T < T_N$ (or T_c for gadolinium). In terbium and dysprosium the c-axis component S_c changes sign in this region and in holmium S_c becomes negative after having passed through a *positive* maximum at lower temperatures. The magnetic transitions are characterized by changes in the slope of the S–T curves rather than by discontinuities such as occur in both the electrical and thermal conductivities. Dysprosium is an exception to this and S_c, S_b and S_a all show abrupt variations at T_c. Since the other metals show no comparable effects this change in dysprosium has been tentatively associated with the crystal structure change rather than with the appearance of superzones.

As with the thermal conductivity, the anisotropy ratio passes through zero in the vicinity of the upper magnetic transition temperature and for $T >$ 300°K (except gadolinium) $|S_b| < |S_c|$ and $(S_b \sim S_c)$ is approximately constant.

With the thermoelectric properties we are again dealing with complex phenomena, the observed variation of which depends on several processes operating simultaneously. Two of these are well known and are responsible for the thermoelectric power of normal metals; these being the diffusion of electrons along the temperature gradient, and the phonon-drag effect. This latter arises from the transfer of momentum from the phonon current to the electrons in the direction of the phonon flow. In the rare earth metals the Seebeck coefficient is further dependent upon the magnetic properties of the materials both through a magnon-drag term and the modification of the Fermi surface by the superzone boundaries. These latter effects have as yet been little studied and although some work has been reported on the role of magnon–phonon scattering [102, 103] the effects of the variable spin wave spectra are still unknown.

To a first approximation the total Seebeck coefficient is given by the sum of the various contributions, so that

$$S_{\mathrm{T}} = S_{\mathrm{e}} + S_{\mathrm{ph}} + S_{\mathrm{m}} \tag{5.25}$$

where S_{m} is the sum of all possible magnetic contributions to the observed S_{T} value.

The electronic contribution S_{e} can be written

$$S_{ij} = \frac{\pi^2 kT}{3e}\left[\frac{\partial \log \sigma_{ij}(E)}{\partial E}\right]_{E = E_{\mathrm{f}}}, \tag{5.26}$$

where σ_{ij} is given by equation (5.8). Sill and Legvold [101] write the total relaxation time τ in terms of the impurity, lattice and magnetic relaxation times as

$$\frac{1}{\tau} = \frac{1}{\tau_i} + \frac{1}{\tau_l} + \frac{1}{\tau_m} \tag{5.27}$$

and allow τ_{mag} to have an energy dependence given by

$$\frac{1}{\tau_m} = \alpha E^n \tag{5.28}$$

This then gives,

$$\begin{aligned} S_z &= (LeT/2E_{\mathrm{f}})[3/(1-\delta) - 2n\tau/\tau_{\mathrm{s}}] \\ \text{and} \quad S_x &= (LeT/2E_{\mathrm{f}})[3 - 2n\tau/\tau_{\mathrm{s}}], \end{aligned} \tag{5.29}$$

where τ_{s} is due to spin disorder scattering, δ is approximately $\Delta E/E_{\mathrm{f}}$, and for a simple model $n = \frac{1}{2}$. Using this approach Sill and Legvold were able to account qualitatively for the observed variations of S_{T} in the region $T > 30$°K.

At low temperatures the phonon and magnon drag terms can be expected to dominate due to the disappearance of S_{e}. In this region S_{ph} is proportional to T^3 up to $\theta_{\mathrm{D}/5} \rightarrow \theta_{\mathrm{D}/10}$ when it passes through a maximum and decreases as T^{-1}. Theoretical indications are that the magnon scattering will result in a similar behaviour with temperature. Since θ_{D} for these metals lies between

152°K for gadolinium and 163°K for erbium, the observed peaks in S_T in the vicinity of 20°K appear to substantiate this view. Further, as the maximum in S_T for the non-magnetic element yttrium is relatively small compared with the magnetically ordered metals Legvold has associated a major part of the observed maximum with the magnon component. As the spin wave energy gap in many of the metals is also of order 20°K a rather rapid onset of the magnon drag component might be expected at about this temperature which will further complicate the analysis. The normal ferromagnetic gadolinium has no such energy gap however, and for this element one finds $S_T \propto T^3$ for $T < 25$°K in agreement with the foregoing but providing no separation of the two drag terms.

References

[1] J. C. McLennan, J. F. Allen and J. O. Wilhelm, *Phil. Mag.*, **10**, 500 (1930).
[2] D. H. Parkinson, F. S. Simon and R. H. Spedding, *Proc. Roy. Soc.*, **A207**, 137 (1951).
[3] W. T. Ziegler, R. A. Young and A. L. Floyd, *J. Am. Chem. Soc.*, **75**, 1215 (1953).
[4] B. T. Matthias, H. Suhl and E. Corenzwit, *Phys. Rev. Letts.*, **1**, 92 (1958).
[5] G. S. Anderson, S. Legvold and F. H. Spedding, *Phys. Rev.*, **109**, 243 (1958).
[6] G. S. Anderson and S. Legvold, *Phys. Rev. Letts.*, **1**, 322 (1958).
[7] N. R. James, S. Legvold and F. H. Spedding, *Phys. Rev.*, **88**, 1092 (1952).
[8] J. K. Alstad, R. V. Colvin, S. Legvold and F. H. Spedding, *Phys. Rev.*, **121**, 1637 (1961).
[9] R. V. Colvin, S. Legvold and F. H. Spedding, *Phys. Rev.*, **120**, 741 (1960).
[10] M. A. Curry, S. Legvold and F. H. Spedding, *Phys. Rev.*, **117**, 953 (1960).
[11] S. Legvold, F. H. Spedding, F. Barson and J. F. Elliott, *Rev. Mod. Phys.*, **25**, 129 (1953).
[12] A. Watabe and T. Kasuya, *J. Phys. Soc. Japan*, **26**, 64 (1969).
[13] H. J. Born, S. Legvold and F. H. Spedding, *J. Appl. Phys.*, **32**, 2543 (1961).
[14] E. M. Zavitskiy, V. F. Terekhova and O. P. Naumkin, *Usp. Fiz. Nauk*, **79**, 263 (1963).
[15] P. M. Hall, S. Legvold and F. H. Spedding, *Phys. Rev.*, **117**, 971 (1960).
[16] P. W. Green, S. Legvold and F. H. Spedding, *Phys. Rev.*, **122**, 827 (1961).
[17] H. E. Nigh, S. Legvold and F. H. Spedding, *Phys. Rev.*, **132**, 1092 (1963).
[18] D. E. Hegland, S. Legvold and F. H. Spedding, *Phys. Rev.*, **131**, 158 (1963).
[19] D. L. Strandburg, S. Legvold and F. H. Spedding, *Phys. Rev.*, **127**, 2046 (1962).
[20] L. R. Edwards and S. Legvold, *Phys. Rev.*, **176**, 753 (1968).
[21] T. T. Jew, Ph.D. Thesis, Iowa State University, Ames, Iowa (1963).
[22] P. G. de Gennes, *J. Phys. Paris*, **23**, 510 (1962).
[23] A. J. Dekker, *J. Appl. Phys.*, **36**, 906 (1965).
[24] Y. A. Rocher, *Adv. Phys.*, **11**, 233 (1962).
[25] M. I. Darby and K. N. R. Taylor, *J. Appl. Phys.*, **37**, 1442 (1966).
[26] K. Yosida, *Progress in Low Temperature Physics*, **4**, 265 (1964).
[27] P. G. de Gennes and J. Friedel, *J. Phys. Chem. Solids*, **4**, 71 (1958).
[28] I. Mannari, *Progr. Theor. Phys.* (Kyoto), **22**, 335 (1959).
[29] K. Niira, *Phys. Rev.*, **117**, 129 (1960).
[30] A. R. Mackintosh, *Phys. Letts.*, **4**, 140 1(963).
[31] A. Yosimari, *J. Phys. Soc. Japan*, **14**, 807 (1959).
[32] T. A. Kaplan, *Phys. Rev.*, **124**, 329 (1961).
[33] F. M. K. Lodge, Ph.D. Thesis, Durham (1969).
[34] F. M. K. Lodge and K. N. R. Taylor, to be published.
[35] A. R. Mackintosh, *Phys. Rev. Letts.*, **9**, 90 (1962).
[36] H. Miwa, *Prog. Theor. Phys.*, **28**, 208 (1962).

[37] R. J. Elliott and F. A. Wedgwood, *Proc. Phys. Soc.*, **81**, 846 (1963).
[38] F. H. Spedding, A. H. Daane and K. W. Herriman, *J. Metals*, **9**, 895 (1957).
[39] F. H. Spedding, J. J. Hanak and A. H. Daane, *J. Less Common Metals*, **3**, 110 (1961).
[40] T. L. Loucks and S. H. Liu, unpublished, see for example reference [41].
[41] D. W. Boys and S. Legvold, *Phys. Rev.*, **174**, 377 (1968).
[42] T. Kasuya, *Magnetism*, eds. G. T. Rado and H. Suhl, Vol. IIB, Academic Press (1966) p. 215.
[43] F. A. Smidt and A. H. Daane, *J. Phys. Chem. Solids*, **24**, 361 (1963).
[44] J. Popplewell, P. G. Arnold and P. M. Davies, *Proc. Phys. Soc.*, **92**, 177 (1967).
[45] A. J. Dekker, *Phys. Letts.*, **11**, 274 (1964).
[46] T. van Peskitinbergen and A. J. Dekker, *Physica*, **29**, 917 (1963).
[47] J. Kondo, *Progr. Theor. Phys.* (Kyoto), **32**, 37 (1964).
[48] T. Sugawara, *J. Phys. Soc. Japan*, **20**, 2252 (1965).
[49] T. Sugawara and H. Eguchi, *J. Phys. Soc. Japan*, **21**, 725 (1966).
[50] A. A. Abrikosov and L. P. Sor'kov, *Sov. Phys. JETP*, **12**, 1243 (1961).
[51] H. Suhl and B. T. Matthias, *Phys. Rev.*, **114**, 977 (1959).
[52] S. S. Kalski, O. Berbeder-Matitiet and P. C. Weiss, *Phys. Rev.*, **36**, 1500 (1964).
[53] T. Sugawara and H. Eguchi, *J. Phys. Soc. Japan*, **23**, 965 (1967).
[54] T. Sugawara and H. Eguchi, *Phys. Letts.*, **25A**, 668 (1967).
[55] M. A. Jensen, A. J. Heeger, L. B. Welsh and G. Gladstone, *Phys. Rev. Letts.*, **18**, 997 (1967).
[56] A. J. Heeger and M. A. Jensen, *Phys. Rev. Letts.*, **18**, 488 (1967).
[57] A. S. Edelstein, *Phys. Rev. Letts.*, **19**, 1184 (1967).
[58] A. S. Edelstein, *Phys. Rev. Letts.*, **20**, 1348 (1968).
[59] H. H. Hill, W. N. Miner and R. O. Elliott, *Phys. Letts.*, **28A**, 588 (1969).
[60] T. F. Smith, *Phys. Rev. Letts.*, **17**, 386 (1966).
[61] P. W. Bridgman, *Proc. Am. Acad. Arts Sci.*, **81**, 1652 (1952).
[62] P. W. Bridgman, *Proc. Am. Acad. Arts Sci.*, **83**, 1 (1954).
[63] L. F. Vereshchagin, A. A. Semerchan and S. V. Popova, *Sov. Phys.-Doklady*, **6**, 609 (1962).
[64] Private communication.
[65] P. C. Souers and G. Jura, *Science*, **145**, 575 (1964).
[66] R. A. Stager and H. G. Drickamer, *Phys. Rev.*, **133**, A830 (1964).
[67] A. R. Wazzan, R. S. Vitt and L. B. Robinson, *Phys. Rev.*, **159**, 400 (1967).
[68] R. Karplus and J. M. Luttinger, *Phys. Rev.*, **95**, 1154 (1954).
[69] J. Kondo, *Progr. Theor. Phys.* (Kyoto), **27**, 772 (1962).
[70] F. E. Maranzano, *Phys. Rev.*, **160**, 421 (1967).
[71] N. V. Volkenshtein, I. K. Grigorova and G. V. Federov, *Sov. Phys. JETP*, **23**, 1003 (1966).
[72] N. A. Babushkina, *Sov. Phys. Solid State*, **7**, 2450 (1966).
[73] R. S. Lee and S. Legvold, *Phys. Rev.*, **162**, 431 (1967).
[74] C. J. Kevane, S. Legvold and F. H. Spedding, *Phys. Rev.*, **91**, 1372 (1953).
[75] J. J. Rhyne, *J. Appl. Phys.*, **40**, 1001 (1969).
[76] J. J. Rhyne, *Phys. Rev.*, **172**, 523 (1968).
[77] N. V. Volkenshtein, I. K. Grigorova and G. V. Federov, *Sov. Phys. JETP*, **24**, 519 (1967).
[78] Yu. P. Irkhin and Su. Sh. Abelskii, *Sov. Phys. Solid State*, **6**, 1283 (1964).
[79] T. Kasuya, *Progr. Theor. Phys.* (Kyoto), **22**, 227 (1959).
[80] J. Mannari, *Progr. Theor. Phys.* (Kyoto), **22**, 335 (1959).
[81] N. V. Volkenshtein and G. V. Federov, *Phys. Lett. Mat.* (*U.S.S.R.*), **18**, 26 (1964).
[82] S. Arajs and R. V. Colvin, *J. Appl. Phys.*, **35**, 1043 (1964).
[83] S. Arajs and R. V. Colvin, *Phys. Rev.*, **136**, A439 (1961).
[84] R. V. Colvin and S. Arajs, *Phys. Rev.*, **133**, A1076 (1964).
[85] S. Arajs and B. R. Dunmyre, *Physica*, **31**, 1466 (1965).
[86] R. W. Powell and B. W. Jolliffe, *Phys. Letts.*, **14**, 171 (1965).
[87] C. F. Gallo, *J. Appl. Phys.*, **36**, 3410 (1965).

[88] R. W. Powell, *J. Appl. Phys.*, **37**, 2518 (1966).
[89] S. Arajs and G. R. Dunmyre, *Z. Naturforsh*, **21**, 1856 (1966).
[90] N. G. Aliev and N. V. Volkenshtein, *Sov. Phys. JETP*, **22**, 17 (1966).
[91] N. G. Aliev and N. V. Volkenshtein, *Sov. Phys. Solid State*, **7**, 2068 (1966).
[92] N. G. Aliev and N. V. Volkenshtein, *Sov. Phys. JETP*, **22**, 997 (1966).
[93] B. W. Jolliffe, R. P. Tye and R. W. Powell, *J. Less Common Metals*, **11**, 388 (1966).
[94] G. S. Nokolskii and V. V. Gremenko, *Phys. Stat. Sol.*, **18**, K123 (1962).
[95] O. V. Lounasmaa and L. J. Sundstrom, *Phys. Rev.*, **158**, 591 (1967).
[96] O. V. Lounasmaa and L. J. Sundstrom, *Phys. Rev.*, **150**, 399 (1966).
[97] L. R. Edwards and S. Legvold, *Phys. Rev.*, **176**, 753 (1968).
[98] W. J. Nellis and S. Legvold, *Phys. Rev.*, **180**, 551 (1969).
[99] J. M. Ziman, *Electrons and Phonons*, O.U.P. (1962) p. 397.
[100] E. I. Kondorskii, O. S. Galkina, P. A. Markov and Yu. M. Borovikov, *Zh. Eksp. Teor. Fiz.*, **57**, 130 (1969).
[101] L. R. Sill and S. Legvold, *Phys. Rev.*, **137**, A1139 (1965).
[102] M. Bailyn, *Phys. Rev.*, **126**, 2040 (1962).
[103] L. E. Gurevich and G. M. Nedlin, *Sov. Phys. JETP*, **18**, 396 (1964).
[104] Y. Kagan and L. A. Maksimov, *Sov. Phys. Solid State*, **7**, 422 (1965).

6

Rare Earth Compounds

The rare earth elements form a prolific number of both metallic and non-metallic compounds [1] and much has been written in the past about their physical properties. From a fundamental viewpoint their behaviour is equally as challenging as that of the pure metals, and the scope for investigation is many times larger as may be proved by a simple count of activities, both past and present, for the two types of material.

With the exception of rare earth–rare gas combinations, simple two element compounds are formed with all the elements to the right of the Cr–Mo–W column in the periodic table. More complex, three component systems involving one other metal and either a group VIB element or a halide are also common and provide us with the ferrites, garnets and pseudo-perovskites which have been the attention of much application work in the past.

The often large number of possible stoichiometric compositions occurring within any one binary or ternary phase diagram, and the extensive substitution which is possible within a single compound, makes a detailed analysis of all the rare earth compounds an impossibly large task. Detailed discussion of the garnets, ferrites and other ternary compounds has been given by Schieber [2] in his book devoted to the experimental magnetochemistry of non-metallic magnetic materials, and the reader is referred to this text for descriptions of the phenomena associated with these compounds. Within this chapter we will restrict ourselves to certain binary (or pseudobinary) materials which are currently of interest. These include the oxides, pnictides, chalcogenides and transition metal compounds, which possess properties typical of insulating, semi-metallic, and metallic materials respectively.

In the insulating oxides and chalcogenides, the effects of exchange are normally small and the crystalline electric field plays a dominant role in determining their behaviour. On the other hand in the metallic materials the role of the conduction electrons is important and in many cases the exchange energy exceeds the crystal field effects, making a complete analysis essential if we are to reliably interpret the many phenomena which have been observed for these systems. The pnictides fall between these two types of material, their semi-metallic properties leading to exchange interactions which are sufficiently small to allow their introduction as a perturbation on the crystal field effects. The detailed properties of each of these series will be described in the following.

6.1 Rare earth oxides

6.1.1 *Trivalent oxides*

These compounds can be observed in hexagonal (A), monoclinic (B) or cubic (C) modifications depending upon the temperature of observation. The A phases are restricted to the light rare earths with larger ionic radii [3]. The low temperature cubic phase of the heavy metal oxides is found to be anti-ferromagnetic [4–10] and the presence of these oxides in the rare earth metals is known to lead to anomalies in the specific heat data in the vicinity of 4°K [11–13]. Only a limited amount of data exists concerning the spin structures [10, 14] but that which is has revealed non-collinear structures with different moment values on the two types of atomic site in the crystals.

The magnetic properties both above and below the Néel point are given in Table 6.1. The low Curie temperatures of these compounds results in the

Table 6.1 Magnetic data for the rare earth sesquioxides [15–20].

	Effective moment μ_{eff} (μ_B)	Paramagnetic Curie point θ_p (°K)	Néel point T_N (°K)
Ce_2O_3			
Pr_2O_3	3·59	−73	
Nd_2O_3	3·66	−27	
Sm_2O_3	1·50	−150	
Gd_2O_3	7·85	−17	1·5
Tb_2O_3	9·67	−13	2·4
Dy_2O_3	9·14	−18	1·2
Ho_2O_3	10·40	−14	
Er_2O_3	10·50	−11	3·4
Tm_2O_3	9·43	−25	
Yb_2O_3	7·52	−90	2·3

magnetic exchange interaction appearing as a small perturbation on the crystal field interaction and the low temperature behaviour can be expected to be strongly influenced by the crystal field. In addition it is possible that dipole–dipole interactions could be the origin of the magnetic order, although Moon *et al* [10] have shown that this is not the case for Er_2O_3 and Yb_2O_3. Its magnitude, however, is still sufficiently large that it may have some influence on the magnetic behaviour in some of the compounds.

The influence of the crystal field on the magnetic moments have been discussed by several authors in the past [10, 18, 21, 22] and the predicted results in the paramagnetic region agree well with experimental values. Direct observation of optical transitions between the levels of the ground state multiplet has been made for Er_2O_3 [23], and the results confirm the existence of two non-equivalent sites. Derived g-values, however, are inconsistent with those obtained directly [24]. Non-equivalent sites were also evident in the neutron diffraction studies of Moon *et al* [10].

6.1.2 *Higher oxides*

Limited measurements have also been made on tetravalent oxides, although the only stable example of these is CeO_2. By growth under high oxygen pressure, however, PrO_2 has been prepared as well as various compositions intermediate between TbO_2 and Tb_2O_3. Néel temperatures of 14°K and 3°K have been reported for PrO_2 and TbO_2 respectively [4, 25] but there has been little discussion of the origin of the physical properties of this class of oxide.

6.1.3 *Divalent oxides*

Oxides of this stoichiometry are known to exist for the elements neodymium, samarium, europium and ytterbium and have a rock salt structure. The most extensively studied of these is the semiconducting compound EuO, which is ferromagnetic [26] with paramagnetic and ferromagnetic Curie temperatures of 76 and 69·4°K respectively. The saturation magnetization at 4·2°K corresponds to 6·8 μ_B per ion, close to the free ion value. Neutron diffraction data [27] shows that in the ordered state the moments are aligned parallel.

The magnetocrystalline anisotropy is quite large with the value of the first order constant $K_1(0) = 4{\cdot}36 \times 10^5$ eg cm^{-3} [28]. The variation of K_1 with both field and temperature is adequately described by the single ion theory of Wolf [29]. Assuming that the interaction of the crystal field with the ionic levels provides the dominant mechanism of the anisotropy a value of $b_4 = 21{\cdot}2 \times 10^{-4}$ cm^{-1}/ion was obtained for the cubic crystal field parameter. Magnetostriction measurements reveal a lattice contraction on ordering.

The simplicity of the magnetic structure of this compound, in which the highly localized, spin-only, moments (Eu^{2+}) are situated in an insulating, face-centred cubic, lattice, make it ideal for treatment as a Heisenberg ferromagnet. This approach was first adopted by McGuire *et al* [30] and a value for the sum of the first and second-neighbour exchange interactions obtained using the molecular field relation [31]

$$\theta = T_c = \tfrac{2}{3}S(S+1)[12\mathscr{J}_1 + 6\mathscr{J}_2]$$

where S is the atomic spin ($\frac{7}{2}$ for Eu^{2+}), $\mathscr{J}_1$ and $\mathscr{J}_2$ are the exchange interactions in °K, and T_c the ferromagnetic Curie temperature. Values of individual exchange interactions were obtained from paramagnetic resonance studies [32] on the assumption that the observed line width originated from exchange narrowing and that the theoretical treatments [33–35] for a simple cubic lattice could be used for this material. The values obtained in this way are given in Table 6.2 along with the magnetic parameters for EuO and the other europium chalcogenides which will be discussed later.

These results show a ferromagnetic coupling to the nearest neighbour Eu^{2+} ions and a much weaker antiferromagnetic coupling to the next nearest neighbours which is normally thought to arise from a superexchange via the intervening oxygen ion. The nature of the ferromagnetic coupling has

Table 6.2 The transition temperatures, magnetic moments and exchange integrals of the europium chalcogenides [32, 36].

	θ_p (°K)	T_N (°K)	T_c (°K)	Moment at $T = 0$°K (μ_B)	$\mathscr{J}_1$ (°K)	$\mathscr{J}_2$ (°K)
EuO	76		69·4	6·80	0·65	−0·06
EuS	19		16·3	6·87	0·20	−0·08
EuSe	4·6	4·58	3·8	6·70	0·13	−0·11
EuTe	−6	9·64			0·03	−0·17

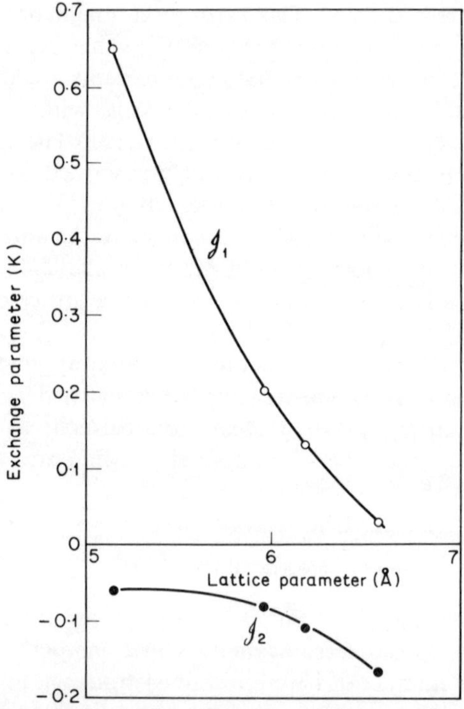

Fig. 6.1 The variation of the exchange integrals of the rare earth chalcogenides with lattice spacing [32].

been the subject of considerable interest and since the original work of Matthias *et al* [26] it has been attributed to an indirect interaction between the localized $4f$ electrons. Several proposals have been made about the nature of this interaction and Methfessel and Mattis [36] have discussed them at length in their review dealing with 'Magnetic Semiconductors'. These models have been developed for EuS, EuSe and EuTe in addition to EuO, and consequently we shall examine them further in Section 6.2.

The variation of the exchange integral $\mathscr{J}_1$ with lattice spacing has been obtained directly from high pressure measurements [37–40], from the magnetic properties of non-stoichiometric films [41] and from the behaviour of films evaporated at low substrate temperatures [42]. The results of these investigations indicate that $\mathscr{J}_1$ decreases continuously with increasing distance between the ions as shown in Fig. 6.1.

The semiconducting nature of EuO extends through the magnetic ordering temperature although optical measurements have shown [43, 44] an unusual behaviour of the absorption edge with temperature as given in Fig. 6.2. The usual increase of the energy gap with decreasing temperature is replaced below 90°K by a shift of the absorption edge to lower energies. This red shift, of more than 0·2 eV, is sensitive to applied magnetic fields through the induced magnetization below 76°K and Busch *et al* [43] have shown that it is

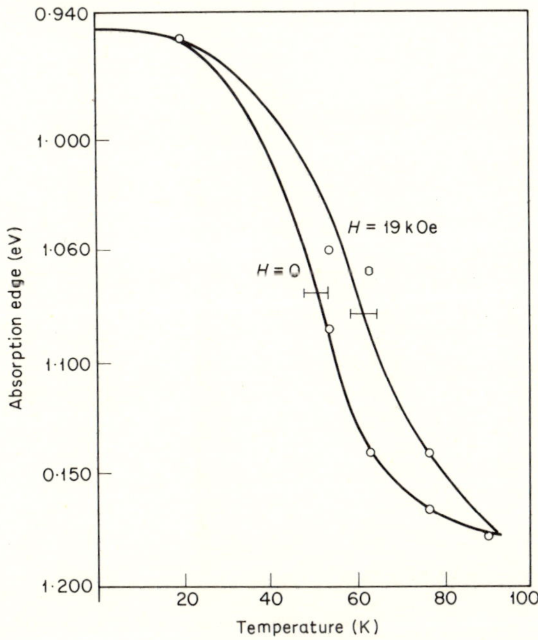

Fig. 6.2 The change in the position of the optical absorption edge in EuO showing the 'red-shift' below 90°K [43].

associated with the ordered magnetic state. The persistence of the red shift to temperatures well in excess of the Curie temperature removes the dependence on magnetization and Methfessel *et al* [45] have shown that the change in the absorption edge is proportional to the nearest neighbour spin correlation function $\langle S_1 S_2 \rangle / S^2$, suggesting a high degree of localization of the electron states responsible for the shift. The reflection and absorption spectra in the vicinity of the absorption edge exhibit marked maxima at about 2·0 eV (300°K) which have been ascribed to a $4f^7$–$4f^6\,5d$ transition. If this transition is taken as that responsible for the absorption edge then the red shift is associated with the excited state and may be attributed to the variation of the overlap of the Γ_5 electrons caused by changes in the relative ionic spin orientation due to the intra-atomic *f–d* exchange at neighbouring europium sites [46–49]. The nature and effectiveness of such 'magnetic excitons' (as they were named by Kasuya) were discussed at length in connection with the electronic and magnetic behaviour of the europium chalcogenides [50–52].

In addition to the reflection maximum at 1·5 eV (80°K) Feinleib *et al* [53] have observed a further peak at 4·7 eV (see Fig. 6.3). The 1·5 eV peak was observed to be a triplet whose shape depended on the state of polarization

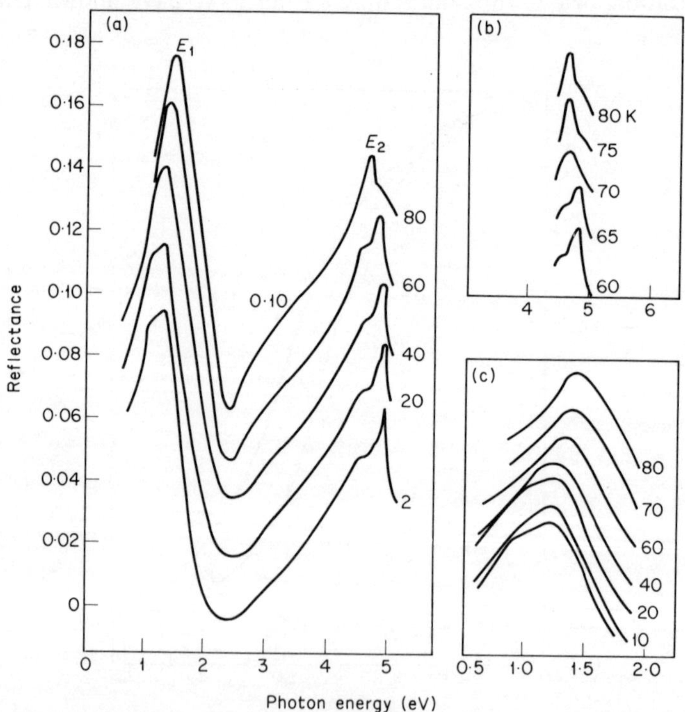

Fig. 6.3 Reflection spectrum of EuO showing the transition to the Γ_3 and Γ_5 levels. (a) Entire spectrum at various temperatures; (b) the 4–6μ region near to T_C; (c) the peak at 1·4μ [53].

of the incident beam, in keeping with the behaviour to be expected from the $4f^7$–$4f^6 5d$ transition. The splitting of the Γ_5 band due to exchange was estimated to be 0·25 eV at low temperatures. Structure was also observed in the higher energy peak although no red shift was evident, and these transitions were associated with the well-known staircase structure of the $4f^7$–$4f^6 5d$ (Γ_3) absorption spectrum of EuF_2. Spin–orbit broadening prevents this staircase structure being visible in the transition to the Γ_5 levels. The schematic band structure obtained in this work is indicated in Fig. 6.4a with a qualitative level scheme for the 5*d* levels being shown in Fig. 6·4b [54]. Further support for this origin of the absorption edge comes from the observed blue-shift with lattice dilation [42]. As the atomic spacing increases the effects of the crystal field decrease and the energy gap between the relatively stable 4*f* level and the Γ_5 level increases.

The onset of magnetic ordering in EuO has also been shown to be accompanied by changes in the refractive index and photoconductive sensitivity [55, 56]. Since photoconduction was found both above and below the Curie temperature it was suggested that optical absorption at the band edge (1·5 eV) occurs by electron excitation into spin polarized conduction bands, since no photoconduction was to be expected from the exciton model [47]. This conduction band model was introduced by Rys [57] and Cho [58] as an alternative explanation for the observed red shift of the absorption edge at the Curie temperature. The dramatic increase in the sensitivity below T_c, by several orders of magnitude, appears to be consistent with the theoretical calculations of the electron mobility in magnetic semiconductors under exchange scattering conditions [59].

The Faraday [60] and Kerr [61] magneto-optical effects are also marked in this material and Suits [62] has observed effects on the transmission of radiation through single crystals of EuO which he attributed to optical diffraction by the magnetic domains. This diffraction was shown to arise from magneto-optical birefringence rather than magneto-optical dichroism, and the linear (Cotton Mouton) birefringence was preferred to the circular (Faraday) birefringence.

Specific heat measurements have been made by Maruzzi [63], and Henderson *et al* [64]. The latter results, which were taken between 0·37 and 4·4°K, were fitted above 0·7°K using spin wave theory developed for EuS by Charap and Boyd [65]. The exchange integrals obtained in this way were $\mathcal{J}_1 = 0{\cdot}76°K$ and $\mathcal{J}_2 = 0{\cdot}08°K$, in good agreement with those obtained from magnetization [66] and resonance [32] data. Below 0·7°K the dominant contribution to the specific heat comes from the hyperfine splitting and is given by

$$C_N = R\sum_i \tfrac{1}{3} I_i(I_i+1)(g_N\mu_N H_{eff}/kT)$$

where $\mu_N I_i$ is the nuclear moment of the europium isotope i and H_{eff} is the nuclear hyperfine field. The derived value of H_{eff} giving the best fit to the

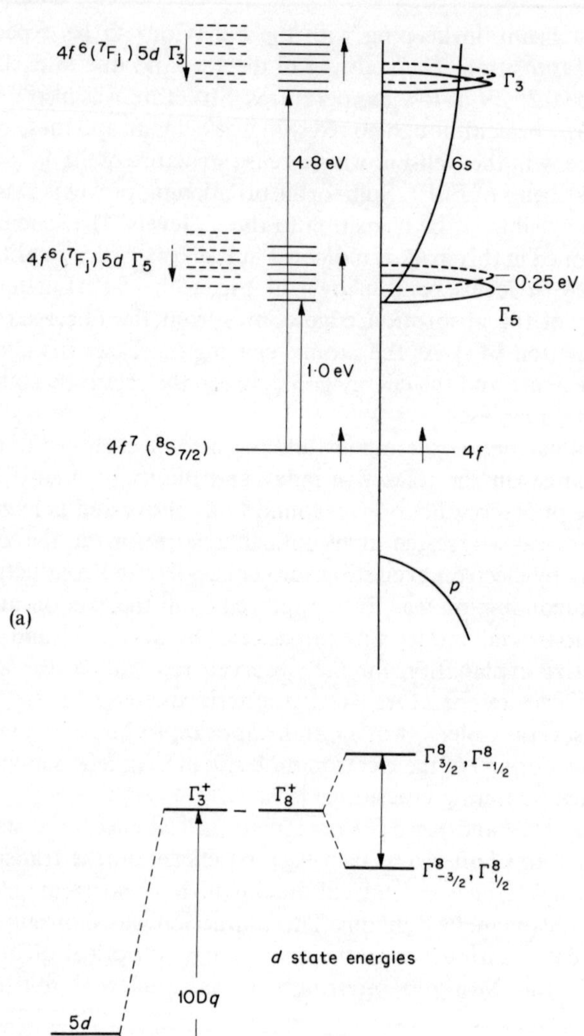

Fig. 6.4 (a) The schematic band structure and (b) the level scheme for the 5*d* states in EuO derived from the reflectivity measurements [53, 54].

experimental data is $H_{eff} = 300 \pm 5$ kOe in good agreement with the value determined from Mössbauer [67] and nmr [66, 68] measurements.

An intermediate oxide also exists for europium (Eu_3O_4) which is metamagnetic with a Néel temperature of 6·1°K and a critical field of 3·5 kOe at 4·2°K [69].

6.2 Compounds formed with sulphur, selenium and tellurium

Since the discovery of ferromagnetism in EuO, considerable attention has been given to the other europium chalcogenides, which also crystallize in the NaCl structure. EuS is found to be ferromagnetic with a Curie temperature of 16·3°K [70–74] while at low temperatures both EuSe and EuTe are antiferromagnetic with Néel points of 4·58 and 9·64°K respectively [70, 72, 74–77]. The values of these transition temperatures given by various authors show a considerable variation which is probably due, in part, to the magnetic field dependence of the spin structure which has been reported [75, 78] for EuSe. Schwob and Vogt [79] found evidence for a spontaneous magnetization in EuSe below 3·8°K, which supports the suggestion made by Busch *et al* [80] that this material becomes ferromagnetic at very low temperatures.

As in the case of EuO these materials may be treated as Heisenberg ferromagnets, allowing the values of the exchange parameters given in Table 6.2 to be obtained. These are given as a function of the lattice parameter in Fig. 6.1 [81]. Also included in Table 6.2 are the results for EuO and the magnetic parameters of all the europium chalcogenides. The variation of the exchange integrals with interatomic spacing inherent in this diagram (Fig. 6.1) is confirmed by measurements of the system Eu(S, Se) [82] and by the direct observation of the increase in Curie temperature with pressure [83, 37–40] for the pure compounds. The form of this dependence is appreciably different from that expected from superexchange in cubic materials [84].

Specific heat and thermal conductivity measurements at low temperatures have been successfully interpreted using spin wave theory and in some cases have given values for the exchange integrals as discussed for EuO [85–87, 89–91]. In the case of EuS these exchange parameters have been used by Wojtowicz [92] to derive the magnetic contributions to the specific heat.

The optical absorption curves of the europium compounds with sulphur selenium and tellurium [44, 70, 88, 93] are similar to those of EuO as may be seen from Fig. 6.5. Unlike normal semiconductors, the onset of absorption shifts to higher energies with decreasing electron affinity of the anion (i.e. from O^{2+} to Te^{2+}). The photon energy at the absorption edges of all the compounds are listed in Table 6.3. Attempts have been made to attribute this variation to changes in the crystal field interaction [49], comparable to those proposed for EuO in connection with the edge shift accompanying lattice dilation [42]. Magnetic ordering has been shown to be accompanied by a

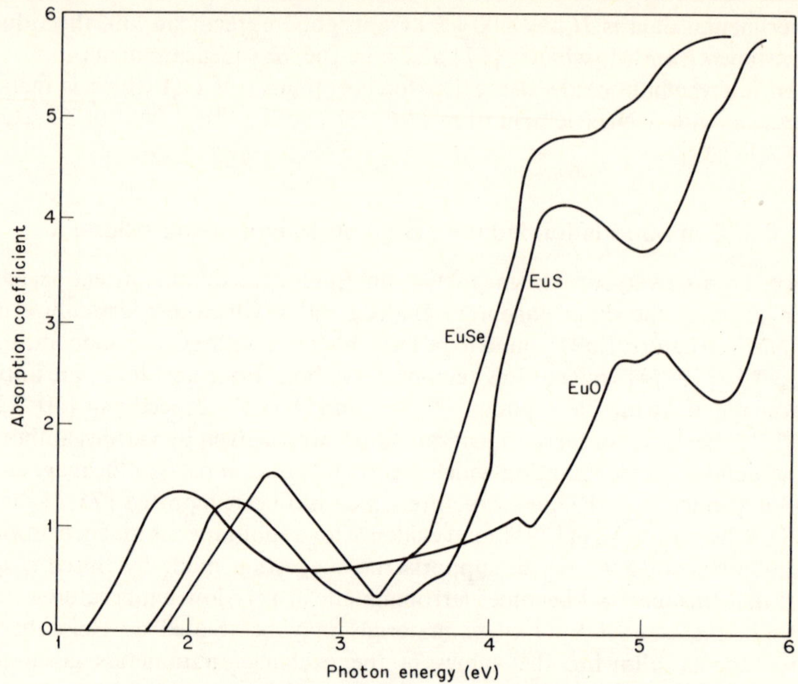

Fig. 6.5 Absorption spectra of europium chalcogenides [88].

Table 6.3 The forbidden energy gap width, and the magnitude of the 'red shift' in EuO, EuS, EuSe and EuTe [44].

	EuO	EuS	EuSe	EuTe
E_g (ev) (300°K)	1·12	1·64	1·85	2·0
Red Shift ΔE_g (ev)	0·26	0·18	0·335	—

red shift of the absorption edge [44, 80, 94] and a resolvable splitting of the low and high energy absorption peaks into a triplet and a doublet respectively [94]. This is comparable to the results for EuO discussed in the previous section and the interpretation of the observations is similar.

Other changes in the optical properties at the Curie temperature have been found for the magnetic birefringence [95] and dichroism [46], luminescence emission [96] and the photoconductivity [55, 56].

As mentioned previously, the mechanism of the ferromagnetic exchange interaction $\mathscr{J}_1$ depends on the band structure adopted to represent these compounds. Methfessel and Mattis [36] have outlined the differences in the various proposals in relation to the position of the 4*f* levels relative to the conduction and valence bands. For 4*f* levels lying below the valence band, as suggested by the energy band model for EuS derived by Cho [58] (see Fig. 6.6) a Bloembergen–Rowland type of indirect exchange is possible by way of the polarized valence electrons [97–100].

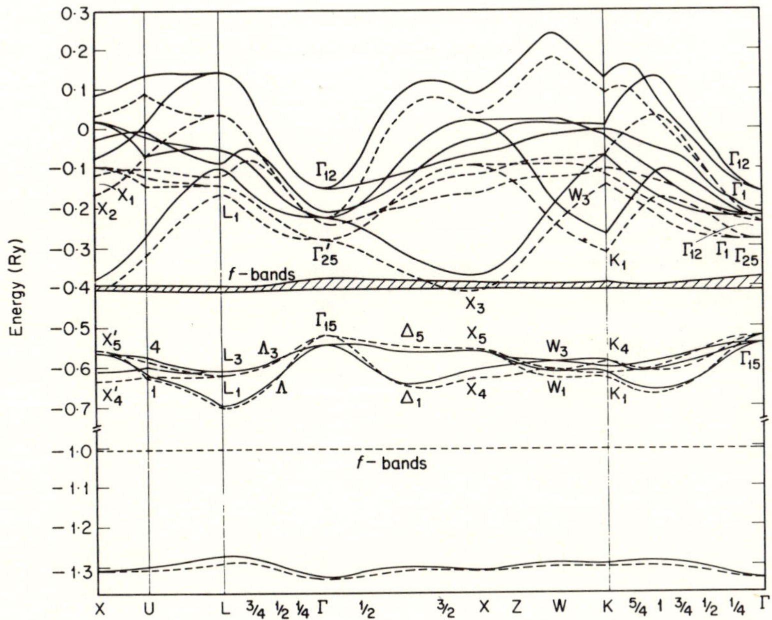

Fig. 6.6 The energy band structure of EuS derived by Cho for spin up (– – –) and spin down (——) states [58].

When the $4f$ levels are situated between the valence and conduction bands the exchange can occur via the d- or s-electrons. Goodenough [101] has proposed such a model based on the mixing of $4f$ and $5d$ states at different cations, which gives for the ferromagnetic exchange

$$\mathscr{J}_1 = +2b^2\mathscr{J}_{fd}/4S^2U^2$$

where b is the $4f$–$5d$ transfer integral, $\mathscr{J}_{fd}$ the interatomic $4f$–$5d$ exchange integral and U the excitation energy. This relation gives reasonable agreement with the experimental variation in $\mathscr{J}_1$ across the series of compounds using values of U derived from the optical measurements.

The effects of changing the conduction electron concentration in the chalcogenides have been studied by alloying the divalent europium compounds with a similar compound formed with a trivalent rare earth ion [102–105]. In the systems (Eu, Gd)Se and (Eu, Gd)S, increasing the gadolinium concentration and hence the conduction electron concentration, causes the Curie temperature to rise rapidly to a maximum at the europium-rich end of the system and subsequently fall monotomically to the negative value corresponding to the terminal gadolinium compound. This behaviour is illustrated in Fig. 6.7, and its similarity to the oscillatory nature of the RKKY interaction [106] has been taken as an indication of an indirect coupling via the conduction electrons in these materials. As the conduction band in these compounds is almost certainly non-parabolic, care must be taken when attempting to draw quantitative conclusions from this similarity.

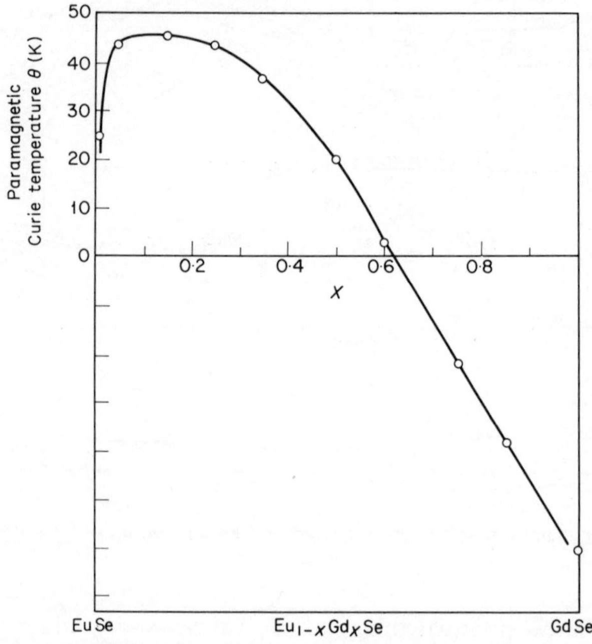

Fig. 6.7 The variation of the Curie temperature in (Eu, Gd)S with increasing conduction electron concentration, showing the behaviour comparable to that to be expected from RKKY theory.

The use of temperature-dependent indirect interactions [153, 154] via the conduction electrons is also of doubtful value since many of these rely on the *f–s* interaction being small enough for perturbation theory to be employed. This of course, is frequently not the case in these materials.

For the low gadolinium concentrations the model of Kasuya and Yanase [50–52] has been very successful in accounting for much of the observed behaviour. In this model the excess electron remains bound to the impurity ion and can interact with the 4*f* spin lattice. The changes in Curie temperature are then taken to be an indication of the energy differences between the impurity complex (consisting of a trivalent rare earth ion surrounded by twelve Eu^{2+} ions) being situated in a disordered and an ordered lattice.

Before leaving the europium chalcogenides it is worth noting that the electrical properties of systems such as (Eu, Gd)Se, investigated by Von

Molnar and Methfessel [107] show anomalously large values of the magneto-resistance, for some compounds, in the vicinity of the ordering temperature (see Fig. 6.8). This is accompanied by a large resistivity in the absence of a field. This behaviour has been ascribed to the reduction of the activation energy for electron motion due to the effect of the magnetic field on empty impurity sites. These effects have been discussed at length by Methfessel and Mattis [36]. More recently measurements on both EuO(Gd) [108] and EuS(Gd) [109] have shown the presence of zero field resistivity anomalies at the Curie temperature. These have been satisfactorily accounted for by critical scattering from correlated spins [110] and from the $(Gd^{3+} + 12Eu^{2+})$ impurity complex with its trapped electron. On this basis estimates of the *s–f* exchange interaction have been made giving $\mathscr{J}_{sf} = 4{\cdot}9 \times 10^{-2}$ eV (EuO) and $4{\cdot}3 \times 10^{-2}$ eV (EuS).

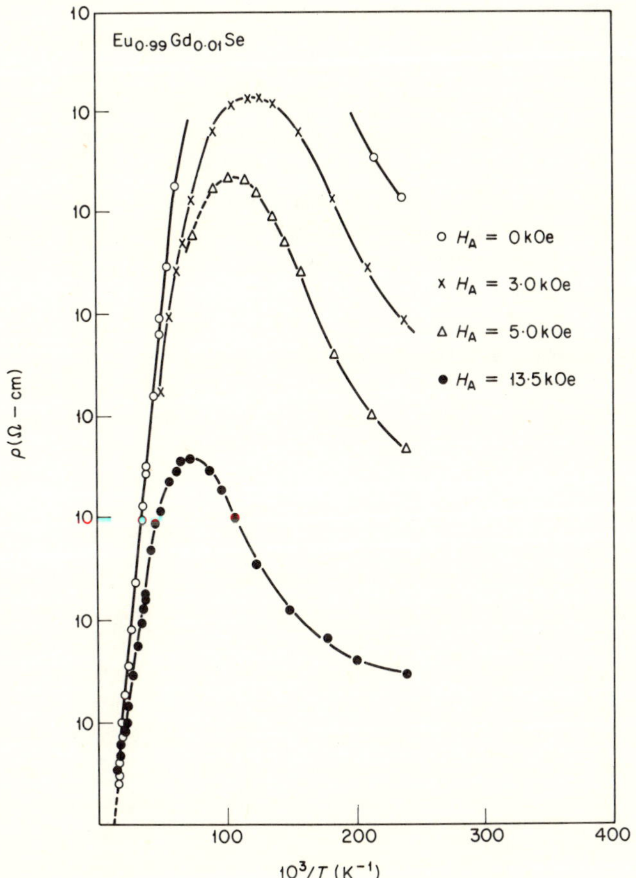

Fig. 6.8 Iso-field resistivity-temperature variation of (Eu, Gd)Se showing the large magneto-resistivity in the vicinity of the magnetic ordering temperature [107].

The equiatomic compounds formed with the trivalent elements have been relatively little studied by comparison. Electrical measurements [111–120] have been interpreted in terms of highly degenerate semiconducting behaviour or metallic conductivity, although the majority of workers appear to favour the latter. Many of the discrepancies of the earlier work may be associated with uncertainties in composition and the effects of the high chemical activity of these materials. The temperature dependence of the magnetic susceptibility may be interpreted in terms of ionic moments which are close to the free ion values [121–127] and the onset of magnetic ordering, for rare earth ions with an odd number of electrons, at low temperatures [125–127]. The magnetic interaction in these materials is taken to be an RKKY type of exchange via the conduction electrons. In the compounds with the lighter elements the crystal fields also play an important part in determining the final magnetic behaviour.

The compounds of stoichiometry A_2B_3 have been examined both in the pure (e.g. Gd_2Se_3) and mixed (e.g. Nd_2O_2Te) form. The structure of these compounds is defect Th_3P_4 type [128–129] which is extremely tolerant of the addition of an excess of the tripositive ion. With the resultant increase in electron concentration which can occur in going from A_2B_3 to A_3B_4 there is a continuous change from semiconducting to metallic conductivity [116, 117, 130–133] and simultaneously the magnetic behaviour shows a tendency to change from antiferromagnetic to ferromagnetic ordering. For example Gd_2Se_3 has a paramagnetic Curie point of $-10°K$ while $Gd_{2\cdot1}Se_{2\cdot9}$ is ferromagnetic with $T \simeq 80°K$.

While it is clear that these changes in the electrical conductivity and magnetic properties are related, a detailed mechanism of the behaviour has yet to be developed. Methfessel and Mattis [36] have attempted to account qualitatively for the correlation in terms of the nature and behaviour of the *5d* orbitals of the cations.

6.3 Compounds with Group V elements

Interesting families of compounds are formed between the rare earth metals and elements of the nitrogen group. The 1:1 compounds which occur all have the simple NaCl structure and as such simplify theoretical analysis of their behaviour. Compounds having the Th_3P_4 structure are formed with either an A_2B_3 or A_3B_4 stoichiometry for a limited number of element combinations. As with the chalcogenides, the A_2B_3 compounds have 4/3 vacant atomic sites per unit cell and as a result a large number of cations can be introduced into the lattice giving a homogeneity range extending to the 3:4 composition.

6.3.1 1:1 *Stoichiometry* (Pnictides)

The pnictides have been the subject of much systematic study [32, 124,

134–142] using various techniques. As Table 6.4 shows, most of the compounds exhibit some kind of magnetic ordering with transition temperatures usually less than 60°K. The magnetic susceptibilities obey the Curie–Weiss law with derived effective moment values agreeing closely with the theoretical value of $g_J(J(J+1))^{\frac{1}{2}}$. The corresponding paramagnetic Curie temperatures are predominantly positive for the nitrides and phosphides and with few exceptions negative for the remainder of the compounds. GdN is the only ferromagnetic compound, although ferrimagnetism is observed in nine of the remaining materials, the rest of which become antiferromagnetic or do not order. With the exception of the gadolinium compounds all these antiferromagnetics show field-induced transitions to ferromagnetism with their critical fields increasing from the phosphides to the antimonides. Fig. 6.9 shows the magnetization-field results for the holmium compounds [138] where these transitions are clearly visible. The magnetic characteristics of all the compounds are given in Table 6.4.

Neutron diffraction measurements at Oak Ridge [134] have revealed complex spin structures in the heavy rare earth nitrides. The pronounced small angle scattering and diffuse scattering arise from magnetic reflections and were originally taken as an indication of short range ferromagnetic correlations occurring well below the Curie temperature. The small angle scattering, however, does not pass through the maximum at the transition temperature which is to be expected from critical scattering. Observations in

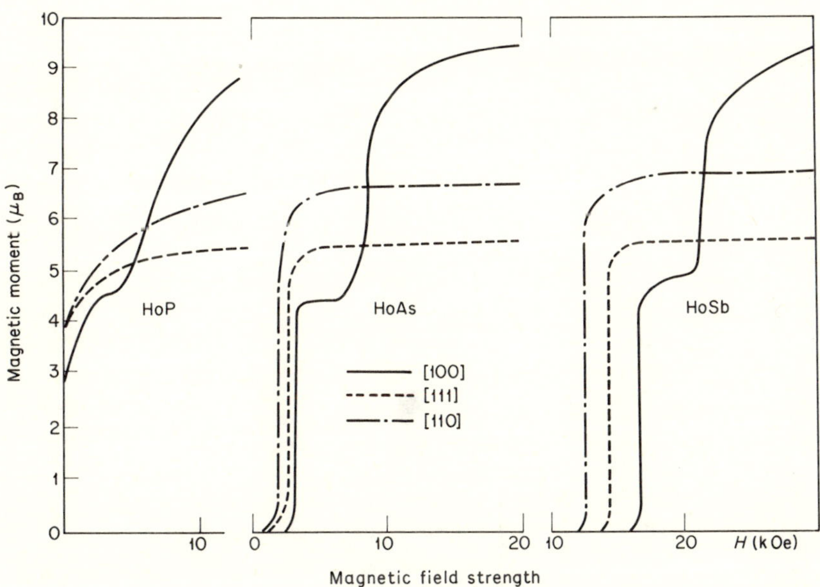

Fig. 6.9 Magnetic moment versus internal field measured at 1·7°K for different crystallographic directions in HoP, HoAs, and HoSb [138].

Table 6.4 Details of the magnetic properties of the compounds formed between the rare earth metals and the group VA elements of the periodic table, after Busch [80]. n.o. = no order, μ_s is in Bohr magnetons.

		Ce	Pr	Nd	Sm	Eu	Gd	Tb	Dy	Ho	Er	Tm	Yb
N	Order	anti	no	ferri(?)	anti(?)	n.o.	ferro	ferri	ferri	ferri	ferri	n.o.	anti(?)
	θ_p	...	0	+24	...	−200	+69	+34	+20	+12	+4	−18	−116
	T_c			32	Van Vleck		72	40	21	13	6		
	$\mu_s(0°K)$			3·1	paramagnetism		6·6	6·3	6·3	9·2	5·5		
P	Order	anti	n.o.	ferri(?)			anti	anti	ferri(?)	ferri	anti	n.o.	n.o.
	θ_p, T_c, T_N	−8 8	−2	+11 +11			+2 +15	+3 +8	+8 +8	+6 +6	+2 +4	−2 ...	−55 ...
As	Order	anti	n.o.	anti			anti	anti	ferri(?)	anti	anti	n.o.	n.o.
	θ_p, T_c, T_N	−5 +7	−6	+4 +11			−12 +25	−4 +10·5	+2 +8·5	+1 +4·8	−1·5 +3·5	−2 ...	−25 ...
Sb	Order	anti		anti			anti	anti	anti	anti	anti	n.o.	n.o.
	θ_p, T_c, T_N	+8 +18		−3 +16			−42 +28	−14 +16·5	−4 +9·5	−2·5 +5·5	−3 +3·5	−1 ...	−60(?)

an applied field show that the magnetic easy direction is parallel to either the (100) or (111) axis depending on the rare earths involved and the results could only be fully interpreted on the basis of a spin structure more complex than a simple ferromagnet. The structure proposed for HoN is shown in Fig. 6.10(a) in which the spins are arranged in ferromagnetic layers which are perpendicular to the 111 direction, the moment direction always being parallel to a 100 axis but changing from one axis to another, every seven atomic layers, in a spiral manner. A similar arrangement of the spins was found for all the other heavy rare earth nitrides except those with thulium and ytterbium. Somewhat different results have been reported by McGuire

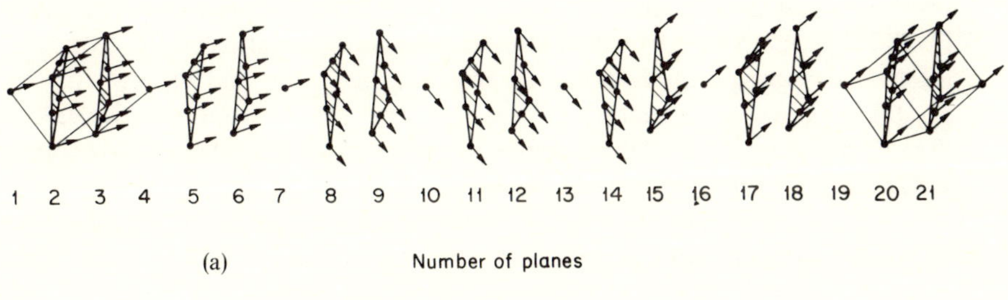

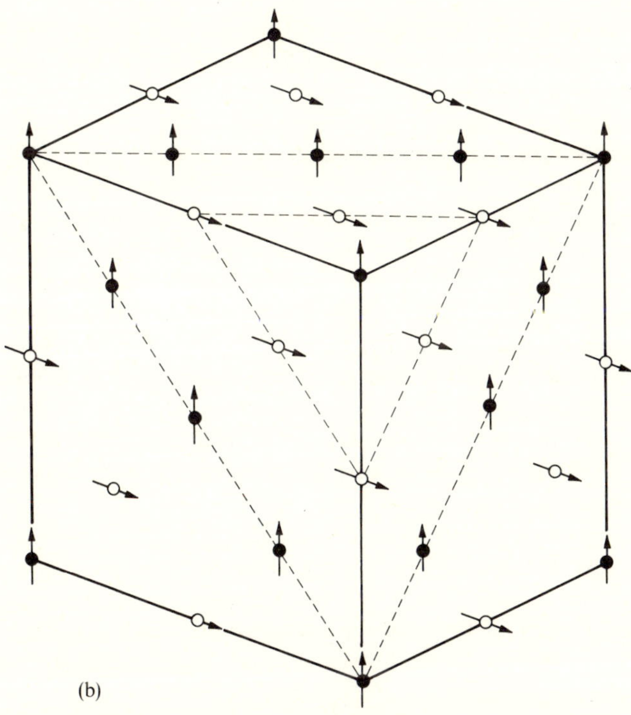

Fig. 6.10 (a) The magnetic structure of HoN. (b) The HoP 'flip-flop' structure [134]

et al [143, 144] who found GdN to be a pure ferromagnet, and GdSb and GdBi to be antiferromagnets of the second kind.

Structure determinations on the remaining compounds give moment configurations, in the antiferromagnetic compounds, of the MnO type. The so-called 'flip-flop' structure of ferromagnetic HoP is shown in Fig. 6.10(b) and it may be seen that there are ferromagnetic planes perpendicular to the cube diagonal and that the moment orientations in adjacent planes are at 90° to one another. This situation results in a ferromagnetic arrangement parallel to the (110) direction and antiferromagnetic at right angles to this.

Determination of the ionic moments from neutron diffraction observations show that in the ordered state the moment values are appreciably less than those associated with the free ions. Crystal field calculations of Trammell [145, 146] are consistent with these findings and predict reduced moments for the compounds of terbium, dysprosium, holmium and erbium. In the crystal field, ions with a Kramers' doublet ground state will not order unless the exchange energy is large enough to cause mixing of the ground and excited states. For singlet ground states, such as those found in Pr and Tm compounds, no magnetic ordering is observed, as shown in Table 6.4. For terbium and holmium cations, however, the exchange field is sufficiently strong to cause appreciable mixing of the higher terms.

Calculations [145, 147] also indicated that the crystal field interactions will give rise to a large magnetic anisotropy leaving the moments directed along the cube edge in dysprosium and holmium compounds and along the cube diagonal in terbium and thulium compounds. These spin orientations have been observed in neutron diffraction studies and the anisotropies have been observed directly using magnetization measurements [138, 148].

Since the magnetic results show that the exchange interaction between the rare earth ions tends to change from ferromagnetic (i.e. positive) to antiferromagnetic in going from the nitrides (small lattice spacing) to the antimonides, Child *et al* [134] have proposed that there would be a critical lattice spacing at which these predominant interactions balanced, and that this was probably the case in the phosphides. Under these conditions the dipole interactions can be sufficient to stabilize a given spin configuration. This was considered to be the case for HoP, and later for HoAs, for which the antiferro-, ferri or ferromagnetic spin structures become stable with increasing field. The magnetic energies required for these transitions are equal to the dipole energy difference of the various spin arrangements.

The destruction of the non-magnetic singlet ground state in the compounds of thulium by the application of large external fields has been examined theoretically by Cooper [149, 150]. This process results in the appearance of an induced moment through the polarization of the ground state wave function by Zeeman mixing with excited states. With a sufficiently large field applied in an easy direction, the magnetization should be equal to 7 μ_B, while for fields applied parallel to the hard (100) direction the magnetization is expected to increase in a series of steps, since with increasing fields the Γ_1 state is crossed by one of the Γ_4 levels and subsequently the Γ_4 state is crossed in turn by a $\Gamma_5^{(2)}$ level. Associated with these increases, at fields below those for which the levels cross, is a magnetization inversion caused by the thermal population of the approaching level, giving rise to a higher magnetization in the hard direction at 20°K then at 10°K and so on. These predicted effects are shown in Fig. 6.11 for TmSb, and the Γ_4 crossing is observed to occur at about 400 kOe. Experimental observations in pulsed fields of up to 265 kOe [151] show a moment of 4·5 μ_B per molecule at 4·2°K. This is in excess of the polarized singlet state and may be associated with admixing of the Γ_4 level.

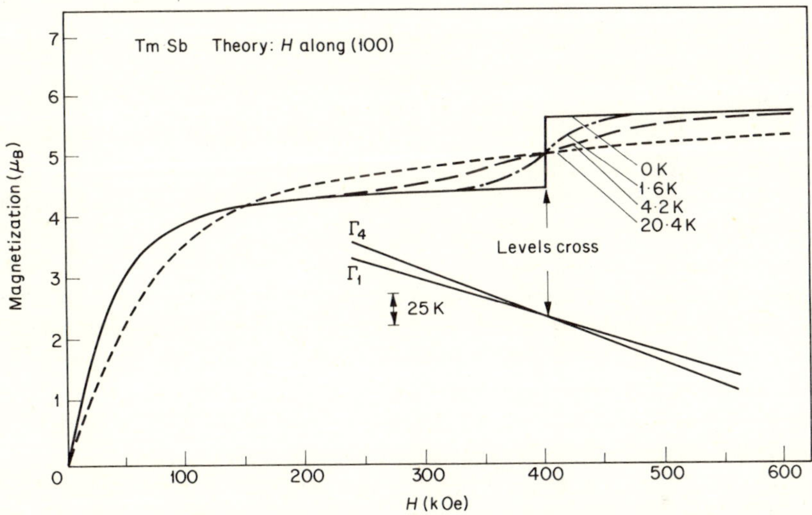

Fig. 6.11 The predicted magnetization of TmSb for the magnetic field applied parallel to the hard 110 direction. The variation of the Γ_1 and the lowest Γ_4 levels are shown.

Obviously the ionic moment development will be accompanied by the growth of exchange forces and a change in the anisotropy of the magnetization. The experimental indications are that the exchange is probably antiferromagnetic and increases rapidly with magnetization.

Paramagnetic resonance has been observed in the thermally populated Γ_5 level of TmN by Cooper *et al* [152], and the temperature dependence of the *g*-shift is attributed to the presence of a small ferromagnetic exchange interaction in addition to the dominant crystal field effects. A small anomaly at $T = 120$°K was observed in the *g*-shift in addition to the overall crystal field–exchange variation. This appeared as a minimum in *g* and its origin is as yet unclear, although attempts were made to associate it with the RKKY interaction, relaxation effects, and anisotropy.

Paramagnetic resonance studies have also been made on the europium and gadolinium compounds [32, 140]. The line widths in these measurements could be accounted for on the basis of exchange narrowing, so allowing approximate values of the nearest and next nearest neighbour exchange parameters $\mathscr{J}_1$ and $\mathscr{J}_2$ to be evaluated. In the gadolinium compounds the resulting temperature dependence of $\mathscr{J}_1$ was associated with the work of Berdyshev and Karpenko [153, 154] on magnetically ordered semiconductors to give a value for possible energy gaps in these materials. These were all approximately 10^{-2} eV but increased slightly in going from GdP to GdBi.

In general, however, direct measurements of electrical conductivity in the nitrides [136, 155, 156] show metallic behaviour and recent self-consistent band structure calculations for the related compounds ScN, ScP and ScAs [157] show the latter two to be metallic and the former to be either a narrow gap semiconductor or semimetallic. The situation is still extremely unclear though, as the nitrides are difficult to prepare and frequently have as much as 5% nitrogen deficiency, and the observation of metallic conductivity is somewhat ambiguous. Further, although it has a similar structure to the rare earth pnictides, ScN is formed with a cation whose diameter is appreciably less than that of the rare earth ions, consequently the atomic spacing may lead to a band structure bearing only a slight resemblance to the rare earth compounds.

Direct measurement of optical energy gaps in the nitrides have been made by Sclar [136] and compared with empirically predicted values obtained by comparison with the III-V semiconductors [158]. These in general showed fair agreement with gap widths of order 2–3 eV. Sclar attempted to reconcile these results with the observations of metallic conductivity behaviour made by himself and others in terms of deviations from stoichiometry causing a conversion of a 2 eV energy gap semiconductor into a defect-type semiconductor with the Fermi level lying in the conduction band. There is of course the further complication that with the normally rather low purity of the rare earth metals used to provide the compounds, the impurity concentration even in 'stoichiometric' compounds is probably already high enough for the impurity levels to form a band and even to overlap the conduction band, again leaving the Fermi level in the conduction band.

Nuclear magnetic resonance Knight shifts have been reported by Jones [142, 159] and have been used by him to obtain information about the $4f$ electron component of the total paramagnetic susceptibility and the associated paramagnetic Curie temperatures. He has also associated the hyperfine field at the non-magnetic site in these compounds with the spin component of the angular momentum of the rare earth ion rather than with the magnetization at the rare earth site. On the basis of the conduction electron spin–polarization exchange model, the magnitude of the Knight shift has been used to obtain values of the exchange energies (see Section 6.4). These are all found to be negative (antiferromagnetic) for the phosphides, arsenides, antimonides and bismuthides; the magnitudes are given in Table 6.5.

The exchange constants have been interpreted by most workers in terms

Table 6.5 The exchange energies of the rare earth pnictides as derived from Knight shift observations and the exchange integrals $\mathscr{J}_1$ and $\mathscr{J}_2$ obtained by a molecular field analysis of the magnetization data [80, 142, 144].

	Nitrides			Phosphides			Arsenides			Antimonides		
	A_0(eV)	$\mathscr{J}_1$ (°K)	$\mathscr{J}_2$ (°K)	A_0	$\mathscr{J}_1$	$\mathscr{J}_2$	A_0	$\mathscr{J}_1$	$\mathscr{J}_2$	A_0	$\mathscr{J}_1$	$\mathscr{J}_2$
Ce				−1·1			−0·74					
Pr				−0·68			−0·50					
Nd		9·6		−0·56	0·5	0·4	−0·42	0·48	0·70		0·43	−1·05
Sm				−0·35			−0·28			−0·22		
Eu				−0·41								
Gd		6·57		−0·37	0·12	−0·25	−0·28	0·1	−0·4		−0·1	−0·45
Tb		4·26		−0·31	0·1	−0·18		0·07	−0·22		0·1	−0·34
Dy		3·43		−0·27	0·23	−0·28		0·1	−0·27		0·08	−0·27
Ho		3·0		−0·24	0·23	−0·28		0·12	−0·20		0·06	−0·23
Er		1·6		−0·23	0·11	−0·26		0·07	−0·23		$\simeq$0	−0·23
Tm				−0·19			−0·16			−0·15		
Yb				−0·27								

of a molecular field model using nearest and next nearest neighbours only. The values of the two exchange parameters $\mathscr{J}_1$ and $\mathscr{J}_2$ obtained in this way are also shown in Table 6.5. Darby and Taylor [160], however, have attempted to associate the magnetic behaviour with a long range indirect exchange between the rare earth ions. On this basis and assuming metallic conductivity, a valence electron concentration of 1·42 per ion was found for all compounds. Their results included the nitrides as well as the other group V elements and consequently did not take account of the change in sign of the paramagnetic Curie temperatures. It would appear therefore, that while this result may apply to either the nitrides or the remaining compounds it is something of a coincidence that the same value of electron concentration appears in both.

6.3.2 4:3 *Stoichiometry*

Gadolinium and dysprosium also form, with the group VA elements antimony and bismuth, the metallic 4:3 compounds which have an inverted Th_3P_4 structure. The rare earth ions occupy the 16 six-fold coordinated sites of this structure. The gadolinium compounds are ferromagnetic with Curie temperatures of 260°K (Sb_3) and 340°K (Bi_3) while Dy_4Sb_3 becomes antiferromagnetic between 67°K and ferromagnetic below 21°K with a very large anisotropy [128]. The effective moments in the series $Gd_4(Sb_{1-x}Bi_x)_3$ are always in excess of 8 μ_B, being as high as 8·8 μ_B in Gd_4Bi_3 compared with the ionic value for gadolinium of 7·94 μ_B. The ordered moments, however, are found to be slightly less than 7 μ_B. The Curie temperatures show a continuous change between the terminal values, and Holtzberg *et al* [128] have discussed the exchange in terms of the RKKY model.

6.4 Rare earth-aluminium compounds

The rare earth–aluminium phase diagrams [161] contain a number of compounds ranging from A_3Al to A_3Al_{11}. Some stoichiometries are found for all the trivalent rare earth elements, whereas others occur for a limited number of elements. Compounds such as the A Al and A Al_3 intermetallics belong to the former category but show changes in structural type in passing from the light to the heavy end of the rare earth series. These differences have been used in the past as a means of studying the influence of size in binary compound formation, and close correlations have been found between the observed stoichiometries of aluminium compounds and the atomic radius of the rare earth element concerned. For example, Buschow and van Vucht [161] have shown that the compound richest in the lanthanide element is A_3Al for the large radius atoms and A_2Al for smaller radii. A compound A_3Al_2 is also found with the heavy elements. Similarly in the aluminium-rich compounds A_3Al_{11} is the last stable compound for the light rare earths and A Al_3 for the heavy rare earths. This latter composition is found for all the elements but passes through a variety of crystallographic structures with decreasing ionic radius of the rare earth part.

In addition to the A Al_3 compounds, the compositions A Al and A Al_2 are found for each of the trivalent rare earth metals and their magnetic behaviour has attracted considerable attention. The dialuminides all possess the cubic Laves phase (C15) structure and have been shown to be ferromagnetic by Williams *et al* [162, 163] with a maximum Curie temperature of 176°K for $GdAl_2$. The variation of the Curie temperature across the series is proportional to the de Gennes factor, $G = (g-1)^2J(J+1)$, for elements to the right of gadolinium. The magnetic ordering of the compounds formed with the elements cerium to samarium, however, persists to much higher temperatures than might be anticipated from any simple theory; the Curie temperatures of 122°K and 65°K for $SmAl_2$ and $NdAl_2$ being well in excess of those for the pure rare earth metals.

The saturation moment values are generally less than those corresponding to the free trivalent ions, a property which is typical not only of the other aluminium compounds, but also of the rare earth intermetallic compounds in general. These values are listed in Table 6.6 along with the transition temperatures and effective paramagnetic moments of those compounds.

Neutron diffraction measurements of $DyAl_2$ and $NdAl_2$ [164] confirm the general features described above and show further that while $NdAl_2$ is a pure ferromagnetic, $DyAl_2$ has a weak antiferromagnetic contribution to the ordering. This behaviour is proposed to account for the appearance of superlattice lines in the neutron diffraction pattern at temperatures below 24°K. The ordered moment of 9·1 μ_B, indicated by this work, is within 10% of the free ion value and it is concluded that crystal field effects play little part in determining the dysprosium moment in $DyAl_2$. It is generally believed,

Table 6.6 The magnetic moments and transition temperatures of the rare earth aluminium compounds [163, 187, 207, 214].

		La	Ce	Pr	Nd	Sm	Eu	Gd	Tb	Dy	Ho	Er	Tm	Yb	Lu	Y
A_3Al_2	θ							285	125	31	10	−3	−10			
	μ_{eff}							8·2	9·6	10·8	10·9	9·6	7·8			
	T_c							282	190	76	33					
	T_N											9	3			
	μ							7·1								
A Al	θ		4	11	−4			85	24	17	22	28	−2			
	μ_{eff}		2·34					8·52	10·11	11·15	11·25	10·05	7·70			
	T_c										26					
	T_N		9	20	29			42	72	20		13	10			
	μ										7·1					
$A\ Al_2$	θ				61			180	108	68		16				
	μ_{eff}				3·1			7·92	9·82	9·7		9·56				
	T_c			33	65	123		170	114	58	27	14·5				
	T_N															
	μ				2·5			7·0	8·1	9·1	7·86	7·6				
$A\ Al_3$	θ		−46	−14	+5			−89	−64	−51	−26	+6	−19	−300		
	μ_{eff}		2·63	3·74	4·11			8·29	10·0	10·85	10·89	9·87	7·88	4·62		
	T_c											21				
	T_N							17	21	23	9					
	μ											> 6·2				

however, that crystal field quenching is an important factor in these compounds [163, 165–167]. Whether this is the only effect is still open to doubt, and the reduced moment values which are observed in magnetization studies can be accounted for in part by a small antiferromagnetic contribution to the exchange interaction. Relatively small crystal field effects are suggested by the Mössbauer studies of the hyperfine interactions of Nowik *et al* [168] and Wiedemann and Zinn [169] and also by the neutron diffraction studies of Olsen *et al* [164].

In their early work, Williams *et al* [162] investigated some pseudobinary compounds formed between the light and heavy rare earth compounds. The magnetization of these systems was consistent with a ferrimagnetic coupling between the different rare earth moments and was indicative of a parallel coupling of the spins of the two types of ion. This follows since for the light rare earth elements with a less than half full 4*f* shell, $J = L-S$, while for the heavy elements with seven or more 4*f* electrons $J = L+S$. If the ionic spins couple by an indirect exchange interaction involving the conduction electrons, as seems likely, then the sign of the net exchange will be independent of the sign of the rare earth conduction electron interaction.

The results of electron spin [170] and nuclear magnetic resonance [171, 172] experiments have shown that the *s*–*f* interaction is negative in these compounds, and that the amplitude of the spin polarization at the aluminium sites is also negative. It would appear that this is also likely to be the case at the neighbouring rare earth sites since the A–A and A–Al distances are comparable. As a result the spins of these neighbour ions will be aligned parallel to one another and to the central ion as anticipated in the previous paragraph. Further evidence for this negative interaction has been provided by the observation of magnetic dilution effects caused by the substitution of yttrium and lanthanum into the magnetic dialuminides [173].

The early resonance measurements [170–172] were interpreted by using a uniform conduction electron polarization model to give values for the Knight shift (K) and the fractional change in the g value ($\Delta g/g$).

Using the usual exchange interaction between the localized *f*-electrons and the conduction electrons, i.e.

$$H_{\mathrm{ex}} = \mathscr{J}_{sf}\mathbf{S}\cdot\mathbf{s} \tag{6.1}$$

the Knight shift may be written [171, 172]

$$\begin{aligned} K &= K_0\{1+(\mathscr{J}_{sf}\chi_{\mathrm{ion}}/2g_J\mu_{\mathrm{B}}^2)[\langle\bar{S}\cdot\bar{J}\rangle/J(J+1)]\} \\ &= K_0\{1+[\mathscr{J}_{sf}(g_J{}^{-1})\chi_{\mathrm{ion}}/2g_J\mu_{\mathrm{B}}^2]\} \end{aligned} \tag{6.2}$$

where K_0 is the Knight shift due to Pauli paramagnetism, χ_{ion} is the spin susceptibility of the tripositive ions and the remaining terms have their usual meaning.

The shift in the electronic g value due to this conduction electron polarization was shown by Yosida [174] to be

$$\frac{\Delta g}{g} = 3n\mathscr{J}_{sf}/2n_{\mathrm{f}}g_J E_{\mathrm{f}} = \frac{3}{4}\frac{z\mathscr{J}_{sf}}{g_J E_{\mathrm{f}}} \tag{6.3}$$

where $z(=2n/n_{\mathrm{f}})$ is the atomic conduction electron density and E_{f} is the Fermi energy.

While these two results provide methods of determining $\mathscr{J}_{sf}$, it is unlikely that the values obtained will agree with one another since the former detects the polarization at the Al site and the latter that at the Gd site. However, the sign of the interaction and the relative strengths across the rare earth series should not be affected by these differences and as Fig. 6.12 shows, the variations of the two sets of experimental results are remarkably similar. Since

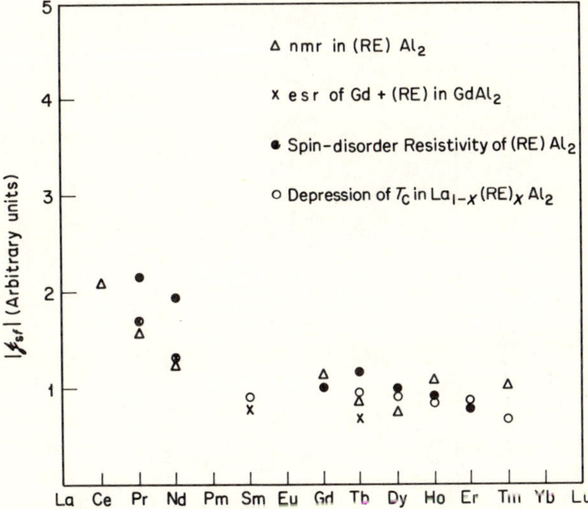

Fig. 6.12 The variation of the exchange parameter $\mathscr{J}_{sf}$ with atomic number for the rare earth dialuminides [179].

the original work, the magnitude of $\mathscr{J}_{sf}$, as determined from the Knight shift observations, has been corrected by several workers [175–177] but the overall form of the variation has remained the same. The latest values are given in Table 6.7 [142].

If a non-uniform polarization model is adopted, such as arises from the RKKY theory, the exchange constant $\mathscr{J}_{sf}$ is replaced by

$$\mathscr{J}_{sf} = -6\pi Z A_0 \sum_n \phi(2k_{\mathrm{f}} R_{mn}) \tag{6.4}$$

with A_0 the exchange integral of the RKKY s–f interaction. On this basis, and making assumptions concerning the values of Z and k_{f}, the values of A_0

Table 6.7 The exchange constants $\mathscr{J}_{sf}$ for the rare earth aluminium compounds obtained using the uniform conduction electron polarization model [166, 167, 170, 171, 175, 176, 192, 214].

	Ce	Pr	Nd	Sm	Gd	Tb	Dy	Ho	Er	Tm	Yb	Mean† A_0
A Al					−0·16	−0·15	−0·15	−0·16	−0·14			−0·39 eV
A Al_2	−0·63	−0·48	−0·38	—	−0·31	−0·28	−0·25	−0·38	−0·30	−0·29		−0·9 eV
A Al_3	−0·41	−0·35	−0·22	−0·21	−0·23	−0·22	−0·24	−0·20	{−0·18*(α) −0·24*(β)	−0·24	−0·23	−1·7 eV

* $ErAl_3$ exists in two modifications.

† Given in [214], and based on assumptions concerning the electronic structure in equation (6.4).

have been obtained for the compounds across the series and are also shown in Table 6.7. As may be seen the exchange parameter of either interpretation decreases in magnitude with increasing atomic number, the variation for compounds with the light rare earths being more pronounced than for those with the heavy elements, for which there is little obvious change in $\mathscr{J}_{sf}$ or A_0 for elements beyond gadolinium.

Other confirmatory evidence for the relative values of the exchange parameters, and for the validity of the RKKY model comes from the experimental determination of the spin disorder resistivity contribution to the total resistivity [178] and from the depression of the superconducting transition temperature of $LaAl_2$ doped by other rare earth elements [179]. The latter property depends on the square of the exchange parameter and consequently gives no indication of its sign. This is perhaps unfortunate since the results of magnetization and conductivity studies suggest a positive interaction, contrary to the indications of the resonance techniques. In addition recent studies by Coles *et al* [180] have shown positive g shifts, and hence positive $\mathscr{J}_{sf}$, in gadolinium doped $LaAl_2$.

Magnetic dilution of the rare earth dialuminides by lanthanum, yttrium and thorium show concentration dependences of the Curie temperatures [173] and hyperfine fields [181] which can also be interpreted using a negative A_0 in conjunction with the RKKY interaction. Although the basic features of spin echo measurements [181] were in keeping with those of the terminal compound $GdAl_2$ obtained by other workers [182, 183], the variation of the Al^{27} resonance is based on rather limited evidence and is somewhat inconclusive. Particularly worrying in this context is the observation of only one of the two possible resonances from the different aluminium sites in these compounds. These are clearly visible at 4·2°K for up to 30% yttrium addition [184]. Other observations of the hyperfine field strength in these compounds [168, 169, 185] show that the rare earth nuclear field is close to the free ion value.

In the RKKY treatment of the exchange interaction no account is taken of contributions from the spin–orbit coupling of the conduction electrons taking part in the exchange. Levy [186] has recently explored the effects of including pair interactions of the type $L_A \cdot S_B$. Although both isotropic and anisotropic pair terms exist, a close agreement between theory and experiment has been possible for the aluminium compounds without the use of the higher degree terms.

The magnetic properties of the $A\,Al_3$ compounds have been investigated by Buschow and Fast [187] and the observed transition temperatures and moment values are listed in Table 6.6. In contrast to the dialuminides, only $NdAl_3$ and $ErAl_3$ are found to be ferromagnetic, the compounds with gadolinium, terbium, dysprosium and holmium are antiferromagnetic. The remainder show no evidence of ordering to the lowest temperatures investigated. Interpretation of the results is complicated by three changes in the equilibrium crystal structure across the series. Compounds to the left of $GdAl_3$ crystallize in the hexagonal Ni_3Sn structure and those to the right of $ErAl_3$ in the cubic Cu_3Au structure, the intermediate compounds possessing a mixture of the two types of atomic stacking in which the cubic character becomes more prominent for the smaller rare earth ions.

The ordering temperatures probably arise from the relatively low rare earth concentration in these compounds and it is assumed that the exchange interaction is appreciably less than the crystal field energy. In the only ferromagnetic heavy rare earth compound, $ErAl_3$, the observed moment is appreciably less than the free ion value $g_J J = 9\ \mu_B$, and in spite of a reported difficulty in obtaining saturation the reduced moment was attributed entirely to crystal field quenching of the orbital moment. As shown by Bleaney [188] the ground state of the Er^{3+} ion, in the absence of sixth order terms, is a Γ_8 quartet of moment 5·5 μ_B and it is necessary to assume that the next highest level (also Γ_8) is sufficiently close to allow mixing to occur between the two states if the observed moment of 6·2 μ_B is to be understood.

The variation in crystal structures makes a comparison of the crystal field effects across the series a difficult task and to date complete calculations have only been attempted for $TmAl_3$ [189] which crystallizes in the cubic, Cu_3Au, structure. The exchange contribution to the final splitting was introduced as a perturbation on the crystal field solutions by introducing a term in the exchange field H_{ex}. Using the molecular field approximation

$$H_{ex} = -\mathscr{J}_{sf}[(g-1)/2\ \mu_B]\langle J_z \rangle_{\text{average}}$$

allows $\mathscr{J}_{sf}$ to be used as a fitting variable. The value of the exchange constant obtained in this way is negative but rather lower than that derived from nuclear magnetic resonance observations on the compounds of this series [189–194]. These values, obtained from a uniform conduction electron polarization model are listed in Table 6.7.

The only other stoichiometries which have received any considerable attention are the A Al and A_3Al_2 compounds, which crystallize in the orthorhombic (DyAl) [195, 196] and the tetragonal (Zr_3Al_2) structures [197] respectively.

Magnetization studies [198–203] show that with the exception of HoAl the equiatomic compounds are antiferromagnetic, but that many of them show evidence of a metamagnetic type of transition at low temperatures. The magnetization curves of these compounds, obtained at 4·2°K, are shown in Fig. 6.13 and the transition temperatures are listed in Table 6.6. The metamagnetic type of transitions are clearly visible but it is worth noting that these are observed for only a small range of temperatures near to 4·2°K and do not persist to the ordering temperature.

Neutron diffraction studies [200–203] of the antiferromagnetic compounds have been interpreted in terms of a non-colinear magnetic structure. The stability of this structure has been attributed by Becle *et al* [202, 204] to

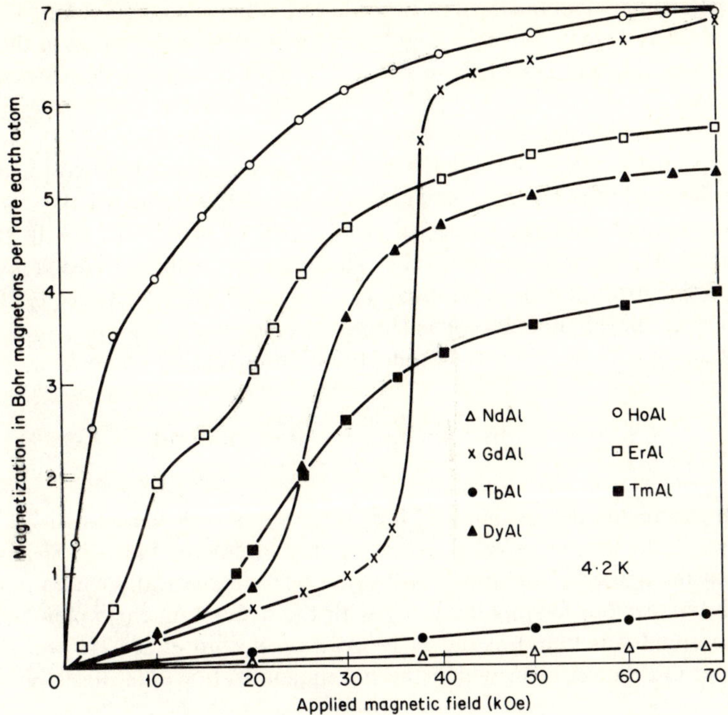

Fig. 6.13 A Al magnetization versus field observations [202].

highly anisotropic magnetic interactions between the rare earth ions and their possible origin has been mentioned briefly in terms of (i) pseudodipolar interactions such as those introduced by Nhung *et al* [205] for the ASi compounds and (ii) distortion of the Fermi surface as suggested by Blandin [206].

The observations of Barbara *et al* [207, 208] on the A_3Al_2 compounds, which are found only with the heavy rare earths, suggest that Er_3Al_2 and Tm_3Al_2 are antiferromagnetic at all temperatures below the ordering temperatures, but that the compounds with gadolinium, terbium, dysprosium and holmium show a transition from ferromagnetism to antiferromagnetism with decreasing temperature. This behaviour is suggested by the observation of metamagnetic-like transitions below some characteristic temperature in much the same way as those found for the equiatomic compounds. In this case, however, Buschow [209] has managed to show that these 'metamagnetic' transitions in Dy_3Al_2 are associated with the onset of a high magnetocrystalline anisotropy and a subsequent reinterpretation of the observed behaviour by Barbara *et al* [210] was carried out in terms of the motion of monatomic domain walls. Observations by Taylor *et al* [211–213] on these and other compounds showing similar behaviour, suggest that the low temperature magnetization shows marked time dependence and as a consequence the form of the experimental magnetization-field curves in these materials can depend markedly on the rate of field change during the measurement.

Knight shift measurements on the equiatomic compounds [214] have been interpreted in terms of the RKKY theory and a value obtained for the exchange integral A_0. The results showed $A_0 = -0{\cdot}39$ eV in all the compounds, compared with the values of $A_0 = -0{\cdot}9$ eV and $-1{\cdot}7$ eV in the AAl_2 and AAl_3 systems respectively. The increase with increasing aluminium concentration was attributed to interband mixing effects [215]. The values of the phenomenological exchange constant $\mathcal{J}_{sf}$ are listed in Table 6.7 along with those for the other stoichiometries.

6.5 Rare earth-transition metal compounds

6.5.1 *Compounds with the 3d-elements*

The magnetic properties of many of the compounds with the 3*d*-transition metals appear to be related, and extensive measurements [216–235] have shown that manganese, iron and cobalt order ferromagnetically with a light rare earth partner but ferrimagnetically with the heavy rare earth elements. This latter trend was found in the early work of Nesbitt *et al* [216] on the Gd–Co and Gd–Fe series in which the net magnetization was observed to

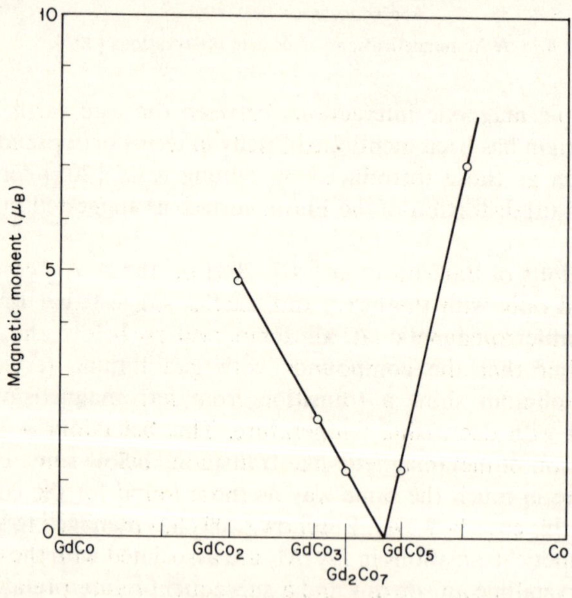

Fig. 6.14 The molecular moments of the compounds in the gadolinium–cobalt system showing the deep minimum arising from the antiparallel moment alignment (data taken from [234, 235]).

decrease rapidly with increasing gadolinium concentration, pass through a minimum in the vicinity of 20 and 25 atomic % gadolinium for the cobalt and iron compounds respectively, and finally increase towards the gadolinium value of 7·55 μ_B ion. Fig. 6.14 shows the results of more recent measurements by Lemaire [234, 235] on the compounds in the Gd–Co system, from which this behaviour is clearly visible. The position of the minimum in this variation follows from a simple antiparallel alignment of gadolinium and cobalt moments with approximate magnitudes of 7 μ_B and 1·7 μ_B respectively. In contrast, the bulk magnetization of the nickel compounds can be interpreted as arising from the ferromagnetic ordering of the rare earth ions only [222, 226, 236–239], the nickel ions apparently being non-magnetic for the majority of stoichiometries. The only exceptions to this are compounds with the heavy elements near to the ANi_3 composition [240–243].

The phase diagrams for the nickel and cobalt series are very similar for all the trivalent rare earth elements, and as Fig. 6.15 shows, reported stoichiometries exist between A_3B and A_2B_{17} for these elements. The phase diagrams of the systems containing iron show a rather different distribution of compounds, the only compositions common to iron, cobalt and nickel being AB_2, AB_3 and A_2B_{17}.

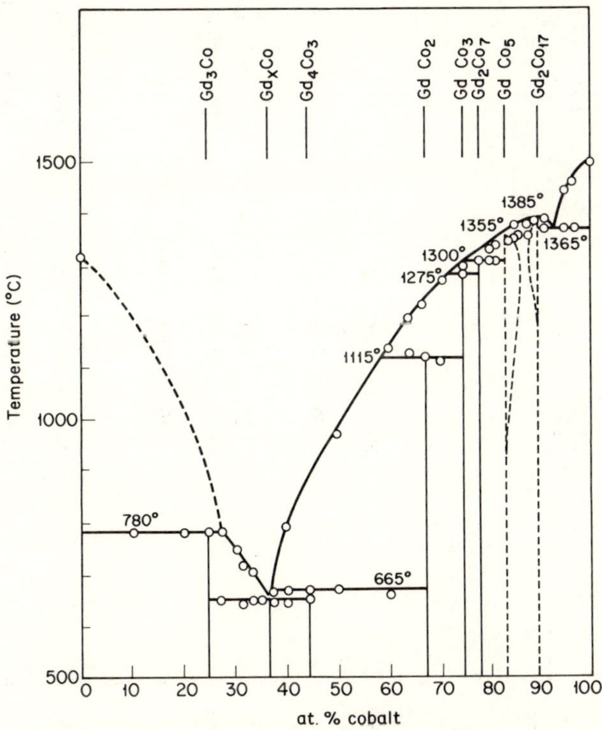

Fig. 6.15 The gadolinium–cobalt phase diagram [334].

Table 6.8 Magnetic data for the rare earth nickel compounds [240–242, 226, 280 , 283].

		La	Ce	Pr	Nd	Sm	Eu	Gd	Tb	Dy	Ho	Er	Tm	Yb	Lu	Y
	θ			−24	0			60	−5	29	−6	−5	0			
	μ_{eff}			3·7	3·6			8·1	10·0	11·1	9·8	7·4				
A_3Ni	T_c												12			
	T_N			2	15			100	62	35	20	9				
	μ															
	θ			23	24			77	40	64	36	13				
	μ_{eff}			3·9	3·7			8·1	9·7	10·7	10·7	9·8				
A Ni	T_c			20	35/28	45		72	51	48/62	31	10	4/8			
	T_N															
	μ			2·26	2·74	0·23		7·4	7·74	8·75	8·65	8·1	5·15			
	θ			4	10			78	35	23	12	10	0			
	μ_{eff}			3·57	3·74			7·82	9·82	10·4	10·5	9·37	7·28			
$A Ni_2$	T_c			15	16	21		85	45	30	22	21	14			
	T_N															
	μ			0·86	1·80	0·25		7·1	7·8	9·2	8·4	6·8	3·2			
	θ															
	μ_{eff}															
$A Ni_3$	T_c			20	27	85		116	98	69	66	62	43	< 20		33
	T_N															
	μ			1·57	1·88	0·33		6·55	6·84	7·0	7·84	5·77	3·86	2·0		0·16

Table 6.8 (continued)

		La	Ce	Pr	Nd	Sm	Eu	Gd	Tb	Dy	Ho	Er	Tm	Yb	Lu	Y
	θ															
	μ_{eff}															
A_2Ni_7	T_c		48	85	87			118	101	81	70	67				58
	T_N															
	μ		0·26	4·36	4·14			12·65	10·72	13·30	12·57	12·28				0·41
	θ		−90					30			8	12	60			
	μ_{eff}		2·65					7·87		9·3	10·5		7·88			
$A\,Ni_5$	T_c				9	25		27	27	15	10	12	7			
	T_N			14												
	μ		0·2	0·4	2·25	0·7		6·8	7·0	9·05	8·1	7·7	6·7			
	θ															
	μ_{eff}															
A_2Ni_{17}	T_c					186		205	178	168	162	166	152			160
	T_N															
	μ					4·5		8·8	8·5	8·05	12·2	9·1	—			5

The molecular moments of the nickel-rich compounds are always found to be close to the theoretical $g_J J$ value of the rare earth ions and in the 1:2 cubic Laves phase compounds YNi_2 and $GdNi_2$ the observed moments are 0·9 μ_B and 7·1 μ_B. In the remaining 1:2 compounds, however, the magnetization is frequently less than that corresponding to the rare earth ions although the effective moments in the paramagnetic state agree well with $g(J(J+1)^{\frac{1}{2}}$ for the rare earth ions (see Table 6.8 for the magnetic parameters of the nickel compounds). Consequently, it appears that in addition to the absence of a nickel moment in the ordered state, the moment of the rare earth atoms is also reduced. Both Bleaney [188] and Skrabek and Wallace [223, 236] have interpreted this decrease in ordered rare earth moment as due to crystal field quenching of the orbital contribution to the total moment. In this work, Bleaney assumed only a B_4 contribution to the crystal field and calculated the resultant moments of the ground state levels for the ions in this crystal field. In many cases these moments were less than the observed moment and it was necessary to admix some of the excited state levels to provide moments which were comparable to the experimental values. The extent to which this mixing occurs depends on the relative strengths of the crystal and molecular fields, and is complete for molecular field energies appreciably greater than the ground and excited state level splitting due to the crystal field. Table 6.9 shows the Curie temperatures and estimated excited state energies for the ions. Using the former as a guide to the molecular field strength it is clear that for terbium, dysprosium and holmium, mixing will be extensive. The theoretical moment will then be that associated with those lower levels, i.e. the values given under 'maximum moment' in Table 6.9. For the remaining compounds, however, the resultant moment will lie somewhere between this value and the ground state moment, the exact value depending on the details of the crystal and molecular field energies. This is found to be the case in practice and it seems certain that this basic model is the correct one to use in

Table 6.9 Estimated and observed data for the ANi_2 series after Bleaney [188, 246].

Compound	Ground state moment	Maximum moment (M_3)	Measured moment at 4·2°K	Low excited states at (°K)	Curie temperature (°K)
$PrNi_2$	0	2·07	1·04	77	15
$NdNi_2$	1·33	3·25	1·89	47	16
$SmNi_2$	0·24	0·71	0·25	112	21
$GdNi_2$	7·0	7·0	7·1	—	85
$TbNi_2$	0	5·62	7·8	27	45
$DyNi_2$	3·35	8·02	9·2	12	30
$HoNi_2$	0	10	8·4	6, 10	22
$ErNi_2$	5·50	6·23	6·8	36	21
$TmNi_2$	0	4·37		37	12

interpreting the magnitudes of the rare earth moments in these compounds. These calculations are of course open to the criticism that they are based on the normal point charge model (see Chapter 1) and consequently are founded on rather drastic assumptions. This is overcome in practice by allowing considerable latitude on the overall splitting rather than holding rigorously to the B_4 and B_6 values of the point charge model.

The compounds formed with cerium are further complicated by the relative ease with which this element becomes quadrivalent in the solid state. In all cases the ordered moment of the cerium ion in the $3d$-transition metal compounds [223–226] is either extremely small or zero, in this latter case Pauli paramagnetism is usually found. Where they have been observed, the paramagnetic moments are in agreement with those calculated from the saturation magnetization. For those compounds in which the cerium moment is greater than zero, the results of both magnetic and specific heat studies [244] may be interpreted in terms of a small residual amount of Ce^{3+} in a matrix of non-magnetic Ce^{4+} ions.

The absence of a nickel moment in these compounds with rare earth metals has generally been attributed to a transfer of electrons from the rare earth atoms, resulting in a filling of localized $3d$ states; leaving neutral nickel atoms with a configuration $3d^{10}$. Recent measurements in the series $Gd(Co_{1-x}Ni_x)_2$ [245], however, show that there is a sudden loss of moment with increasing nickel content at $x \simeq 0{\cdot}6$ and it appears more likely that the non-magnetic state of nickel is the result of a more complex mechanism. It is possible that as the magnetization of the transition metal sublattice decreases, due to the gradual addition of electrons into a $3d$-band, a point is reached at which the exchange interaction is no longer able to support the splitting of the $3d$-sub-bands and the observed moment disappears. This will be a function of both the shape of the $3d$-band and the position of the Fermi level in the band. Unfortunately, it is likely to be some considerable time before band structure details of these compounds are available and the detailed reasons for the absence of a moment on the nickel ions is still uncertain.

The magnetic moments and transition temperatures of the compounds formed with cobalt and iron are listed in Tables 6.10 and 6.11. In contrast to the nickel compounds it is generally much more difficult to ascribe a definite moment to either ion in these cases since the observed magnetization is due to the ferro- or ferrimagnetic coupling of the two magnetic sublattices. It is to be expected that the moments of the rare earth ions will still be quenched to some extent and the observed magnetizations of the gadolinium and yttrium compounds suggest that the cobalt and iron moments may vary over a considerable range. In YCo_2 the cobalt moment is either extremely small ($\simeq 0{\cdot}05\ \mu_B$) or zero, whereas in $GdCo_2$ it is approximately $1{\cdot}1\ \mu_B$ and in Gd_2Co_{17} is $1{\cdot}7\ \mu_B$. Similarly, in the iron compounds with gadolinium, the moment associated with the iron ion varies from $1{\cdot}6\ \mu_B$ to $2{\cdot}1\ \mu_B$.

Table 6.10 The transition temperatures and magnetic moments of the compounds formed with cobalt [226, 234–236, 284].

		La	Ce	Pr	Nd	Sm	Eu	Gd	Tb	Dy	Ho	Er	Tm	Yb	Lu	Y
	θ			18	35			159	85	65	44	20	0			
	μ_{eff}															
A_3Co	T_c			7								7	5			
	T_N				14			127	82	45	24					
	μ			1·1				metamagnetic				6	3·5			
	θ							242		62	50	28				
	μ_{eff}															
A_4Co_3	T_c							230		55	44	25				13
	T_N															
	μ							6·5		5·2	7·4	5·75				0·11
	θ				107				228	129		28				
	μ_{eff}															
ACo_2	T_c			50	116	259		408	256	159	95	36	18			
	T_N															
	μ			3·2	3·8	2·0		4·9	6·7	7·6	7·8	7·0	4·7			
	θ															
	μ_{eff}															
ACo_3	T_c		78	349	395			612	506	450	418	401	370			
	T_N															
	μ		0·2	3·8	5·6			2·2	3·4	4·3	5·6	3·9	3·0			

Table 6.10 (continued)

		La	Ce	Pr	Nd	Sm	Eu	Gd	Tb	Dy	Ho	Er	Tm	Yb	Lu	Y
	θ															
	μ_{eff}															
A_2Co_7	T_c	490	123	574	609	713		775	717	647	670	644			453	639
	T_N															
	μ	6·6						2·4	5·3		6·0					7·4
	θ															
	μ_{eff}															
ACo_5	T_c	840	737	912	910											977
	T_N					1020		1008	980	966	1000	986	1020			
	μ	7·1	5·5	9·9	10·5	7·8		1·3	0·7	1·6	1·9	1·4	1·6			8·1
	θ															
	μ_{eff}															
A_2Co_{17}	T_c		1083	1171	1150	1190		1209	1180	1152	1173	1180	1182		1157	1176
	T_N															
	μ		26·1	32·8	31·0	23·6		14·4	10·8	8·3	7·8	10·2	13·6		27·4	27·8

Table 6.11 The transition temperatures and magnetic moments of the iron compounds [231, 250, 251, 337].

		La	Ce	Pr	Nd	Sm	Eu	Gd	Tb	Dy	Ho	Er	Tm	Yb	Lu	Y
	θ															
	μ_{eff}															
AFe_2	T_c		235/878			700		782	705	638	614	593	613		610	550
	T_N															
	μ		2·38			2·5		3·35	3·68	4·91	5·1/6·1	4·75	2·52		2·97	2·91
	θ															
	μ_{eff}															
AFe_3	T_c					657		728 •	648	600	567	553	539			
	T_N															
	μ							1·6	3·6	4·6	4·6	3·5	1·6			
	θ							220								
	μ_{eff}							15·3								7·0
A_6Fe_{23}	T_c				492			659	574	524	501	493	475		471	484
	T_N															
	μ											7·2				37·8
	θ															
	μ_{eff}															
A_2Fe_{17}	T_c			282				466	408	363	325	299				244
	T_N															
	μ							21·2	16·8	15·6	15				33·8	34·2

In attempting to attribute ionic moments to these materials it is usually essential to make some estimate of the likely rare earth moment by extrapolation from the nickel compounds. In this way a reasonably consistent picture evolves in which the cobalt moments increase from a value of around 1·0 μ_B for the 1:2 compounds (with the exception of YCo_2), to 1·7 μ_B for those rich in cobalt, and the iron moments vary from $1{\cdot}8 \pm 0{\cdot}4$ μ_B for the 1:2 compounds to 2·0 μ_B in the 2:17 stoichiometry. In addition to this general increase in the magnitude of the cobalt moment towards 1·7 μ_B there is a tendency, with some rare earth partners, for a maximum to appear at a stoichiometry in the region of 1:3·5 [234, 235].

As with the nickel compounds, the reduced moments associated with the transition metal ions were originally thought to arise from the filling of *localized* 3*d* states by electrons originating from the rare earth ions. In his treatment of the cobalt compounds, Bleaney [246] further proposed that the observed transition metal moment was induced by the rare earth moment through the exchange interaction. Predicted values for these induced moments were then obtained by scaling from the $GdCo_2$ results. Unfortunately, the results on which this analysis was based were subject to considerable revision in the following years and the agreement between the predicted molecular moments and observed cobalt moments is now less convincing.

More recently there has been a tendency to associate the transition metal moment in these compounds with a 3*d*-band, to which the concepts of transition metal ferromagnetism may be applied. Some evidence of the correctness of this approach has been furnished by Taylor and his co-workers [247, 248] from their observations on various pseudobinary series. Fig. 6.16 shows the net observed moment for the $Y(Fe_{1-x}Co_x)_2$ series, the initial increase in the moment with increasing cobalt concentration occurring at a rate of 1 μ_B per added 3*d* electron. The maximum may then be considered to occur when the spin-up sub-band is full and the subsequent decrease (also at 1 μ_B per added electron) due to the filling of the spin-down sub-band. This is of course analogous to the behaviour of the classical Fe–Co, Co–Ni, Ni–Cu alloy systems. The collapse of the transition metal moment near to YCo_2 appears to be related to the collapse in the $Gd(Co_{1-x}Ni_x)_2$ series mentioned earlier, and to similar systems involving different rare earth partners, which have been studied by Slanicka and Taylor [248]. On this basis one may view the increase in the cobalt and iron moments which occur in the transition metal-rich compounds, such as the 2:7, 1:5 and 2:17 compositions, as being due jointly to the strength of the exchange interaction, the shape of the 3*d*-band and the position of the Fermi level relative to this band.

The nature of the exchange interactions in these materials is still open to speculation. In most of the compounds there are three couplings to be considered, namely the rare earth–rare earth, rare earth–transition metal and transition metal–transition metal interactions. The evidence we have for the

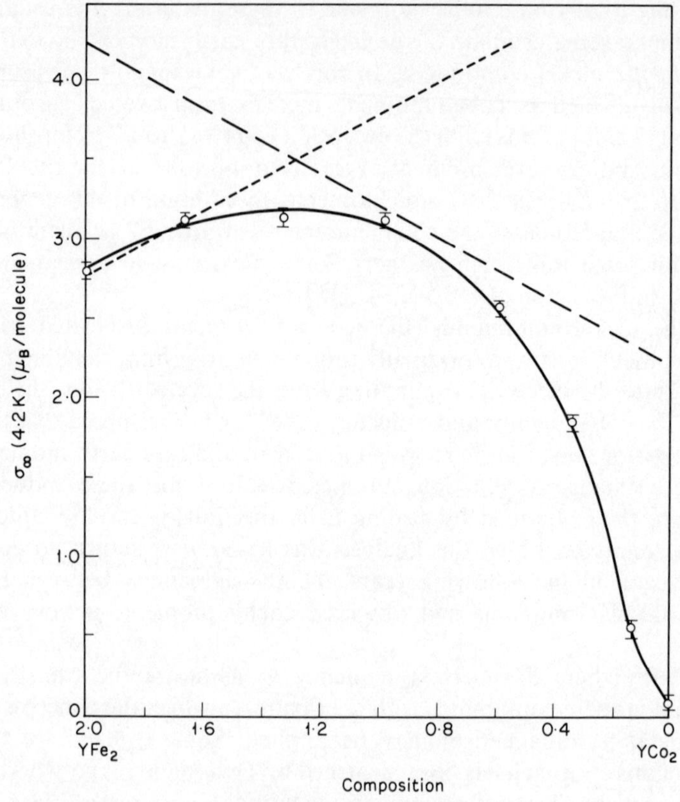

Fig. 6.16 The variation of the molecular moment in Y $(Fe_{1-x}Co_x)_2$. The broken lines have gradients of ± 2 μ_B/3d-electron [247].

total strength of the interactions and also for the relative strengths of the components, comes largely from the Curie temperatures of the compounds. In the 1:1, 1:2 and 1:5 series with nickel the Curie points are low, with a maximum of 77°K for $GdNi_2$. As Fig. 6.17 shows, for any one series these vary linearly with G although for some of the compounds a better fit may be obtained with the function $S(S+1)$. It will be recalled that the results for the isostructural $A\,Al_2$ compounds also follow the factor G and as a result it has been generally accepted that in compounds in which only the rare earth ion carries a moment then the indirect interaction between the rare earth ions is the only one to be considered. The interaction of the rare earth ion with the potentially magnetic nickel ion appears to be small or zero, except in the case of the $A\,Ni_3$, A_2Ni_7 and A_2Ni_{17} compounds in which a small nickel moment is observed [240–243] and for which the extrapolation of the de Gennes function to $G = 0$ intercepts the temperature axis at $T > 0$. The strength of the interaction in these cases will depend on the nickel moment

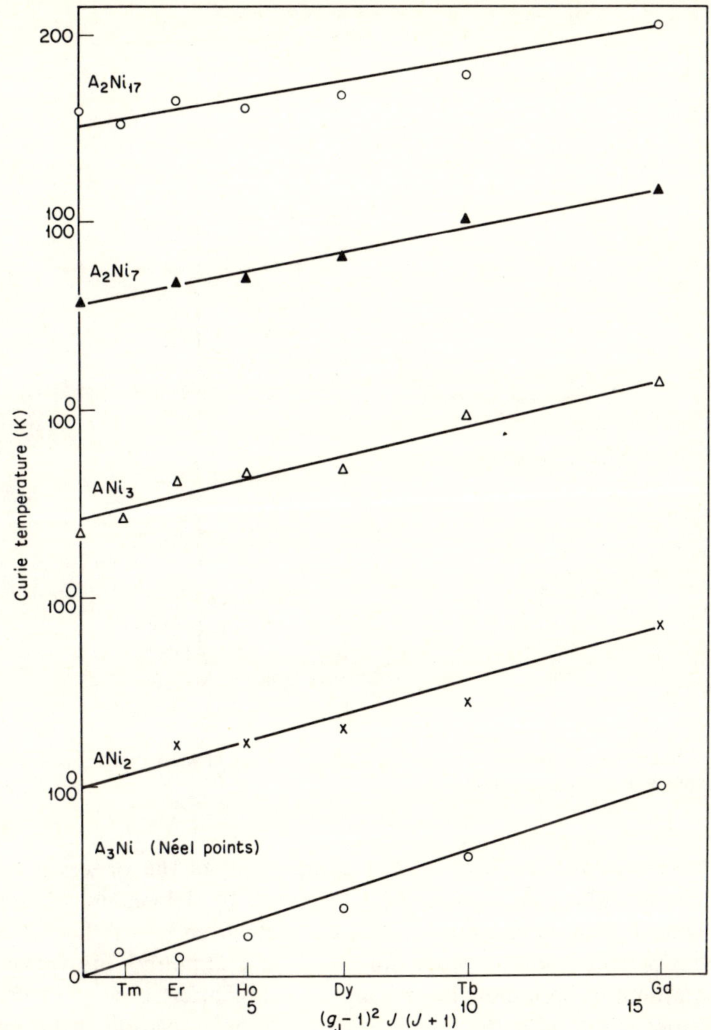

Fig. 6.17 The Curie temperatures of the nickel compounds plotted as a function of the de Gennes factor $G = (g_J - 1)^2 J(J+1)$.

implies, at least in these nickel compounds, that the $3d$ band is not completely filled. The variation of the ordering temperature for compounds of varying nickel composition are shown in Fig. 6.18, from which may be seen the marked increase in transition temperature, for compounds in which the magnetization of the nickel sublattice is greater than zero. It would appear then that the presence of a magnetized transition metal sublattice causes either one or both of the remaining exchange mechanisms to become

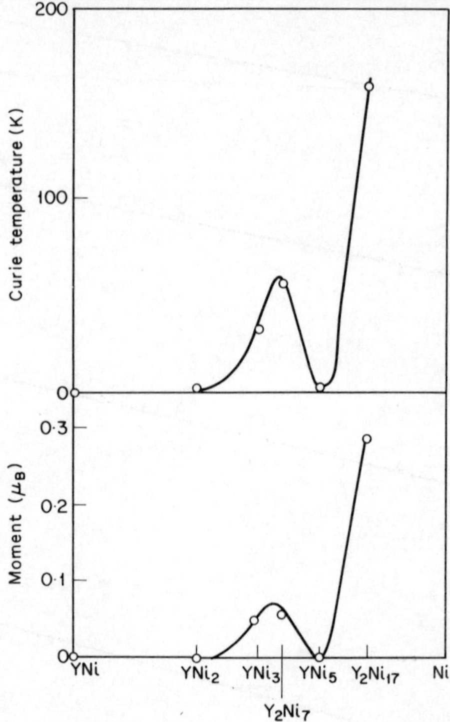

Fig. 6.18 The molecular moments and Curie temperatures of the nickel-rich compounds with yttrium, as a function of nickel content.

important. This is also evident from a comparison of the ordering temperatures of the AFe_2, ACo_2 and ANi_2 systems. In the second of these, ordering occurs at a temperature approximately four times that of the equivalent nickel compound, while for the iron compound the factor is nearer to ten. In addition, the YFe_2–YCo_2 pseudobinaries show an increase in Curie temperature with the magnetization of the transition metal sublattice [247]. The nature of the exchange interactions involving the 3*d* elements is not yet established, however, but it would seem from the value of the interatomic distance between the 3*d*-ions that the transition metal–transition metal interaction is probably a direct interaction comparable to that in the pure metals. The nature of the rare earth–transition metal exchange is less clear as this could proceed either by a direct interaction between the 3*d*- and 4*f*-electrons or by the polarization of the electrons in the 3*d*- or conduction bands. As the Curie points for any stoichiometry in the iron and nickel series are dependent upon the rare earth partner it would seem that it is this latter exchange which is dominant in determining the transition temperature. Further, the increase in ordering temperatures above those for the yttrium compounds agrees reasonably well with the de Gennes factor so that the indirect exchange appears more probable. Bloch and Lemaire [249] have recently proposed a model in which the transition metal moment in the cobalt compounds originates from an exchange enhanced susceptibility and suggest the involvement of the 3*d*-electrons in the rare earth–rare earth interaction.

Two points require mentioning, however, in connection with the nature of the exchange interactions. First, in the iron series the Curie temperatures of compounds formed with a given rare earth element decrease with increasing iron concentration [250] while in the cobalt series increasing the fraction of transition metal raises the ordering temperature; see Fig. 6.19 (a and b). Second, in the cobalt-rich compounds, for which the cobalt ionic moment is comparable to that in cobalt metal, the Curie temperature becomes essentially independent of the rare earth ion involved and rapidly approaches the value for cobalt metal. It would seem then that in the cobalt series, the rare earth sublattice ceases to have a controlling effect on the magnetic behaviour of the compounds and the governing exchange mechanism is likely to be a direct interaction between the cobalt ions. The decrease in Curie temperature in the iron series, with increasing iron concentration is something of an anomaly and has been discussed by Buschow and van der Goot for the erbium compounds [232]. Since the moment of iron appears to be reasonably constant, at approximately 2 μ_B/ion, the fall in Curie temperature with increasing iron content can not be caused by a reduced iron moment, as appears to be the case for the cobalt compounds. Nor can it be associated with a decreasing exchange interaction due to the erbium ions, since a comparable behaviour is also found in the compounds formed with yttrium and lutetium. They propose, therefore, that the observed variation may arise from some other effects such as coordination number, interatomic distance or a change in the degree of localization of the 3*d*-electrons.

A simple model of the ferrimagnetic coupling has been proposed by Wallace [251] and follows from the known sign of the interaction between the *f* and *d* electrons of the two types of ion and the conduction electrons. Following the analysis of the work by Peter [170] and Jaccarino *et al* [171, 172] on the rare earth dialuminides, discussed in Section 6.4, we know that the conduction electron polarization due to the rare earth ion is negative with respect to the spin of the ion at the ion site and also negative at the aluminium site. Since, in the AB_2 compounds the interatomic AB and AA distances are comparable the polarization is then still negative at the neighbouring rare earth ions, so leading to a ferromagnetic rare earth sublattice. Assuming that in the AFe_2 and ACo_2 compounds the interaction of the *d*-electrons with the conduction electrons is positive, as it appears to be for other systems involving iron [252], then the spins of the iron and cobalt ions, which lie at sites equivalent to the aluminium ions in RAl_2, will be aligned antiparallel to the spins of the rare earth ions. Consequently, for those com-

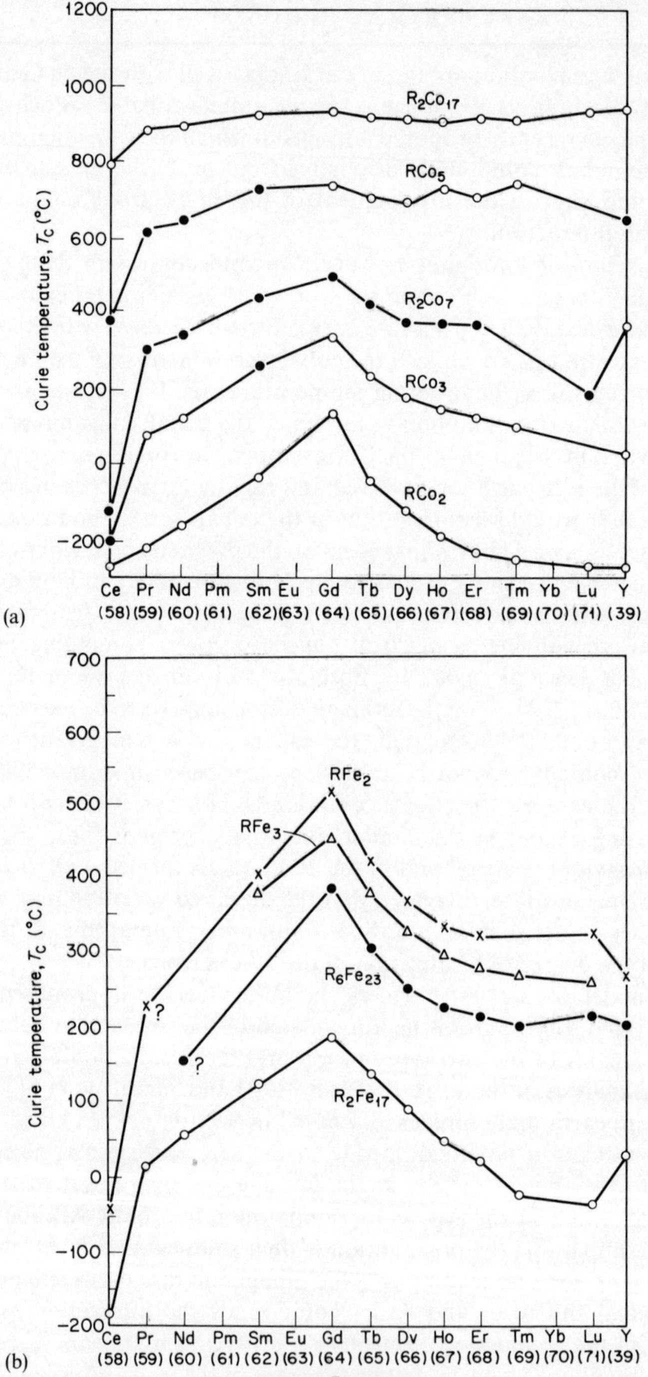

Fig. 6.19 Curie temperatures versus Rare Earth atomic number for (a) the A–Co and (b) the A–Fe series [250].

pounds formed between the rare earth ions from the second half of the series and iron or cobalt, ferrimagnetic coupling will occur, since for these lanthanides $J = L+S$, while for compounds formed with the light rare earths ferromagnetism will result as in this case $J = L-S$. A diagrammatic view of this situation is shown in Fig. 6.20 for both types of ion.

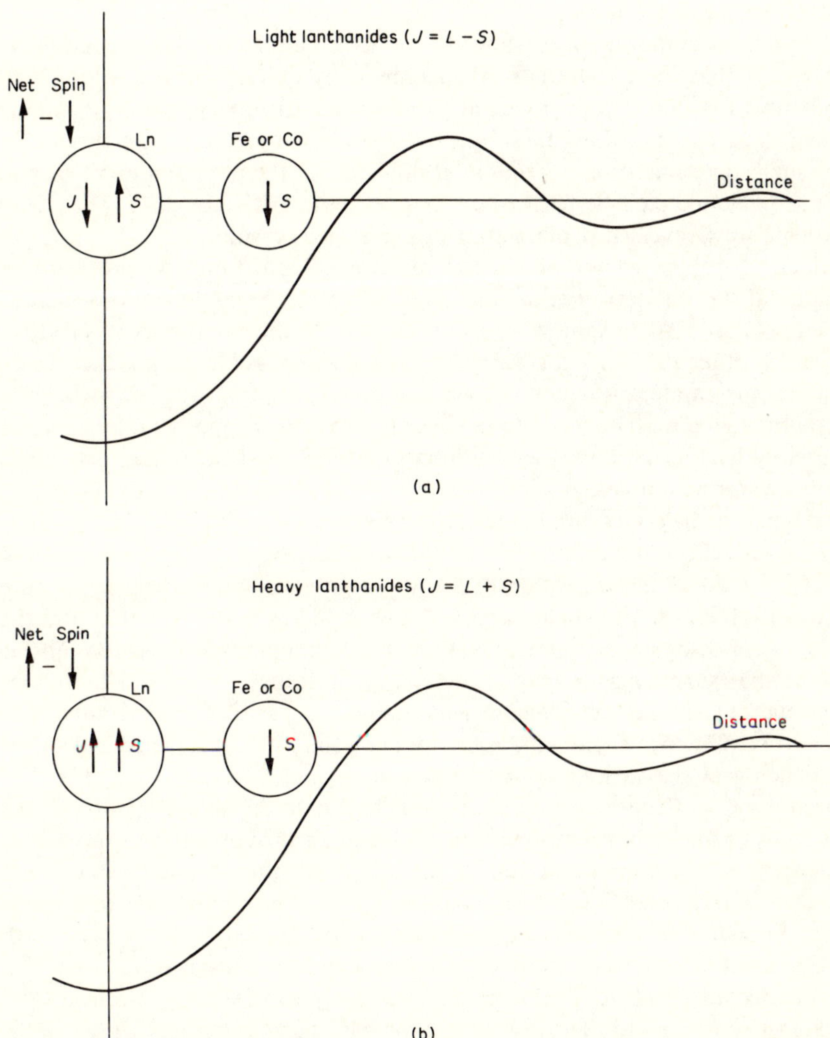

Fig. 6.20 A simplified representation of the interaction between the rare earth and transition metal ions which occurs through the polarization of the conduction electrons [251]. (a) For the light rare earth compounds. (b) For the heavy rare earth compounds.

It has been possible to determine the easy direction, in some of the compounds, directly both by magnetic measurement [253–255] and neutron diffraction studies [256–258], and indirectly through the observation of the Mössbauer spectra [259, 260]. In the cubic Laves phase AB_2 compounds with iron, Bowden [261] has shown that the observed easy directions are in agreement with those predicted theoretically on the assumption that the crystal field is the controlling factor in determining the direction of magnetization relative to the lattice.

Magnetocrystalline anisotropy measurements on the AB_5 compounds have revealed that these materials are magnetically extremely hard [262–265], having a first anisotropy constant greater than 10^7 erg/cc, the *c*-axis giving the lowest energy. Using these observations in conjunction with the earlier moment determinations, Strnat *et al* pointed out the potential of YCo_5 and $SmCo_5$ as fine particle permanent magnet materials [262, 264]. The initial calculations revealed a maximum possible energy storage of 28·1 MGOe, which is well in excess of normal materials. Unfortunately, problems of material stability have limited the value which can be obtained for practical magnet materials to somewhat less than this. By the simultaneous substitution of other elements for cobalt, it has been possible to produce more satisfactory materials which consist of a fine dispersion of RCo_5 phase in a non-magnetic matrix. Reviews of these permanent magnetic materials have recently been given by several authors [266–268] and the reader is referred to these for a complete bibliography.

Hyperfine field measurements in the AFe_2 compounds [223, 225, 261, 269] and spin echo observations in the cobalt compounds [270] show that the nuclear field at the transition metal ion is almost independent of the rare earth partner, the Fe^{57} field varying between 188 kOe and 232 kOe, and the Co^{59} field being approximately 60 kOe for all compounds measured. This is somewhat surprising in view of the change in ionic moments at both sites throughout the series. From measurements on pseudobinary compounds based on $GdCo_2$, Taylor and Christopher [160] have proposed a detailed balancing of the various contributions to the net hyperfine field. This has been done on the assumption that the field due to the cobalt ions (both the central and neighbouring ions) can be taken as proportional to the cobalt moment in the compounds under investigation, the absolute value being scaled from the known Co^{59} field in pure cobalt for which the ionic moment is also known. On this basis the contribution to the field from the rare earth ions, assumed to arise from conduction electron polarization, was also evaluated and leads to a value of approximately 10 kOe per $4f$ electron spin. This value is close to that observed at the Al^{27} nucleus in $GdAl_2$ [181–183]. The hyperfine field at the rare earth nucleus [185, 260, 269, 271–275] is usually close to the free ion and metal values discussed in Chapter 4, the differences generally being attributed to conduction electron polarization effects.

Limited studies have also been made of the electrical conductivity of these compounds [276–279] with some of the most interesting results being those on $CeCo_2$ in which superconductivity is observed below 0·84°K.

In addition to the cobalt- and nickel-rich compounds discussed above, magnetic studies have been made on the equatomic ANi compounds [280–282] and the rare earth-rich A_3Co and A_3Ni compounds [212, 213, 283–286]. In the first of these series the observed, room temperature, structure changes from the CrB type (LaNi to TbNi) to the FeB type (DyNi to TmNi) [282]. This change has been shown to be reflected in the magnetic structure with HoNi and ErNi possessing a non-collinear spin structure while in NdNi the magnetic moments are observed to be parallel [282]. The magnetic moments in both the ferromagnetic and paramagnetic states are listed in Table 6.8 along with the transition temperatures. The ordered moments are appreciably less than the free ion $g_J J$ values, and since the neutron diffraction observations suggest that nickel is again non-magnetic, this must be associated with crystal field quenching of the rare earth moment for ions other than gadolinium.

The A_3B compounds have been found to possess a rather varied magnetic behaviour. In static field measurements, Feron *et al* [283, 284] have reported the observation of ferromagnetism in Tm_3Ni, Pr_3Co, Tm_3Co and Er_3Co and metamagnetism in the remainder. The metamagnetic transitions are similar to those observed in the equiatomic rare earth–aluminium compounds (see Section 6.4) and do not persist to the magnetic ordering temperature [284, 213]. The values of the moments and transition temperatures of these compounds are listed in Tables 6.8 and 6.10. In addition to a metamagnetic transition on the initial magnetization at 4·2°K, the terbium compounds show an open hysteresis loop for subsequent reversals of the applied magnetic field. Using pulsed fields Taylor and Primavesi [212, 213] have observed critical field strengths for the metamagnetic transitions which are appreciably greater than those reported by Feron *et al* [282, 284], see Fig. 6.21. Further, Dy_3Co also displayed an open hysteresis loop after the initial magnetization. In their work, Taylor and Primavesi suggested that the differences may be due to a time-dependent magnetization associated with the onset of a high magnetocrystalline anisotropy at low temperatures. The competition between the applied field energy, the exchange interactions and the anisotropy were emphasized by Feron *et al* as being the origin of the complex spin structures which are observed for these compounds [286].

Manganese compounds have attracted appreciably less attention than those with the 3*d*-elements to the right of manganese in the periodic table. Phase diagram studies of the systems involving the heavy rare earths have been carried out by Kirchmayr [287] and some results obtained for the enthalpy of melting. These reveal that in the systems examined, there are three stable compounds with stoichiometries AMn_2, A_6Mn_{23} and AMn_{12}. Magnetization studies are rather limited for all but the A_6Mn_{23} composition,

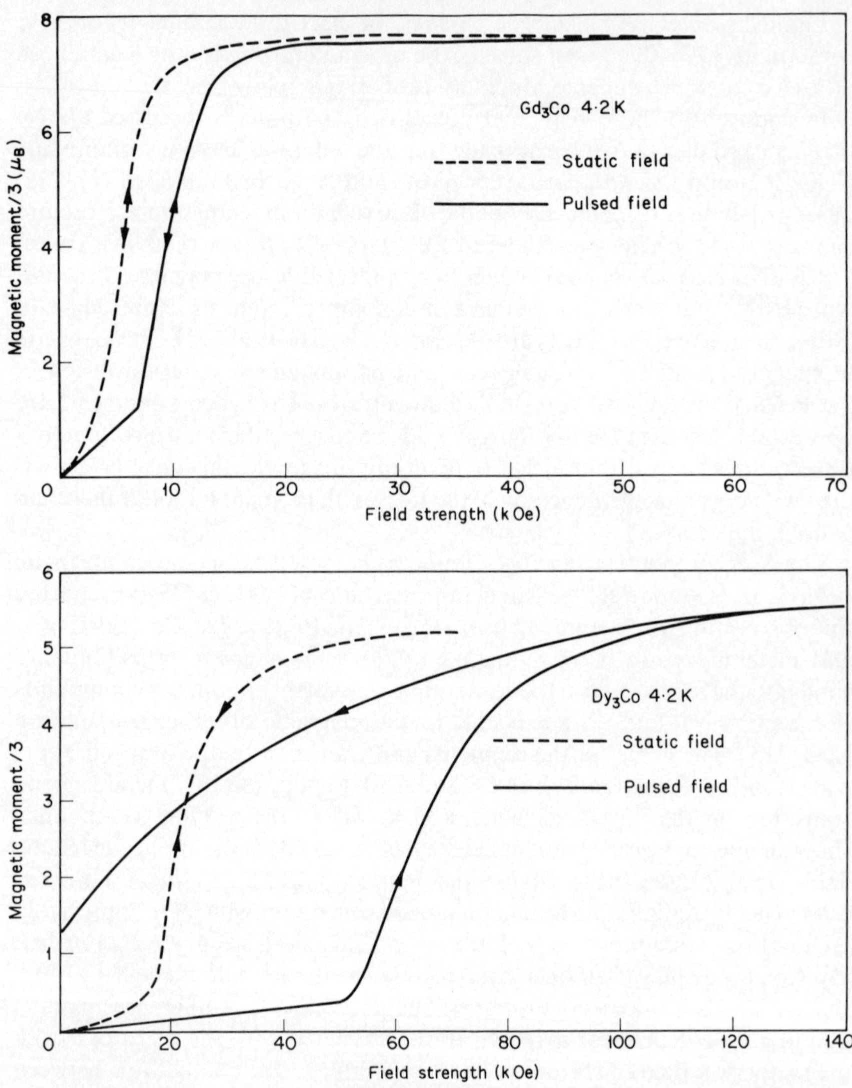

Fig. 6.21 A comparison of the isothermal magnetization curves for Gd_3Co and Dy_3Co obtained using static [282, 284] and 100μsec pulsed [212, 213] fields.

the published results having many omissions as indicated in Table 6.12 where some of the magnetic parameters are listed [228, 230, 287–292]. The interpretation of the observed results appears to be a little difficult as the magnetic structure changes from antiferromagnetic to ferrimagnetic with increasing field strength [288] and the magnetic moment of the manganese ion varies over a wide range of values. Kirchmayr [287] has summed up the behaviour in terms of a mutual interaction of different ferrimagnetic states,

Table 6.12 The magnetic transition temperatures and moment values of some rare earth manganese compounds [287].

		La	Ce	Pr	Nd	Sm	Eu	Gd	Tb	Dy	Ho	Er	Tm	Yb	Lu	Y
	θ				−90			45		−15		−90				
	μ_{eff}				4·9			8·86		10·9		10·1				
AMn_2	T_c							86								
	T_N															
	μ											7·8	5·44			
	θ					85		−90		25	−5	30				475
	μ_{eff}					5·04		10·04		11·47	11·83	10·56				2·35
A_6Mn_{23}	T_c					439		468		443	434	415				486
	T_N															
	μ					0·5		8·4		8·3	8·2	7·6				2·06
	θ							−25		−398	−65					
	μ_{eff}							17·2		16·3		13·2				
AMn_{12}	T_c															
	T_N									110						
	μ															

Table 6.13 The magnetic properties of the ordered phase of the Cubic Laves phase compounds formed between the rare earth elements and ruthenium, rheunium, osmium and iridium [294, 300].

		La	Ce	Pr	Nd	Sm	Eu	Gd	Tb	Dy	Ho	Er	Tm	Yb	Lu	Y
	θ															
	μ_{eff}															
A Ru_2	T_c			~38				~83								
	T_N															
	μ			0·61				6·35								
	θ															
	μ_{eff}															
A Rh_2	T_c			8·6	~6	~22		~75	~40	~35	~16	≃7				
	T_N															
	μ			2·24	1·67	0·53		6·80	7·07	8·16	7·74	7·26				
	θ															
	μ_{eff}															
A Os_2	T_c			~28	23			66	34	15	9	3				
	T_N															
	μ			1·46	1·34			6·71	7·31	6·73	6·0	5·31				
	θ															
	μ_{eff}															
A Ir_2	T_c			16	12	37		90	45	23	12	3	1	~0		
	T_N															
	μ			1·76	1·47	0·24		6·83	6·95	7·65	7·45	6·10	2·92	1·70		

but it is clear that there is still a considerable amount of work to be done in this area of the rare earth intermetallics.

6.5.2 *Compounds with elements of the other transition series*

Both ferromagnetic and superconducting behaviour have been found in a number of compounds with the 4*d*-(Ru, Rh) and 5*d*-(Os, Ir) metals [293–300]. The magnetic ordering temperatures are usually low, but are generally in accord with those to be expected from an indirect exchange. In most of the compounds the ordered molecular moments are significantly less than those to be expected from the free ions. This has been confirmed for $TbIr_2$ and $HoIr_2$ by neutron diffraction observations [301] and is assumed to arise from a combination of crystal field effects and conduction electron polarization. Table 6.13 gives some of the magnetic parameters of these compounds. Hyperfine field measurements at the Ir^{193} nucleus show that the nuclear

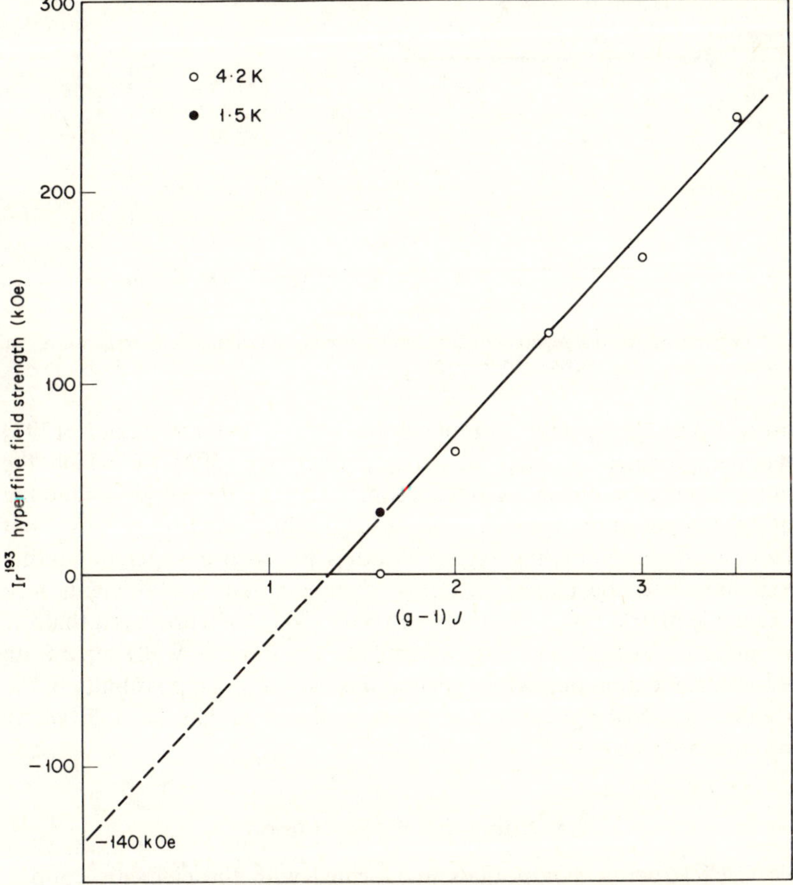

Fig. 6.22 Hyperfine fields at the Ir^{193} nucleus in the AIr_2 compounds.

field varies linearly with $(g_J - 1)J$, extrapolating to a value of $H = -140$ kOe at $(g-1)J = 0$ [338], see Fig. 6.22.

The pseudobinary system $CeRu_2$–$GdRu_2$, passes from superconducting materials at the cerium end of the system to ferromagnetism for the gadolinium compounds. Matthias *et al* [293] and Bozorth *et al* [294–296] showed that for some compositions in the vicinity of 7% $GdRu_2$ a range of overlap exists in which both superconductivity and ferromagnetism can be observed for a single material, see Fig. 6.23. Similar behaviour was also observed in the system $LaOs_2$–$GdOs_2$.

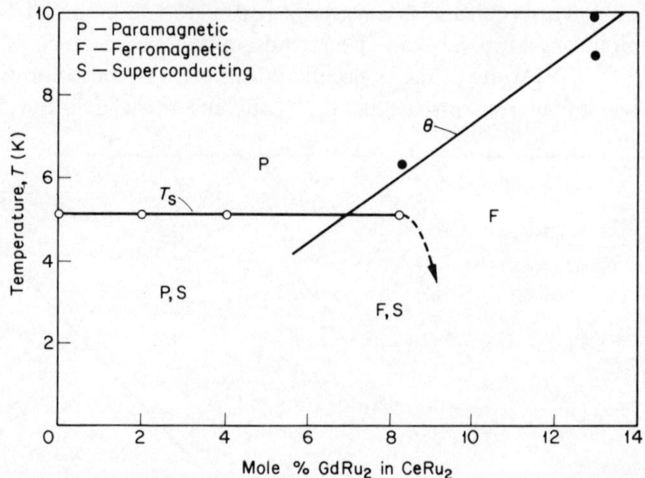

Fig. 6.23 Dependence of the superconducting and ferromagnetic transition temperatures on composition, in the system $GdRu_2$–$CeRu_2$ [295].

Observations on specific compositions in the overlap region [297] showed the presence of incipient superconductivity [302] for which the resistance decreases but does not become zero at the superconducting transition temperature. Further, the susceptibility remains positive but small in the superconducting region. Models for such a superconducting material have been developed in terms of superconducting filaments in a ferromagnetic matrix [303].Matthias and Suhl [304] have proposed that the superconducting regions in these systems are the domain walls separating the ferromagnetic domains. More general discussion of the possibility of the coexistence of ferromagnetism and superconductivity has been given by various authors [305–307].

6.6 Noble metal compounds

Simple CsCl structure compounds are formed with the elements copper, silver and gold, although the entire series is found only in conjunction with silver. With only two exceptions (CeAg and PrAg) investigation of the magnetic properties has shown that all the compounds are antiferromagnetic with Néel temperatures proportional to *G*. The paramagnetic Curie temperatures of the copper and silver compounds are negative [308–312] while those with gold reverse sign from +29°K for GdAu to −5°K for TmAu

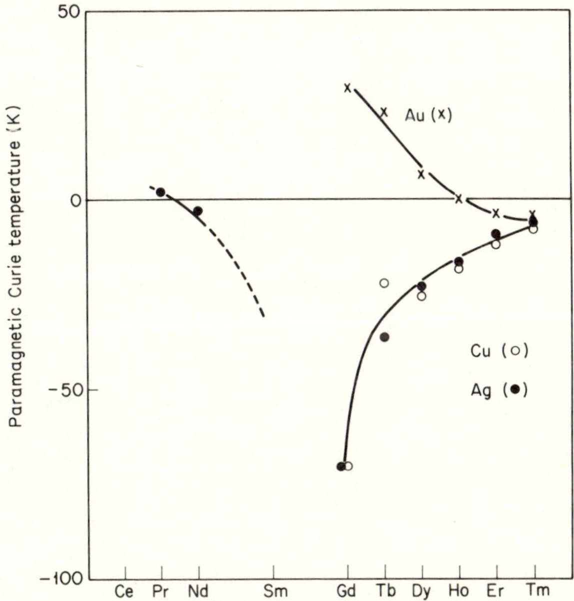

Fig. 6.24 The variation of the paramagnetic Curie temperature across the series for the copper, silver and gold compounds.

(see Fig. 6.24). The paramagnetic moments, listed in Tables 6.14–6.16, along with the transition temperatures of these materials are always rather high compared with the free ion value.

The results of the magnetostatic measurements and the nature of spin structure determined by neutron diffraction techniques [216, 217] have been interpreted in terms of the RKKY interaction. In conjunction with resistivity studies this has been used to obtain values of the exchange parameter A_0 and also various details of the electronic band structure. The RKKY theory was also used by de Wijn *et al* [312] to discuss the results of Knight shift measurements in the copper compounds. They found that the data could best be fitted by an exchange constant $A_0 \simeq 0{\cdot}2$ eV for all the compounds studied. Further the derived variation of A_0 with Fermi momentum was taken as indicating appreciable interband mixing effects of the type discussed by Watson *et al* [215].

Magnetic data is incomplete for both the A B_2 and A B_3 compounds with gold and silver, the published data relating to neutron diffraction studies of

Table 6.14 Some magnetic parameters of the rare earth–copper compounds [309, 311, 322, 323, 335].

		La	Ce	Pr	Nd	Sm	Eu	Gd	Tb	Dy	Ho	Er	Tm	Yb	Lu	Y
ACu	θ							−70	−22	−26 −18	−18 −13	−12 −14	−5 −8			
	μ_{eff}															
	T_c															
	T_N							140	116	64	28	15	10			
	μ															
ACu_2	θ															
	μ_{eff}			3·51	3·56		7·4	8·4	9·8	10·75	10·5	9·35	7·49			
	T_c															
	T_N							41	54	24	9	11				
	μ		0·8	2·3	1·9	0·1	5·8	6·0	7·4	8·7	9·2	5·6	4·2			
ACu_5	θ								2	2	$\simeq 0$	$\simeq 0$	$\simeq 0$			
	μ_{eff}								9·6	10·9	10·8	9·7	7·4			
	T_c															
	T_N								15	7						
	μ									6·5	7·6	7·3	4·9			

Table 6.15 Some magnetic parameters of the rare earth–silver compounds [308, 310, 318].

		La	Ce	Pr	Nd	Sm	Eu	Gd	Tb	Dy	Ho	Er	Tm	Yb	Lu	Y
A Ag	θ			2	−3			−70	−36	−23	−17	−9·5	−7			
	μ_{eff}			3·44	3·64			8·24	10·15	10·45	10·25	9·22	7·15			
	T_c															
	T_N				22			150	106	63	33	21	9·5			
	μ															
$A\,Ag_2$	θ															
	μ_{eff}															
	T_c															
	T_N								35							
	μ								8·95							
$A\,Ag_3$	θ															
	μ_{eff}															
	T_c															
	T_N															
	μ															

Table 6.16 Some magnetic parameters of the rare earth–gold compounds [314, 319].

		La	Ce	Pr	Nd	Sm	Eu	Gd	Tb	Dy	Ho	Er	Tm	Yb	Lu	Y
	θ							29	23	7	0	−4	−5			
	μ_{eff}							7·92	9·54	10·22	10·50	9·42	7·32			
A Au	T_c															
	T_N							31	40	14	10	13/19	8/19			
	μ															
	θ															
	μ_{eff}															
A Au_2	T_c															
	T_N									34						
	μ								9·0	9·2						

terbium, dysprosium and holmium compounds by Atoji [318–320].

The ACu_2 compounds, crystallizing in the orthorhombic, $CeCu_2$, structure [321] are antiferromagnetic [322], most of the compounds studied showing metamagnetic transitions at critical applied fields of up to 80 kOe at 4·2°K (Fig. 6.25). Saturation could not be obtained and consequently there is no

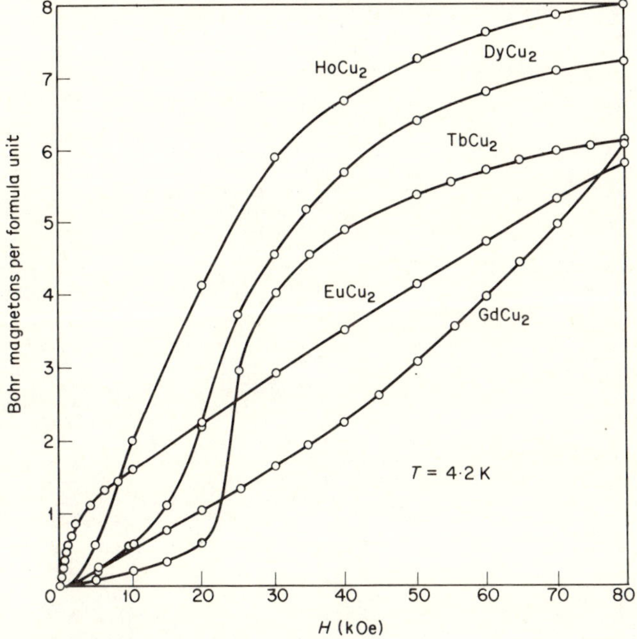

Fig. 6.25 Magnetization curves measured at 4·2°K for some of the ACu_2 compounds [322].

evidence of the magnitude of the ordered moment in these materials. The effective paramagnetic moment is again appreciably in excess of the free ion value as may be seen from Table 6.14 where the published results for these compounds are listed. These excessive moments appear to be characteristic of the noble metal compounds but as yet no explanation seems to have been given to account for their appearance.

Finally, the ACu_5 compounds, investigated by Buschow *et al* [323] show a continual change from antiferromagnetism in $TbCu_5$ to ferromagnetism in the compounds with holmium, erbium and thulium.

6.7 Compounds with elements of Group IV A

6.7.1 *Silicon and germanium*

The rare earth elements form compounds of basic formulae A_5B_4 and AB_2 with both silicon and germanium and additional compositions A_5B_3 and AB with germanium. Not all these compounds are formed with all members

Table 6.17 Some magnetic properties of the A_5Si_4 and ASi_2 compounds [324, 325].

		La	Ce	Pr	Nd	Sm	Eu	Gd	Tb	Dy	Ho	Er	Tm	Yb	Lu	Y
A_5Si_4	θ							349	216	233	69	20				
	μ_{eff}							8·15	9·31	10·3	11·1	9·9				
	T_c															
	T_N															
	μ							7·24	6·48	7·1	7·44	6·12				
$A\,Si_2$	θ															
	μ_{eff}							7·8	9·9	10·4	10·4	9·3				
	T_c															
	T_N							27	17	17	18					
	μ															

Table 6.18 The transition temperatures and effective paramagnetic moments of A_5Ge_4, A_5Ge_3, AGe and AGe_2 [324, 326–328].

		La	Ce	Pr	Nd	Sm	Eu	Gd	Tb	Dy	Ho	Er	Tm	Yb	Lu	Y
A_5Ge_3	θ		−50	10	20			70	93	45	10	35				
	μ_{eff}		2·66	3·80	3·73			8·42	9·98	10·47	11·10	9·48				
	T_c				45											
	T_N		<4·2	12				48	85	40	10	31				
	μ															
A_5Ge_4	θ							94	80	43	16	10				
	μ_{eff}							8·1	9·61	10·5	10·7	9·75				
	T_c															
	T_N							15	30	40	21	7				
	μ															
A Ge	θ		≃0	30	18			−13	−5	≃0	−5	−15				
	μ_{eff}		2·54	3·85	3·62			7·94	9·72	10·63	10·60	9·60				
	T_c		10	39	28	40		62	48	36	18	7				
	T_N															
	μ															
A Ge_2	θ			22	7			−54	−48	−26	−4					
	μ_{eff}			3·6	3·7			8·1	9·9	10·9	10·7					
	T_c			19	3·6											
	T_N							28	42	28	11					
	μ			2·18												

of the rare earth series, as may be seen from Tables 6.17 and 6.18, in which are listed some of their magnetic parameters.

The A_5B_4 silicon compounds are ferromagnetic, with Curie temperatures proportional to the de Gennes factor *G*, whereas those formed with germanium are antiferromagnetic [324] with a maximum Néel temperature of 40°K for Dy_5Ge_4. The susceptibilities show Curie–Weiss behaviour with effective moment values close to the free ion value. In the ordered silicon compounds the maximum molecular moment observed by Holtzberg *et al* [324] is appreciably less than that corresponding to five rare earth ions, a fact which may be associated with high magnetocrystalline anisotropy.

The ASi_2 and AGe_2 compounds have been investigated by Sekizawa *et al* [325, 326] who report antiferromagnetic behaviour for both series with the exception of $PrGe_2$ and $NdGe_2$, which are ferromagnetic with Curie temperatures of 19°K and 3·6°K respectively. The variation of the transition temperatures across the series deviates appreciably from that to be expected from the simple indirect exchange theory. Sekizawa *et al* [326] have attempted to account for this variation in terms of the changes in the radial extent of the 4*f* wavefunctions of the rare earth elements, and have shown that a correlation exists between this parameter and the type of magnetic ordering observed.

Buschow and Fast have investigated the A_5Ge_3 [327] and A Ge [328] series and find these compounds to be antiferromagnetic with the exceptions of those formed with lanthanum, which are Pauli paramagnetic and PrGe and NdGe which are ferromagnetic. Field-induced transitions to ferromagnetism occur for many of the A_5Ge_3 compounds (with A = Tb, Dy, Ho and Er) and for HoGe.

The variation of the transition temperatures of these materials across the rare earth series have been interpreted in terms of the RKKY theory, using a modified de Gennes factor *CG*, where *C* is defined as an interaction constant which may be assumed to depend on the nature of the 4*f* wavefunction [326]. Alternatively it may be treated as a variable in materials with more than one magnetic sublattice, by assuming that changes occur in the spin structure from one material to another. To date there is no neutron diffraction data with which to explore this possibility further.

6.7.2 *Tin*

Compounds having the Cu_3Au structure are formed between tin and the light rare earth metals. The transition temperatures and effective moments of these materials are given in Table 6.19 [329, 330]. The magnetostatic behaviour of these materials, which show antiferromagnetic ordering for the compounds with Pr, Nd and Sm, is not yet understood. The hyperfine field at the Sn nucleus has been observed to be 67 kOe in $PrSn_3$ [331] and attributed entirely to a transferred field, presumably through conduction electron polarization. Shenoy *et al* [331] point out that specimen deterioration occurs

Table 6.19 Some details of the magnetic properties of ASn_3 [329].

		La	Ce	Pr	Nd	Sm	Eu	Gd	Tb	Dy	Ho	Er	Tm	Yb	Lu	Y
	θ			−8	−22			−73								
	μ_{eff}		2·8	3·42	3·6			8·0								
A Sn_3	T_c		Not					Not								
	T_N		ordered	8·6	4·7	12		ordered								
	μ															

rapidly for these materials, liberating free tin which may be the cause of inconsistencies in the results of other authors. The experimental results, obtained using the Mössbauer effect, suggest only one third of all the nuclei see this hyperfine field and as a result the antiferromagnetic ordering is taken to be of the first kind. Ordering of the second kind would have placed equal numbers of neighbours with opposing spins about each Sn ion and consequently all the hyperfine fields would have been zero.

Experimental results for lead compounds have been reported for a few materials but there is insufficient data to be useful.

6.8 Some final comments

The materials discussed briefly in the previous pages have in general been pure compounds which are presently receiving, or have received in the past, a considerable amount of attention for one reason or another. In addition to these materials, there exists an even bigger and fast increasing literature on pseudobinary compounds formed between various series and used to provide evidence for the models developed to account for both structural and physical properties. In order to keep this present chapter within reasonable limits, little mention has been given to such work.

Magnetic and structural information exists for many other compounds and may be found by reference to the Bibliography of Magnetic Materials produced by the Research Materials Information Centre at the Oak Ridge National Laboratory [322] and to the data complications of Wallace [251, 333].

In general this information does not exist for all the compounds across the rare earth series and is frequently subject to uncertainty because of conflicting results. The discussions of the observed behaviour is usually in terms of the competing crystal field and exchange effects and follows much the same lines as for many of the materials discussed in this chapter.

In conclusion, the understanding of the physical properties of the rare earth compounds, particularly the intermetallic compounds, is still very far from complete. A great deal of careful experimental work is still to be done in obtaining a clear picture of the behaviour of many of the compounds. This includes the determination of the nature of the magnetic ordering and the magnetic interactions responsible for the ordering, the transitions in the ordering and the effects of the ordering on other physical properties. So far only a few families of compounds have received more than a superficial investigation, but there is probably enough of this preliminary information available to make the choice of worthwhile experiments somewhat easier.

In order to interpret these results there is still much to be learnt of the nature of electrostatic crystal field interactions in the metallic compounds and the detailed form of the exchange interaction. While it is generally accepted that this proceeds through the polarization of the conduction electrons in the well-known RKKY interaction little allowance is made for the anisotropy of this exchange or for realistic band structure details which can affect the interaction appreciably.

Many of these problems may take some considerable time to solve, but one must hope that we will soon be able to decide unambiguously details such as the signs of the magnetic *s–f* interactions and the ferromagnetism, antiferromagnetism or ferrimagnetism of compounds such as A_3Al_2, A_3Co and A_3Ni.

References

[1] J. A. Gibson and G. S. Harvey, Tech. Rept. AFML-TR-65-430 (1966).
[2] M. M. Schieber, *Experimental Magnetochemistry*, North-Holland (1967).
[3] I. Warshaw and R. Roy, *Prog. Sci. Tech. of Rare Earths*, Vol. 1, Pergamon (1964) p. 203.
[4] M. K. Wilkinson, W. C. Koehler, E. O. Wollan and J. W. Cable, *Bull. Am. Phys. Soc.*, **2**, 127 (1957).
[5] E. F. Westrum, *Rare Earth Research*, ed. E. V. Kleber, Macmillan (1961) p. 69.
[6] J. B. MacChesney, H. J. Williams, R. C. Sherwood and J. F. Potter, *J. Chem. Phys.*, **44**, 596 (1966).
[7] R. E. Brown and W. M. Hubbard, *Fifth Rare Earth Conf., Ames, Iowa*, Book 4 (unpublished) (1965) p. 31.
[8] H. Bonrath, K. H. Hellwege, K. Nicolay and G. Weber, *Phys. Kondens, Materie.*, **4**, 382 (1966).
[9] W. E. Henry, *Phys. Rev.*, **98**, 226 (1955).
[10] R. M. Moon, W. E. Koehler, H. R. Child and L. J. Raubenheimer, *Phys. Rev.*, **176**, 722 (1968).
[11] O. V. Lounasmaa and P. R. Roach, *Phys. Rev.*, **128**, 622 (1962).
[12] O. V. Lounasmaa and R. A. Guenther, *Phys. Rev.*, **126**, 1357 (1962).
[13] O. V. Lounasmaa, *Phys. Rev.*, **129**, 2460 (1963).
[14] E. F. Bertaut and R. Chevalier, *Compt. Rend.*, **262**, B1707 (1966).
[15] H. Hacker, M. S. Lin and E. F. Westrum, *Proc. Fourth Conf. on Rare Earth Research*, ed. LeRoy Eyring, Gordon and Breach (1965).
[16] F. Arajs and R. V. Colvin, *J. Appl. Phys.*, **33**, 2517 (1962).
[17] L. T. Crane, *J. Chem. Phys.*, **36**, 10 (1962).
[18] S. Kern, *J. Chem. Phys.*, **40**, 208 (1964).
[19] A. E. Miller, Ph.D. Thesis, Iowa State University, Ames, Iowa (1964).
[20] N. A. Smolkov and N. V. Dobrovol'skaya, *Izv. Akad. Nauk. SSSR. Neorg. Marerialy*, **1**, 1564 (1965).
[21] W. G. Penney, *Phys. Rev.*, **43**, 485 (1933).
[22] J. H. Van Vleck and W. G. Penney, *Phil. Mag.*, **17**, 961 (1934).
[23] D. Bloor, J. R. Dean and G. E. Stedman, *J. Appl. Phys.*, **41**, 1242 (1970).
[24] G. Schafer and S. Scheller, *Phys. Kondens, Materie.*, **5**, 48 (1966).
[25] J. B. MacChesney, H. J. Williams, R. C. Sherwood, *J. Chem. Phys.*, **41**, 3177 (1964).
[26] B. T. Matthias, R. M. Bozorth and J. H. Van Vleck, *Phys. Rev. Letts.*, **7**, 160 (1961).
[27] N. G. Nereson, C. E. Olsen and G. P. Arnold, *Phys. Rev.*, **127**, 2101 (1962).
[28] N. Miyata and B. E. Argyle, *Phys. Rev.*, **157**, 448 (1967).
[29] W. P. Wolf, *Phys. Rev.*, **108**, 1152 (1957).
[30] T. R. McGuire, B. E. Argyle, M. W. Shafer and J. S. Smart, *J. Appl. Phys.*, **34**, 1345 (1963).
[31] J. S. Smart, *Magnetism*, Vol. III, ed. G. T. Rado and H. Suhl, Academic Press (1963) p. 63.

[32] G. Busch, B. Natterer and H. R. Neukomm, *Phys. Letts.*, **23**, 190 (1966).
[33] J. H. Van Vleck, *Phys. Rev.*, **74**, 1168 (1948).
[34] P. W. Anderson and P. R. Weiss, *Rev. Mod. Phys.*, **25**, 269 (1953).
[35] R. Kubo and K. Tomita, *J. Phys. Soc. Japan*, **9**, 888 (1954).
[36] S. Methfessel and D. C. Mattis, *Handbuch der Physik*, ed. H. P. J. Wijn, Springer Verlag (1968) Vol 18(1), p. 389.
[37] G. K. Skolova, K. M. Demchuk, K. P. Rodionov and A. A. Samokhalov, *JETP* (USSR), **49**, 452 (1965).
[38] R. Stevenson and M. C. Robinson, *Can. J. Phys.*, **43**, 1744 (1965).
[39] D. B. McWhan, P. C. Sauers and G. Jura, *Phys. Rev.*, **143**, A385 (1966).
[40] P. Schwob and O. Vogt, *J. Appl. Phys.*, **40**, 1328 (1969).
[41] K. Y. Ahn and M. W. Shafer, *J. Appl. Phys.*, **41**, 1260 (1970).
[42] K. Lee and J. C. Suits, *J. Appl. Phys.*, **41**, 954 (1970).
[43] G. Busch and P. Wachter, *Phys. Kondens, Materie.*, **5**, 232 (1966).
[44] G. Busch, P. Junod and P. Wachter, *Phys. Letts.*, **12**, 11 (1964).
[45] S. Methfessel, M. J. Freiser, G. D. Pettit and J. C. Suits, *J. Appl. Phys.*, **38**, 1500 (1967).
[46] J. C. Suits, B. E. Argyle and M. J. Freiser, *Phys. Rev. Letts.*, **15**, 882 (1965).
[47] S. Methfessel, *Z. angew, Phys.*, **18**, 414 (1965).
[48] J. C. Suits, B. E. Argyle and M. J. Freiser, *J. Appl. Phys.*, **37**, 1391 (1966).
[49] S. Methfessel, F. Holtzberg and T. R. McGuire, *IEEE Trans. Magnetics*, **2**, 305 (1966).
[50] A. Yanase and T. Kasuya, *J. Appl. Phys.*, **39**, 430 (1968).
[51] T. Kasuya and A. Yanase, *J. Phys. Soc. Japan*, **25**, 1025 (1968).
[52] T. Kasuya and A. Yanase, *Rev. Mod. Phys.*, **40**, 684 (1968).
[53] J. Feinleib, W. J. Scouler, J. O. Dimmock, J. Hanus, T. B. Reed and C. R. Pidgeon, *Phys. Rev. Letts.*, **22**, 1385 (1969).
[54] J. O. Dimmock, J. Hanus and J. Feinleib, *J. Appl. Phys.*, **41**, 1088 (1970).
[55] R. Bachmann and P. Wachter, *Phys. Letts.*, **26A**, 478 (1968).
[56] R. Bachmann and P. Wachter, *J. Appl. Phys.*, **40**, 1326 (1969).
[57] F. Rys, J. S. Helman and W. Baltensperger, *Phys. Kondens, Materie.*, **6**, 105 (1967).
[58] S. J. Cho, *Phys. Rev.*, **157**, 632 (1967).
[59] C. Haas, *Solid State Comm.*, **5**, 657 (1967).
[60] K. Y. Ahn and J. C. Suits, *Trans. Mag. IEEE*, MAG-3, 453 (1967).
[61] J. H. Greiner and G. J. Fan, *Appl. Phys. Letts.*, **9**, 27 (1966).
[62] J. C. Suits, *J. Appl. Phys.*, **38**, 1498 (1967).
[63] V. L. Maruzzi, D. T. Teaney and C. F. Guerci, *Bull. Am. Phys. Soc.*, **12**, 133 (1967).
[64] A. J. Henderson, G. R. Brown, T. B. Reed and H. Meyer, *J. Appl. Phys.*, **41**, 946 (1970).
[65] S. H. Charap and E. L. Boyd, *Phys. Rev.*, **133**, A811 (1964).
[66] E. L. Boyd, *Phys. Rev.*, **145**, 174 (1966).
[67] H. H. Wickman and E. Catalano, *J. Appl. Phys.*, **39**, 1248 (1968).
[68] G. A. Uriano and R. L. Streever, *Phys. Letts.*, **17**, 205 (1965).
[69] L. Holmes and M. Schieber, *J. Appl. Phys.*, **37**, 968 (1966).
[70] G. Busch, P. Junod, M. Risi and O. Vogt, *Proc. Int. Conf. Semiconductors*, Exeter (1962) 727.
[71] T. R. McGuire, B. E. Argyle, M. W. Shafer and J. S. Smart, *Appl. Phys. Letts.*, **1**, 17 (1962).
[72] U. Enz, J. F. Fast, S. Van Houten and J. Smit, *Philips Res. Rept.*, **17**, 451 (1962).
[73] V. L. Maruzzi and D. T. Teaney, *Solid State Comm.*, **1**, 127 (1963).
[74] G. Busch, P. Junod, R. G. Morris and J. Munheim, *Helv. Phys. Acta.*, **37**, 637 (1964).
[75] S. J. Pickart and H. A. Alperin, *Bull. Am. Phys. Soc.*, **10**, 32 (1965).
[76] G. Will, S. J. Pickart, H. A. Alperin and R. Nathans, *J. Phys. Chem. Solids*, **24**, 1679 (1963).
[77] G. Busch, P. Junod, R. G. Morris, J. Munheim and W. Statius, *Phys. Letts.*, **11**, 9 (1964).
[78] R. J. Joenk, *Bull. Am. Phys. Soc.*, **11**, 109 (1966).
[79] P. Schwob and O. Vogt, *Phys. Letts.*, **22**, 374 (1966).

[80] G. Busch, *J. Appl. Phys.*, **38**, 1386 (1967).
[81] T. R. McGuire and M. W. Shafer, *J. Appl. Phys.*, **35**, 984 (1964).
[82] T. R. McGuire, M. W. Shafer and W. Palmer, *Proc. Int. Conf. Magnetism*, Nottingham (1964) p. 474.
[83] P. Schwob and O. Vogt, *Phys. Letts.*, **24A**, 242 (1967).
[84] D. Bloch, Ph.D. Thesis, University of Grenoble (1965).
[85] D. C. McCollum and J. Callaway, *Phys. Rev. Letts.*, **9**, 376 (1962).
[86] J. Callaway and D. C. McCollum, *Phys. Rev.*, **130**, 1741 (1963).
[87] S. H. Charap, *J. Appl. Phys.*, **35**, 988 (1964).
[88] G. Guntherodt, J. Schoenes and P. Wachter, *J. Appl. Phys.*, **41**, 1083 (1970).
[89] D. C. McCollum, R. L. Wild and J. Callaway, *Phys. Rev.*, **136**, A426 (1964).
[90] H. B. Callen and E. Callen, *Phys. Rev.*, **136**, A1675 (1964).
[91] G. G. Low, *Proc. Phys. Soc.*, **82**, 992 (1963).
[92] T. J. Wojtowicz, *J. Appl. Phys.*, **35**, 991 (1964).
[93] G. Busch, P. Junod and O. Vogt, *Thermodynamical, Physical and Structural Properties of Semi-Metallic Compounds*, CNRS, Paris (1967) p. 325.
[94] S. Methfessel, M. J. Freiser, G. D. Pettit and J. C. Suits, *J. Appl. Phys.*, **38**, 1500 (1967).
[95] J. C. Suits and B. E. Argyle, *Phys. Rev. Letts.*, **14**, 687 (1965).
[96] G. Busch and P. Wachter, *Phys. Letts.*, **20**, 617 (1966).
[97] A. M. de Graaf and S. Strassler, *Phys. Kondens, Materie.*, **1**, 13 (1963).
[98] A. M. de Graaf and R. M. Xavier, *Phys. Letts.*, **18**, 225 (1965).
[99] J. Danon and A. M. de Graaf, *J. Phys. Chem. Solids*, **27**, 1953 (1966).
[100] R. M. Xavier, *Phys. Letts.*, **25A**, 244 (1967).
[101] J. B. Goodenough, *Magnetism and the Chemical Bond*, J. Wiley, New York (1963) p. 149.
[102] F. Holtzberg, T. R. McGuire, S. Methfessel and J. C. Suits, *Phys. Rev. Letts.*, **13**, 18 (1964).
[103] F. Holtzberg, T. R. McGuire, S. Methfessel and J. C. Suits, *Proc. Int. Conf. Magnetism*, Nottingham (1964) p. 470.
[104] S. Methfessel, *Z. angew Phys.*, **18**, 414 (1965).
[105] F. Holtzberg, T. R. McGuire and S. Methfessel, *J. Appl. Phys.*, **37**, 976 (1966).
[106] D. C. Mattis, N. Anthony and L. Horowitz, *I.B.M. Report*, RC 945 (1963).
[107] S. von Molnar and S. Methfessel, *J. Appl. Phys.*, **38**, 959 (1967).
[108] S. von Molnar and M. W. Shafer, *J. Appl. Phys.*, **41**, 1093 (1970).
[109] S. von Molnar and T. Kasuya, *Phys. Rev. Letts.*, **21**, 1758 (1968).
[110] P. G. de Gennes and J. Friedel, *J. Phys. Chem. Solids*, **4**, 71 (1958).
[111] R. Didchenko and F. P. Gortsema, *J. Phys. Chem. Solids*, **24**, 863 (1963).
[112] J. F. Miller, F. J. Reid and R. C. Heimes, *J. Electrochem. Soc.*, **106**, 1043 (1959).
[113] J. F. Miller, F. J. Reid and R. C. Heimes, *J. Phys. Chem. Solids*, **25**, 969 (1964).
[114] J. W. McClure, *J. Phys. Chem. Solids*, **24**, 871 (1963).
[115] M. D. Houston, *Ceramic Age* (1961) p. 49.
[116] D. J. Haase and H. Steinfink, *J. Appl. Phys.*, **36**, 3490 (1965).
[117] D. J. Haase and H. Steinfink, *J. Appl. Phys.*, **37**, 2246 (1966).
[118] R. C. Vickery and H. M. Muir, *Rare Earth Research*, ed. E. V. Kleber, MacMillam (1961) p. 223.
[119] M. D. Houston, *ibid*, p. 255.
[120] V. P. Zhuze, A. V. Golubkov, E. V. Goncharova and V. M. Sergeeva, *Sov. Phys. Solid State*, **6**, 208, 213, 343 (1965).
[121] M. C. Picon and M. Palric, *Compt. Rend.*, **242**, 1321 (1956).
[122] A. Benacerraf, L. Domague and J. Flahaut, *Compt. Rend.*, **248**, 1672 (1959).
[123] L. Domague, J. Flahaut and M. Guittard, *Compt. Rend.*, **249**, 697 (1959).
[124] A. Iandelli, *Rare Earth Research*, ed. E. V. Kleber, MacMillan (1961) p. 135.
[125] G. A. Smolenskii, V. E. Adamyan and G. M. Loginov, *Phys. Letts.*, **23**, 16 (1966).
[126] G. A. Smolenskii, V. E. Adamyan and G. M. Loginov, *J. Appl. Phys.*, **39**, 786 (1968).
[127] G. A. Smolenskii, V. E. Adamyan and G. M. Loginov, *Phys. Stat. Sol.*, **18**, 873 (1966).

[128] F. Holtzberg, T. R. McGuire and S. Methfessel, *J. Appl. Phys.*, **35**, 1033 (1964).
[129] F. Holtzberg and S. Methfessel, *J. Appl. Phys.*, **37**, 1433 (1966).
[130] M. Cutler and J. F. Leavy, *Phys. Rev.*, **133**, A1153 (1964).
[131] M. Cutler, J. F. Leavy and R. L. Fitzpatrick, *Phys. Rev.*, **133**, A1143 (1964).
[132] J. R. Henderson, M. Moramoto, E. Loh and J. C. Gruber, *J. Chem. Phys.*, **47**, 3347 (1967).
[133] G. V. Lashkarev and A. V. Savitskii, *Sov. Phys. Solid State*, **9**, 1485 (1968).
[134] H. R. Child, M. K. Wilkinson, J. W. Cable, W. C. Koehler and E. O. Wollan, *Phys. Rev.*, **131**, 922 (1963).
[135] G. Busch, P. Junod, O. Vogt and F. Hulliger, *Phys. Letts.*, **6**, 79 (1963).
[136] N. Sclar, *J. Appl. Phys.*, **35**, 1534 (1964).
[137] D. P. Schumacher and W. E. Wallace, *J. Appl. Phys.*, **36**, 984 (1965).
[138] G. Busch, P. Schwob and O. Vogt, *Phys. Letts.*, **23**, 636 (1966).
[139] F. Levy and O. Vogt, *Phys. Letts.*, **24A**, 444 (1967).
[140] M. B. Allenson and K. N. R. Taylor, *J. Appl. Phys.*, **39**, 1094 (1968).
[141] T. Tsuchida, Y. Nakamura and T. Kaneko, *J. Phys. Soc. Japan*, **26**, 284 (1969).
[142] E. D. Jones, *Phys. Rev.*, **180**, 455 (1969).
[143] T. R. McGuire, R. J. Gambino, S. J. Pickart and H. A. Alperin, *J. Appl. Phys.*, **40**, 1009 (1969).
[144] R. J. Gambino, T. R. McGuire, H. A. Alperin and S. J. Pickart, *J. Appl. Phys.*, **41**, 933 (1970).
[145] G. T. Trammell, *J. Appl. Phys.*, **31**, 3625 (1960).
[146] G. T. Trammell, *Phys. Rev.*, **131**, 932 (1963).
[147] P. Junod and A. Menth, *Phys. Letts.*, **25A**, 602 (1967).
[148] G. Busch and O. Vogt, *Phys. Letts.*, **25A**, 449 (1967).
[149] B. R. Cooper, *Phys. Letts.*, **22**, 24 (1966).
[150] B. R. Cooper, *Phys. Letts.*, **22**, 244 (1966).
[151] B. R. Cooper, I. S. Jacobs, R. C. Fedder, J. S. Kouvel and D. P. Schumacher, *J. Appl. Phys.*, **37**, 1384 (1966).
[152] B. R. Cooper, R. C. Fedder and D. P. Schumacher, *Phys. Rev.*, **163**, 506 (1967).
[153] B. V. Karpenko and A. A. Berdyshev, *Sov. Phys. Solid State*, **5**, 2494 (1964).
[154] A. A. Berdyshev, *Sov. Phys. Solid State*, **8**, 1104 (1966).
[155] R. Didchenko and F. P. Gortsema, *J. Phys. Chem. Solids*, **24**, 863 (1963).
[156] J. J. Veysie and F. Anselin, *Colloq. Int. CNRS sur les derives Semi-metallique*, Orsay (1965) p. 349.
[157] A. C. Switendick and E. D. Jones, *Bull. Am. Phys. Soc.*, **13**, 365 (1968).
[158] N. Sclar, *J. Appl. Phys.*, **33**, 2999 (1962).
[159] E. D. Jones, *Phys. Rev. Letts.*, **19**, 432 (1967).
[160] M. I. Darby and K. N. R. Taylor, *Phys. Letts.*, **14**, 179 (1965).
[161] H. K. J. Buschow and J. H. N. Van Vucht, *Philips Res. Rep.*, **22**, 233 (1967).
[162] H. J. Williams, J. H. Wernick, E. A. Nesbitt and R. C. Sherwood, *J. Phys. Soc. Japan*, **17**, Suppl. B1, 91 (1962).
[163] J. A. White, H. J. Williams, J. H. Wernick and R. C. Sherwood, *Phys. Rev.*, **131**, 1039 (1963).
[164] C. E. Olsen, G. Arnold and N. Nereson, *J. Appl. Phys.*, **38**, 1395 (1967).
[165] M. J. McDermott and K. K. Marklund, *J. Appl. Phys.*, **40**, 1007 (1969).
[166] K. H. J. Buschow, A. M. van Diepen and H. W. de Wijn, *Phys. Letts.*, **24A**, 536 (1967).
[167] H. W. de Wijn, A. M. van Diepen and K. H. J. Buschow, *Phys. Rev.*, **161**, 253 (1967).
[168] I. Nowik, S. Ofer and J. H. Wernick, *Phys. Letts.*, **20**, 232 (1966).
[169] W. Wiedemann and W. Zinn, *Phys. Letts.*, **24A**, 506 (1967).
[170] M. Peter, *J. Appl. Phys.*, **32**, 338 (1961)
[171] V. Jaccarino, B. T. Matthias, M. Peter, H. Suhl and J. H. Wernick, *Phys. Rev. Letts.*, **5**, 251 (1960).
[172] V. Jaccarino, *J. Appl. Phys.*, **32**, 102 (1961).

[173] K. H. J. Buschow, J. F. Fast, A. M. van Diepen and H. W. de Wijn, *Phys. Stat. Sol.*, **24**, 715 (1967).
[174] K. Yosida, *Phys. Rev.*, **106**, 893 (1957).
[175] E. D. Jones and J. I. Budnick, *J. Appl. Phys.*, **37**, 1250 (1966).
[176] R. G. Barnes and E. D. Jones, *Solid State Comm.*, **5**, 285 (1967).
[177] H. J. van Daal and K. H. J. Buschow, *Solid State Comm.*, **7**, 217 (1969).
[178] H. J. van Daal and K. H. J. Buschow, *Phys. Rev. Letts.*, **23**, 408 (1969).
[179] M. B. Maple, *Solid State Comm.*, **8**, 1915 (1970).
[180] B. R. Coles, D. Griffiths, R. J. Lown and R. H. Taylor, *J. Phys. C.*, **3**, L121 (1970).
[181] F. Dintelmann, E. Dormann and K. H. J. Buschow, *Solid State Comm.*, **8**, 1911 (1970).
[182] J. I. Budnick, R. E. Gegenworth and H. Wernick, *Bull. Am. Phys. Soc.*, **10**, 317 (1965).
[183] N. Shamir, N. Kaplan and J. H. Wernick, *J. Phys.* (*Paris*), **32**, C1-902 (1971).
[184] G. Brown, M. A. A. Issa and K. N. R. Taylor, unpublished work.
[185] S. Ofer, M. Rakavy, E. Segal and B. Khurgin, *Phys. Rev.*, **138**, A241 (1965).
[186] P. M. Levy, *J. Appl. Phys.*, **41**, 902 (1970).
[187] K. H. J. Buschow and J. F. Fast, *Z. fur Phys. Chem. Neue Folge*, **50**, 1 (1966).
[188] B. Bleaney, *Proc. Roy. Soc.*, **A276**, 28 (1963).
[189] H. W. de Wijn, A. M. van Diepen and K. H. J. Buschow, *Phys. Rev.*, **B1**, 4203 (1970).
[190] A. M. van Diepen, H. W. de Wijn and K. H. J. Buschow, *J. Chem. Phys.*, **47**, 3489 (1967).
[191] H. W. de Wijn, A. M. van Diepen and K. H. J. Buschow, *Phys. Rev.*, **161**, 253 (1967).
[192] A. M. van Diepen, H. W. de Wijn and K. H. J. Buschow, *Phys. Letts.*, **26A**, 340 (1968).
[193] Private communication.
[194] H. W. de Wijn, K. H. J. Buschow and A. M. van Diepen, *Phys. Stat. Sol.*, **30**, 759 (1968).
[195] C. Becle and R. Lemaire, *Compt. Rend.*, **264**, 543 (1967).
[196] C. Becle and R. Lemaire, *Acta Cryst.*, **23**, 840 (1967).
[197] K. H. J. Buschow, *J. Less Comm. Metals*, **8**, 209 (1965).
[198] F. Kissel and W. E. Wallace, *J. Less Comm. Metals*, **11**, 417 (1966).
[199] B. Barbara, C. Becle, R. Lemaire and R. Pauthenet, *J. Appl. Phys.*, **39**, 1084 (1968).
[200] C. Becle, R. Lemaire and R. Pauthenet, *Compt. Rend.*, **266**, 994 (1968).
[201] C. Becle and R. Lemaire, *Phys. Letts.*, **27A**, 541 (1968).
[202] C. Becle, R. Lemaire and D. Paccard, *J. Appl. Phys.*, **41**, 855 (1970).
[203] C. Becle and R. Lemaire, *Les Elements des Terres Rares* II, CNRS (1969) p. 223.
[204] C. Becle, R. Lemaire and E. Parthe, *Solid State Comm.*, **6**, 115 (1968).
[205] N. V. Nhung, J. Sivardiere and A. Apostolov, *Les Elements des Terres Rares* II, CNRS (1969), p. 261.
[206] A. Blandin, *ibid.*, p. 505.
[207] B. Barbara, C. Becle, R. Lemaire and R. Pauthenet, *Compt. Rend.*, **267**, 309 (1968).
[208] B. Barbara, C. Becle, J. L. Feron, R. Lemaire and R. Pauthenet, *Compt. Rend.*, **267**, 244 (1968).
[209] K. H. J. Buschow, *Phys. Letts.*, **29A**, 12 (1969).
[210] B. Barbara, C. Becle, R. Lemaire and D. Paccard, *Compt· Rend.*, **271**, 880 (1970).
[211] K. N. R. Taylor and G. J. Primavesi, submitted to *J. Phys.*
[212] K. N. R. Taylor and G. J. Primavesi, submitted for publication.
[213] K. N. R. Taylor, submitted for publication.
[214] A. M. van Diepen, H. W. de Wijn and K. H. J. Buschow, *Phys. Stat. Sol.*, **29**, 189 (1968).
[215] R. E. Watson, S. Koide, M. Peter and A. J. Freeman, *Phys. Rev.*, **139**, A167 (1965).
[216] E. A. Nesbitt, J. H. Wernick and E. Corenzwit, *J. Appl. Phys.*, **30**, 265 (1959).
[217] W. H. Hubbard, E. Adams and J. V. Gilfrich, *J. Appl. Phys.*, **31**, 368S (1960).
[218] K. Nassau, L. V. Cherry and W. E. Wallace, *J. Phys. Chem. Solids*, **16**, 123 (1960).
[219] K. Nassau, L. V. Cherry and W. E. Wallace, *J. Phys. Chem. Solids*, **16**, 131 (1960).
[220] R. C. Vickery, W. C. Sexton, V. Novy and E. V. Kleber, *J. Appl. Phys.*, **31**, 366S (1960).
[221] W. H. Hubbard and E. Adams, *J. Phys. Soc. Japan*, **17**, Suppl. B1, 143 (1962).

[222] E. A. Nesbitt, H. J. Williams, J. H. Werwick and R. C. Sherwood, *J. Appl. Phys.*, **33**, 1674 (1962).
[223] W. E. Wallace and E. A. Skrabek, *Rare Earth Research*, II, ed. K. S. Vorres, Gordon and Breach (1963) p. 431.
[224] J. W. Ross and J. Crangle, *Phys. Rev.*, **133**, A509 (1964).
[225] W. E. Wallace, *J. Chem. Phys.*, **41**, 3857 (1964).
[226] J. Farrell and W. E. Wallace, *Inorg. Chem.*, **5**, 105 (1966).
[227] H. Kirchmayr and K. H. Schindle, *Z. angew Phys.*, **19**, 517 (1965).
[228] H. Kirchmayr, *IEEE Trans. Magnetics*, **MAG-2**, 493 (1966).
[229] K. Strnat, G. Hoffer, W. Ostertag and W. C. Olson, *J. Appl. Phys.*, **37**, 1252 (1966).
[230] B. F. de Savage, R. M. Bozorth, F. E. Wang, *J. Appl. Phys.*, **36**, 992 (1965).
[231] K. Strnat, G. Hoffer and A. E. Ray, *IEEE Trans. Magnetics*, **MAG-2**, 489 (1966).
[232] K. H. J. Buschow and A. S. van der Goot, *Phys. Stat. Solids*, **35**, 515 (1969).
[233] A. R. Piercy and K. N. R. Taylor, *J. Appl. Phys.*, **39**, 1096 (1968).
[234] R. Lemaire, *Cobalt*, **32**, 132 (1966).
[235] R. Lemaire, *Cobalt*, **33**, 201 (1966).
[236] E. A. Skrabek and W. E. Wallace, *J. Appl. Phys.*, **34**, 1356 (1963).
[237] S. C. Abrahams, J. L. Bernstein, R. C. Sherwood, *J. Phys. Chem. Solids*, **25**, 1069 (1964).
[238] R. Walline and W. E. Wallace, *J. Chem. Phys.*, **41**, 1587 (1964).
[239] C. Becle, R. Lemaire and D. Paccard, *J. Appl. Phys.*, **41**, 855 (1970).
[240] D. Paccard and R. Pauthenet, *Compt. Rend.*, **264**, 1956 (1967).
[241] R. Lemaire, D. Paccard and R. Pauthenet, *Compt. Rend.*, **265**, 1280 (1967).
[242] J. La Forest, R. Lemaire, D. Paccard and R. Pauthenet, *Compt. Rend.*, **264**, 676 (1967).
[243] R. Lemaire, D. Paccard, R. Pauthenet and J. Schweizer, *J. Appl. Phys.*, **39**, 1092 (1968).
[244] M. Dixon, A. Aoyagi, R. S. Craig and W. E. Wallace, to be published.
[245] K. N. R. Taylor, *Phys. Letts.*, **29A**, 372 (1969).
[246] B. Bleaney, *Rare Earth Research*, II, ed. K. S. Vorres, Gordon and Breach (1963), p. 499.
[247] A. R. Piercy and K. N. R. Taylor, *J. Phys. C.*, **1**, 1112 (1968).
[248] M. I. Slanicka and K. N. R. Taylor, to be published.
[249] D. Bloch and R. Lemaire, *Phys. Rev.* (1970) to be published.
[250] L. R. Salmans, K. Strnat and G. I. Hoffer, *Tech. Rep.*, AFML-TF-68-159 (1968).
[251] W. E. Wallace, *Prog. Sci. Tech. of Rare Earths*, Pergamon (1968) Vol. 3, p. 1.
[252] M. B. Stearns, *J. Appl. Phys.*, **36**, 913 (1965).
[253] G. I. Hoffer and K. Strnat, *IEEE Trans. Mag.*, **2**, 487 (1966).
[254] K. Strnat, G. I. Hoffer, J. Olson, W. Ostertag and J. J. Becker, *J. Appl. Phys.*, **38**, 1001 (1967).
[255] W. A. J. J. Velge and K. H. J. Buschow, *Z. angew. Phys.*, **26**, 157 (1969).
[256] H. Bartholin, B. van Laar, R. Lemaire and J. Schweizer, *J. Phys. Chem. Solids*, **27**, 1287 (1966).
[257] R. Lemaire and J. Schweizer, *J. Phys., Paris*, **28**, 216 (1967).
[258] E. Kren, J. Schweizer and F. Tasset, *Phys. Rev.*, **186**, 479 (1969).
[259] G. K. Wertheim, V. Jaccarino and J. H. Wernick, *Phys. Rev.*, **136**, A151 (1964).
[260] G. J. Bowden, D. St. P. Bunbury, A. P. Guimaraes and R. E. Snyder, *J. Phys. C.*, **1**, 1376 (1968).
[261] G. J. Bowden, Ph.D. Thesis, Manchester (1967).
[262] H. Bartholus, B. van Laar, R. Lemaire and J. Schweizer, *J. Phys. Chem. Solids*, **27**, 1287 (1966).
[263] G. Hoffer and K. J. Strnat, *J. Appl. Phys.*, **38**, 1377 (1967).
[264] W. A. J. J. Velge and K. H. J. Buschow, *J. Appl. Phys.*, **39**, 1717 (1968).
[265] K. H. J. Buschow and W. A. J. J. Velge, *Z. angew. Phys.*, **26**, 157 (1969).
[266] E. A. Nesbitt, *J. Appl. Phys.*, **40**, 1259 (1969).
[267] J. J. Becker, *J. Appl. Phys.*, **41**, 1055 (1970).
[268] K. J. Strnat, *IEEE Trans. Mag.*, **6**, 182 (1970).

[269] R. E. Gegenworth, J. I. Budnick, S. Skalski and J. H. Wernick, *J. Appl. Phys.*, **37**, 1244 (1966).
[270] K. N. R. Taylor and J. T. Christopher, *J. Phys. C.*, **2**, 2237 (1969).
[271] S. Ofer and I. Nowik, *Nucl Phys.*, **A93**, 689 (1967).
[272] U. Aztomy, E. R. Bauminger and S. Ofer, *Nucl. Phys.*, **89**, 433 (1966).
[273] R. L. Cohen and J. H. Wernick, *Phys. Rev.*, **134**, B503 (1964).
[274] R. L. Cohen, *Phys. Rev.*, **134**, A94 (1964).
[275] A. Guimaraes, Ph.D. Thesis, Manchester (1971).
[276] T. F. Smith and I. R. Harris, *J. Phys. Chem. Solids*, **28**, 1846 (1967).
[277] H. L. Luo, M. B. Maple, I. R. Harris and T. F. Smith, *Phys. Letts.*, **27A**, 519 (1968).
[278] M. P. Kawatra, S. Shalski, J. A. Mydosh and J. I. Budnick, *Phys. Rev. Letts.*, **23**, 83 (1969).
[279] M. P. Kawatra, J. A. Mydosh and J. I. Budnick, *Phys. Rev.*, **B2**, 605 (1970).
[280] S. C. Abrahams, J. L. Bernstein, R. C. Sherwood, J. H. Wernick and H. J. Williams, *J. Phys. Chem. Solids*, **25**, 1069 (1964).
[281] R. E. Walline and W. E. Wallace, *J. Chem. Phys.*, **41**, 1587 (1964).
[282] R. Lemaire and D. Paccard, *Les Elements des Terres Rares* II, CNRS (1969), p. 231.
[283] J. L. Feron, R. Lemaire, D. Paccard and R. Pauthenet, *Compt. Rend.*, **267B**, 371 (1968).
[284] J. L. Feron, D. Gignoux, R. Lemaire and D. Paccard, *Les Elements des Terres Rares* II, CNRS (1969) p. 75.
[285] O. A. Strydom and L. Alberts, *J. Less Comm. Metals*, **22**, 503 (1970).
[286] D. Gignoux, R. Lemaire and D. Paccard, *Sol. State Comm.*, **8**, 391 (1970).
[287] H. R. Kirchmayr and F. Lihl, *Tech. Rept. AFML*, TR-66-366 (1966).
[288] E. A. Nesbitt, H. J. Williams, J. H. Wernick and R. C. Sherwood, *J. Appl. Phys.*, **34**, 1347 (1963).
[289] L. M. Corliss and J. M. Hastings, *J. Appl. Phys.*, **35**, 1051 (1964).
[290] G. Felcher, L. M. Corliss and J. M. Hastings, *J. Appl. Phys.*, **36**, 1001 (1965).
[291] H. R. Kirchmayr, Ph.D. Thesis, Vienna (1968).
[292] A. Dwarak, H. R. Kirchmayr and H. Rauch, *Z. angew. Phys.*, **24**, 318 (1968).
[293] B. T. Matthias, H. Suhl and E. Corenzwit, *Phys. Rev. Letts.*, **1**, 449 (1958).
[294] R. M. Bozorth, B. T. Matthias, H. Suhl, E. Corenzwit and D. D. Davis, *Phys. Rev.*, **115**, 1595 (1959).
[295] R. M. Bozorth, D. D. Davis and A. J. Williams, *Phys. Rev.*, **119**, 1570 (1960).
[296] R. M. Bozorth and D. D. Davis, *J. Appl. Phys.*, **31**, 3215 (1960).
[297] J. J. Drautman, C. J. Anderson and R. Del Grosso, *J. Appl. Phys.*, **35**, 974 (1964).
[298] B. T. Matthias, T. H. Geballe and V. B. Compton, *Rev. Mod. Phys.*, **35**, 1 (1963).
[299] A. E. Berkowitz, F. Holtzberg and S. Methfessel, *J. Appl. Phys.*, **35**, 1030 (1964).
[300] J. Crangle and J. W. Ross, *Proc. Int. Conf. Magnetism*, Nottingham (1964) p. 240.
[301] G. P. Felcher and W. C. Koehler, *Phys. Rev.*, **131**, 1518 (1963).
[302] R. T. Webber and J. M. Reynolds, *Phys. Rev.*, **73**, 640 (1948).
[303] H. Koppe, *Ann. Physik.*, **6**, 375 (1949).
[304] B. T. Matthias and H. Suhl, *Phys. Rev. Letts.*, **4**, 51 (1960).
[305] H. Suhl, B. T. Matthias and E. Corenzwit, *J. Phys. Chem. Solids*, **11**, 346 (1959).
[306] L. P. Gorkov and A. I. Rusinov, *Sov. Phys. JETP*, **19**, 922 (1964).
[307] M. A. Jensen and H. Suhl, *Magnetism*, ed. G. T. Rado and H. Suhl, Academic Press (1966) **IIB**, p. 183.
[308] R. E. Walline and W. E. Wallace, *J. Chem. Phys.*, **41**, 3285 (1964).
[309] R. E. Walline and W. E. Wallace, *J. Chem. Phys.*, **42**, 604 (1965).
[310] J. Pierre and R. Pauthenet, *Compt. Rend.*, **260**, 2739 (1965).
[311] J. Pierre, *Les Elements des Terres Rares* II, CNRS (1969) p. 55.
[312] H. W. de Wijn, K. H. J. Buschow and A. M. van Diepen, *Phys. Stat. Sol.*, **30**, 759 (1968).
[313] T. Kaneko, *J. Phys. Soc. Japan*, **25**, 905 (1968).
[314] F. Kissell and W. E. Wallace, *J. Less Comm. Metals*, **11**, 417 (1966).
[315] C. C. Chao, *J. Appl. Phys.*, **37**, 2081 (1966).

[316] J. W. Cable, W. C. Koehler and E. O. Wollan, *Phys. Rev.*, **136**, A240 (1964).
[317] G. Arnold, N. Nereson and C. Olsen, *J. Chem. Phys.*, **46**, 4041 (1967).
[318] M. Atoji, *J. Chem. Phys.*, **48**, 3380 (1968).
[319] M. Atoji, *J. Chem. Phys.*, **48**, 560 (1968).
[320] M. Atoji, quoted by W. E. Wallace (see ref. [251]).
[321] A. R. Storm and K. E. Benson, *Acta. Cryst.*, **16**, 701 (1963).
[322] R. C. Sherwood, H. J. Williams and J. H. Wernick, *J. Appl. Phys.*, **35**, 1049 (1964).
[323] K. H. J. Buschow, A. M. van Diepen and H. W. de Wijn, *Proc. 8th Rare Earth Conf.*, Reno (1970) p. 53.
[324] E. Holtzberg, R. J. Gambino and T. R. McGuire, *J. Phys. Chem. Solids*, **28**, 2283 (1967).
[325] K. Sekizawa and K. Yasukochi, *J. Phys. Soc. Japan*, **21**, 274 (1966).
[326] K. Sekizawa, *J. Phys. Soc. Japan*, **21**, 1137 (1966).
[327] K. H. J. Buschow and J. F. Fast, *Phys. Stat. Sol.*, **21**, 593 (1967).
[328] K. H. J. Buschow and J. F. Fast, *Phys. Stat. Sol.*, **16**, 467 (1966).
[329] T. Tsuchida and W. E. Wallace, *J. Chem. Phys.*, **43**, 3811 (1965).
[330] E. Bucher, K. Andres, J. P. Maita and G. W. Hull, *Helv. Phys. Acta*, **41**, 723 (1968).
[331] G. K. Shenoy, B. D. Dunlap, G. M. Kalvius, A. M. Toxen and R. J. Gambino, *J. Appl. Phys.*, **41**, 1303 (1970).
[332] T. F. Connolly and E. D. Copenhaver, URNL, RMIC-7, (Rev. 2) UC-25 (1970).
[333] W. E. Wallace, *Prog. Solid State Chem.*, in press.
[334] K. H. J. Buschow and A. S. van der Goot, *J. Less Comm. Metals*, **17**, 249 (1969).
[335] J. Pierre, *Compt. Rend.*, **265**, B1169 (1967).
[336] R. Lemaire, C. Berthet Columinas, J. La Forest, R. Pauthenet and J. Schwiezer, *Cobalt*, **39**, 97 (1968).
[337] G. I. Hoffer and L. R. Salmans, *Proc. 7th Rare Earth Conf.*, Coronado, California (1968).
[338] A. Henberger, F. Pobell and P. Kienle, *Z. fur Physik* 205, 503 (1967).

Appendix 1

Units

This text employs (Gaussian) c.g.s. units, atomic units, and relativistic units, together with other special units which are associated with particular branches of physics, e.g. the Angstrom $\equiv 10^{-8}$ cm in optical and X-ray spectroscopy. Our reason for using these many different units, rather than adopting a more coherent set, such as the SI system, is that they have been, and are continuing to be used in most scientific journals. It is assumed that the c.g.s. units need no discussion. The atomic, relativistic, and SI units are briefly described below.

Atomic units

The Schrödinger equation for a single electron of an hydrogen-like atom is

$$\left(-\frac{\hbar}{2m}\nabla^2+\frac{Ze^2}{r}\right)\psi = E\psi.$$

This equation is obviously simplified if units are chosen so that

$$m = \hbar = e = 1 \tag{A.1}$$

i.e. the unit of mass is the electronic mass, and the unit of charge is the electronic charge. With this choice some other units are

length $\quad \dfrac{\hbar^2}{me^2} = a_0,$ radius of the smallest Bohr orbit of a hydrogen atom.

$\qquad\qquad = 0{\cdot}529 \times 10^{-8}$ cm.

energy $\quad \dfrac{e^2}{a_0} =$ twice the ionization energy of normal hydrogen

$\qquad\qquad = 27{\cdot}20$ eV

time $\quad \dfrac{\hbar^3}{me^4} = 2{\cdot}415 \times 10^{-17}$ sec.

In these atomic units it should be noted that the numerical value of the speed of light is $c = 137{\cdot}04$.

Rational relativistic units

In solving the relativistic Dirac equation

$$(c\boldsymbol{\alpha}\cdot\mathbf{p}+\beta mc^2)\psi = E\psi$$

it is convenient to measure velocities in units of the speed of light, and energies in units of the rest mass of the electron. If units are chosen such that

$$c = m = \hbar = 1$$

then the units of some other quantities are:

length $\quad \dfrac{\hbar}{mc}$ = electron Compton wavelength
$= 3{\cdot}857\times10^{-11}$ cm.

energy $\quad mc^2$ = rest mass of an electron
$= 5{\cdot}118\times10^{5}$ eV

time $\quad \dfrac{\hbar}{mc^2} = 1{\cdot}286\times10^{-19}$ sec

momentum $\quad mc$

In these relativistic units the numerical value of the electronic charge is given by $e^2 = 1/137{\cdot}04$.

SI units

The SI system is almost identical with the rationalized m.k.s. system which is universally employed in electromagnetic theory. The basic units are given in Table A.1 and some derived units having special names are given in Table A.2. It is recommended that only powers of 10^3 are used as multiples and sub-

Table A.1

Quantity	SI unit	Unit symbol	c.g.s. unit	Conversion factor from SI to c.g.s.
Length	metre	m	centimeter	10^2
Mass	kilogramme	kg	gramme	10^3
Time	second	s	second	1
Electric current	ampere	A	e.m.u.	10^{-1}
Thermodynamic temperature	degree Kelvin	K	degree Kelvin	1
Luminous intensity	candela	cd		
Plane angle	radian	rad	radian	1
Solid angle	steradian	sr	steradian	1

Table A.2

Quantity	SI unit	Unit symbol	c.g.s. unit	Conversion factor from SI to c.g.s.
Force	Newton	$N = kg\,m\,s^{-2}$	dyne	10^5
Energy	Joule	$J = Nm$	erg	10^7
Power	Watt	$W = J\,s^{-1}$	erg sec^{-1}	10^7
Electric charge	Coulomb	$C = A\,s$	e.m.u.	10^{-1}
Electric potential difference	Volt	$V = W/A$	e.m.u.	10^8
Electric capacitance	Farad	$F = A\,s\,V^{-1}$	e.s.u.	9×10^{11}
Electric resistance	Ohm	$\Omega = V\,A^{-1}$	e.m.u.	10^9
Frequency	Hertz	$Hz = s^{-1}$	cycles sec^{-1}	1
Magnetic flux	Weber	$Wb = V\,s$	maxwell	10^8
Magnetic flux density (**B**)	Tesla	$T = Wb\,m^{-2}$	gauss	10^4
Inductance	Henry	$H = V\,s\,A^{-1}$	e.m.u.	10^9
Luminous flux	Lumen	$lm = cd\,sr$	lumen	1
Illumination	lux	$lx = lm\,m^{-2}$	phot	10^4

Table A.3

Multiplying factor	Prefix	Symbol
10^{12}	tera	T
10^{9}	giga	G
10^{6}	mega	M
10^{3}	kilo	k
10^{-3}	milli	m
10^{-6}	micro	μ
10^{-9}	nano	n
10^{-12}	pico	p
10^{-15}	femto	f
10^{-18}	atto	a

multiples of the primary unit, but in some cases this condition is waived. Table A.3 gives the prefixes corresponding to various multiplying factors.

An important difference between the SI and c.g.s. systems is that the fields **D** and **E**, and similarly **H** and **B**, have different dimensions in SI. Consider the fields **H** and **B**. For an isotropic medium, for which **B**, **H** and **M** are all parallel, one formally has

	c.g.s. (e.m.u.)	*SI*	
(i)	$\mathbf{B} = \mathbf{H}+4\pi\mathbf{M}$	$\mathbf{B} = \mu_0(\mathbf{H}+\mathbf{M})$	
(ii)	$\mathbf{M} = \chi\mathbf{H}$	$\mathbf{M} = \chi\mathbf{H}$	(χ = susceptibility per volume)
(iii)	$\mathbf{B} = \mu\mathbf{H}$	$\mathbf{B} = \mu\mu_0\mathbf{H}$	(μ = permeability of the medium)
(iv)	$\mu = 1+4\pi\chi$	$\mu = 1+\chi$	

It is the appearance of μ_0 in the SI system which, through (i), gives rise to the fact that **B** and **H** have different dimensions. μ_0 is the permeability of free space and it is defined so that the unit of current, the ampere, has its usual practical value:

$$\mu_0 = 4\pi \times 10^{-7} \quad \mathrm{Hm^{-1}\,(JA^2m^{-1})}.$$

The permittivity of free space ε_0 is then defined by $\varepsilon_0\mu_0 = c^{-2}$. From (iv) it is clear that the volume susceptibility χ in SI units is 4π times the magnitude of χ in e.m.u. Some other useful conversions are listed in Table A.4.

Table A.4

Quantity	SI unit	c.g.s. unit	Conversion factor from SI to c.g.s.
Magnetic induction **B**	T	gauss	10^4
Magnetic field **H**	$\mathrm{A\,m^{-1}}$	oersted	$4\pi \times 10^{-3}$
Intensity of magnetization **M**	$\mathrm{A\,m^{-1}}$	e.m.u.	10^{-3}
Magnetic moment **m**	Wb m	e.m.u.	10^3
Volume susceptibility χ	$\mathrm{SI\,m^{-3}}$	$\mathrm{e.m.u.\,cm^{-3}}$	$1/4\pi$
Mass susceptibility	$\mathrm{SI\,kg^{-1}}$	$\mathrm{e.m.u.\,g^{-1}}$	$10^3/4\pi$
Gramme-molar susceptibility	$\mathrm{SI\,(g\text{-}mole)^{-1}}$	$\mathrm{e.m.u.\,(g\text{-}mole)^{-1}}$	$10^6/4\pi$

The magnetic field **H** is also often defined in m.k.s. units through the relation $\mathbf{B} = \mu_0\mathbf{H}+\mathbf{M}$ rather than by (i) above. The intensity of magnetization **M** then has units $\mathrm{Wb\,m^{-2}}$, and equation (ii) becomes $\mathbf{M} = \chi\mu_0\mathbf{H}$. Expressions (iii) and (iv) are unchanged. For **M** the conversion factor from m.k.s. to e.m.u. is $10^4/4\pi$.

Some miscellaneous units of energy

Table A.5 Some miscellaneous units of energy

Unit	Symbol	Conversion factors
Electronvolt	eV	$1{\cdot}6019 \times 10^{-12}$ erg sec
Rydberg	Ry	13·60 eV
Atomic Unit	a.u.	27·20 eV ≡ 2 Ry
Energy of an electron with wavenumber one $\mathrm{cm^{-1}}$ (hc)	$\mathrm{cm^{-1}}$	$1{\cdot}240 \times 10^{-4}$ eV
Energy associated with a temperature of one °K (Boltzmann's constant × 1°K)	°K	$8{\cdot}62 \times 10^{-5}\,\mathrm{eV} \equiv 1{\cdot}439\,\mathrm{cm^{-1}}$

Appendix 2

The Determination of Crystal Field Splittings

As a very simple illustration of crystal field splittings consider the effect of the cubic potential

$$V_c = B_4^0[O_4^0 + 5O_4^4]$$

upon a multiplet ground state with $J = \frac{5}{2}$. The free ion is six-fold degenerate but in the presence of the cubic potential above the degeneracy will be partially removed leaving a Γ_8 quartet and a Γ_7 doublet.

To calculate the splittings, we require the matrix elements of V_c between states $|J = \frac{5}{2}, M_J\rangle$. To assist in the evaluation of these elements it is convenient to define the spin raising (J_+) and spin lowering (J_-) operators:

$$J_+ = J_x + iJ_y, \qquad J_- = J_x - iJ_y$$

Then for a given term, the required operations are:

$$J_+|J, M_J\rangle = \{J(J+1) - M_J(M_J+1)\}^{\frac{1}{2}}|J, M_J+1\rangle$$
$$J_-|J, M_J\rangle = \{J(J+1) - M_J(M_J-1)\}^{\frac{1}{2}}|J, M_J-1\rangle$$
$$J_z|J, M_J\rangle = M_J|J, M_J\rangle$$
$$J^2|J, M_J\rangle = J(J+1)|J, M_J\rangle.$$

The $|J, M_J\rangle$ are orthogonal and assumed to be normalized to unity so that

$$\langle J, M_J|J'M'_{J'}\rangle = \delta_{J,J'}\delta_{M_J,M'_{J'}}.$$

Following the above, we can write for a general element

$$\langle J, M_J|V_c|J', M'_{J'}\rangle = B_4^0\{\langle J, M_J|O_4^0|J', M'_{J'}\rangle + 5\langle J, M_J|O_4^4|J'M'_{J'}\rangle\}$$

Considering a specific element, say $\langle -\frac{5}{2}|O_4^4|\frac{3}{2}\rangle$ this then gives (from Table 1.7)

$$\langle -\tfrac{5}{2}|O_4^4|\tfrac{3}{2}\rangle = \tfrac{1}{2}\langle -\tfrac{5}{2}|J_+^4|\tfrac{3}{2}\rangle + \tfrac{1}{2}\langle -\tfrac{5}{2}|J_-^4|\tfrac{3}{2}\rangle$$

where $J = \frac{5}{2}$ is implicitly assumed.

Then

$$J_+|\tfrac{3}{2}\rangle = \{\tfrac{5}{2}(\tfrac{5}{2}+1) - \tfrac{3}{2}(\tfrac{3}{2}+1)\}^{\frac{1}{2}}|\tfrac{3}{2}+1\rangle$$
$$= \sqrt{5}|\tfrac{5}{2}\rangle$$

and

$$J_+^2|\tfrac{3}{2}\rangle = J_+\{\sqrt{5}|\tfrac{5}{2}\rangle\} = 0$$

since no state $|\tfrac{7}{2}\rangle$ exists. Hence $J_+^3|\tfrac{3}{2}\rangle = 0$ and therefore $\langle -\tfrac{5}{2}|J_+^4|\tfrac{3}{2}\rangle = 0$.

Similarly

$$J_-|\tfrac{3}{2}\rangle = \{\tfrac{5}{2}(\tfrac{5}{2}+1)-\tfrac{3}{2}(\tfrac{3}{2}-1)\}^{\frac{1}{2}}|\tfrac{3}{2}-1\rangle\sqrt{8}|\tfrac{1}{2}\rangle$$
$$J_-^2|\tfrac{3}{2}\rangle = J_-\{\sqrt{8}|\tfrac{1}{2}\rangle\} = 3\sqrt{8}|-\tfrac{1}{2}\rangle$$
$$J_-^3|\tfrac{3}{2}\rangle = J_-\{3\sqrt{8}|-\tfrac{1}{2}\rangle\} = 24|-\tfrac{3}{2}\rangle$$
$$J_-^4|\tfrac{3}{2}\rangle = J_-\{24|-\tfrac{3}{2}\rangle\} = 24\sqrt{5}|-\tfrac{5}{2}\rangle.$$

Hence finally

$$\langle -\tfrac{5}{2}|O_4^4|\tfrac{3}{2}\rangle = \tfrac{1}{2}\langle -\tfrac{5}{2}|J_-^4|\tfrac{3}{2}\rangle = 12\sqrt{5}\langle -\tfrac{5}{2}|-\tfrac{5}{2}\rangle = 12\sqrt{5}.$$

The other matrix elements may be evaluated in the same way, or more simply reference may be made to Hutchings (ref. [28] of Chapter 1). The complete 6×6 energy matrix is given below, where the rows and columns are labelled by the free ion states, and the arrangement of elements has been specifically chosen to give the block form.

	$\|\tfrac{5}{2}\rangle$	$\|-\tfrac{3}{2}\rangle$	$\|\tfrac{3}{2}\rangle$	$\|-\tfrac{5}{2}\rangle$	$\|\tfrac{1}{2}\rangle$	$\|-\tfrac{1}{2}\rangle$
$\langle\tfrac{5}{2}\|$	$60\,B_4^0$	$60\sqrt{5}\,B_4^0$	0	0	0	0
$\langle-\tfrac{3}{2}\|$	$60\sqrt{5}\,B_4^0$	$-180\,B_4^0$	0	0	0	0
$\langle\tfrac{3}{2}\|$	0	0	$-180\,B_4^0$	$60\sqrt{5}\,B_4^0$	0	0
$\langle-\tfrac{5}{2}\|$	0	0	$60\sqrt{5}\,B_4^0$	$60\,B_4^0$	0	0
$\langle\tfrac{1}{2}\|$	0	0	0	0	$120\,B_4^0$	0
$\langle-\tfrac{1}{2}\|$	0	0	0	0	0	$120\,B_4^0$

The new eigenvalues λ are calculated by subtracting λ from each diagonal element, and solving the secular equation which is obtained by setting the determinant of the resulting matrix to zero. In our example the determinant factorizes into two identical quadratics and two equalities, viz:

$$\begin{vmatrix} 60\,B_4^0-\lambda & 60\sqrt{5}\,B_4^0 \\ 60\sqrt{5}\,B_4^0 & -180\,B_4^0-\lambda \end{vmatrix} = 0 \text{ and } 120\,B_4^0-\lambda = 0$$

The solutions of the quadratic are $\lambda = -240\,B_4^0$ and $\lambda = 120\,B_4^0$. The effect of V_c is therefore to split the six-fold degenerate free ion state into a four-fold degenerate level, designated Γ_8, and a doublet level, designated Γ_7, which respectively have energies $120\,B_4^0$ and $-240\,B_4^0$ relative to the original level. In the simple example above it can be seen from the block form of the energy matrix that the V_c considered mixes states $|\frac{3}{2}\rangle$ and $|-\frac{5}{2}\rangle$, and $|\frac{5}{2}\rangle$ and $|-\frac{3}{2}\rangle$. As mentioned in Chapter 1, Lea *et al* (ref. [29] of that chapter) have tabulated the level splittings for $2 < J < 8$ in a general way as a function of B_4^0 and the fractional value of B_4^0/B_6^0. From their work the complete set of wavefunctions for the split levels are immediately found to be:

$$\Gamma_7: \quad \begin{array}{l} 0{\cdot}4083|+\frac{5}{2}\rangle - 0{\cdot}9129|-\frac{3}{2}\rangle \\ 0{\cdot}4083|-\frac{5}{2}\rangle - 0{\cdot}9129|+\frac{3}{2}\rangle \end{array}$$

$$\Gamma_8: \quad \begin{array}{l} 0{\cdot}9129|+\frac{5}{2}\rangle + 0{\cdot}4083|-\frac{3}{2}\rangle \\ 0{\cdot}9129|-\frac{5}{2}\rangle + 0{\cdot}4083|+\frac{3}{2}\rangle \\ |+\frac{1}{2}\rangle \\ |-\frac{1}{2}\rangle \end{array}$$

Usually one is interested in the perturbation of these states by a magnetic field, and the new energy levels can be determined by setting up and solving another secular equation which contains the matrix elements of the magnetic interaction $g_J\mu_B\mathbf{J}\cdot\mathbf{H}$ between the wavefunctions of Γ_7 and Γ_8 (ref. [32], Chapter 1). This is not done here, but to illustrate the use of some results given in Section 1.3.4, we consider the splitting of the ground state Γ_7 Kramers' doublet by a magnetic field. It is clear that the upper and lower wavefunctions of the doublet given above have the respective form of $|+\rangle$, $|-\rangle$ in equation (1.49). Hence, for the case of a magnetic field H_z applied parallel to the z-axis, equation (1.50) immediately gives for the splitting

$$\Delta E_{\|} = 2g_J\mu_B H_z\{(0{\cdot}4083)^2(\tfrac{5}{2}) + (0{\cdot}9129)^2(-\tfrac{3}{2})\}.$$

When a magnetic field $H_\perp$ is applied perpendicular to this z-axis it is necessary to evaluate the matrix element $\langle +|J_x|-\rangle$ in equation (1.51) in order to calculate the splittings. Since $J_x = \frac{1}{2}(J_+ + J_-)$ we therefore require the following results:

$$J_+|-\tfrac{5}{2}\rangle = \sqrt{5}|-\tfrac{3}{2}\rangle$$
$$J_-|-\tfrac{5}{2}\rangle = 0$$
$$J_+|\tfrac{3}{2}\rangle = \sqrt{5}|\tfrac{5}{2}\rangle$$
$$J_-|\tfrac{3}{2}\rangle = \sqrt{8}|\tfrac{1}{2}\rangle.$$

Then we have

$$\langle +|J_x|-\rangle = \tfrac{1}{2}[0{\cdot}4083\langle\tfrac{5}{2}| - 0{\cdot}9129\langle -\tfrac{3}{2}|\} \times$$
$$\{0{\cdot}4083\sqrt{5}|-\tfrac{3}{2}\rangle - 0{\cdot}9129[\sqrt{5}|\tfrac{5}{2}\rangle + \sqrt{8}|\tfrac{1}{2}\rangle]\}$$
$$= -(0{\cdot}4083)(0{\cdot}9129\sqrt{5})$$

So that finally

$$\Delta E_{\perp} = 2g_J\mu_B\, 0{\cdot}4083(0{\cdot}9129\sqrt{5})H_{\perp}$$

In general it is more convenient to set up a single secular equation containing both the crystal field and the magnetic interaction, because then the free ion wavefunctions may be used as a basis, which simplifies the evaluation of the matrix elements.

Index

SUBJECT INDEX

AUTHOR INDEX